Business Statist

A multimedia guid
concepts and applic

G000112167

Business Statistics

A multimedia guide to concepts and applications

Chris Robertson
European Institute of Oncology
with software by

Moya McCloskey
Pricewaterhouse Coopers

ARNOLD

A member of the Hodder Headline Group
LONDON
Co-published in the United States of America by
Oxford University Press Inc., New York

First published in Great Britain in 2002 by
Arnold, a member of the Hodder Headline Group,
338 Euston Road, London NW1 3BH

http://www.arnoldpublishers.com

Co-published in the United States of America by
Oxford University Press Inc.,
198 Madison Avenue, New York, NY10016

The advice and information in this book are believed to be true and
accurate at the date of going to press, but neither the author nor the publisher
can accept any legal responsibility or liability for any errors or omissions.

British Library Cataloguing in Publication Data
A catalogue record for this book is available from the British Library

Library of Congress Cataloging-in-Publication Data
A catalog record for this book is available from the Library of Congress

ISBN 0 340 71927 3

1 2 3 4 5 6 7 8 9 10

Production Editor: James Rabson
Production Controller: Bryan Eccleshall
Cover Design: Terry Griffiths

Typeset in Times 10/12 by Replika Press Pvt Ltd, Delhi 110 040, India
Printed and bound in Malta by Gutenberg Press Ltd

What do you think about this book? Or any other Arnold title?
Please send your comments to feedback.arnold@hodder.co.uk

To my wife Clare, and our children, Kieron, Douglas and Hamish

Contents

List of figures

List of tables

Preface

This textbook and the accompanying software are written from different perspectives and with different users in mind. We believe that the main users of the book and software will be business and social science students taking an introductory course in statistics at undergraduate level. It may also be relevant for postgraduate MBA students.

One of the main difficulties in teaching statistics at these levels is coping with the motivation and mathematical ability of the students. We have tried to address the problem of motivation by using a large number of articles and graphs from newspapers and websites within the areas of business, economics, social science and everyday life. Generally each chapter of the book and each software module is application-motivated rather than just having an example to illustrate the technique.

As a minimum we expect that the vast majority of students will have a mathematical qualification from school at the Standard grade, in Scotland, or GCSE grade in England, Wales and Northern Ireland. For some of the chapters this level of mathematical ability is not necessary, but other chapters have more advance mathematics. A large number of students will have an A level, AS level or Scottish Higher qualification in mathematics, and we hope that the book has sufficient mathematical detail for such students.

We believe that the software will be an appropriate way into the textbook for students who have little mathematical background and wish to know why they need to study statistics before delving into the technicalities. It can also be used as a quick refresher for the main topics which are covered in much more detail in the book. Finally, it could be used to reinforce the concepts in the book and in a lecture course as it tries to explain the concepts from a more pictorial point of view.

The textbook is a fairly traditional run through the material that is likely to be covered in an introductory statistics course for business and social science students. It is written in such a way as to try to explain the statistical concepts, together with the required mathematics, rather than just as a discussion of statistical methods and when and how to use them.

There is also a lot of material which is not likely to be covered in a 40- or 50-hour course. For example, we have never taught two-sample t tests or multiple regression in an introductory course for business students. However, much of the more interesting analysis comes through the use of these procedures and we feel that it is important for students to have this information, if only for reference, and to make the book fairly complete up to the level of two-sample paired and unpaired tests.

We foresee that the material in Chapters 1–7 could be used to form the basis of a 25-hour lecture course for students with minimal mathematical abilities. Most of the equations and mathematical manipulations come in Chapters 8–15. Throughout the text there are a number of more detailed case studies. Other case studies are presented on the website for the book. Some extra material covering slightly harder or less important topics is also available on the accompanying website at www.arnoldpublishers.com/support/business statistics.

The software and textbook are loosely linked and can be used independently of each other. The following table gives the links between the six modules and the chapters in the textbook. The software is a very quick introduction to a number of concepts, while the book is a more detailed look at the concepts, methods and interpretation.

Module	Book sections
1. Presenting data graphically	2.3 Location and variability in quantitative variables
	3.2 Histogram
	3.3 Stem and leaf plot
	3.4 Boxplot
	3.6 Bar chart and pie chart
	3.7 Scatter plot
	3.8 Time series plot
2. Basic statistics	2.1 Populations and samples
	2.2 Types of variables
	2.3 Location and variability in quantitative variables
	5.1 Planning a survey
	5.2 Sampling schemes
3. Data models	9.1 Key Ideas of random variables, expectation and variability
	9.2 Binomial distribution
	9.3 Continuous models
	9.4 Normal distribution
	10.2 Random samples, repeated samples and sampling distributions
	10.4 Standard error
	10.5 Central limit theorem
	11.2 Confidence interval for the mean of a population
4. Analysing data	3.7 Scatter plot
	6.2 Common transformations to linearity
	6.4 Correlation
	6.5 Regression lines
	14.1 Correlation
	14.3 Testing
	15.2 Linear regression model
	15.4 Confidence intervals for the estimates
	15.7 Residuals and outliers
	15.8 Confidence intervals for a predicted mean and a predicted single value
5. Time series data	7.1 Introduction
	7.2 Simple time series models
	7.3 Moving averages
	7.4 Exponential smoothing
	7.5 Correcting for seasonality
	7.6 Forecasting
6. Survey data	2.1 Populations and samples
	5.1 Planning a survey
	5.2 Sampling schemes
	5.5 Questionnaire designs
	5.6 Bias and representative studies
	5.7 Data presentation, tables and bar charts

The main chapters of the textbook which are not covered by any part of the software are as follows:

Chapter	Material
1. Introduction	Examples of statistical ideas in statistics and society. Section 1.6 is covered a little in Module 1
4. Index numbers	Weighted averages, simple index numbers, Laspeyres and Paasche index numbers, retail price index
8. Probability	Events, definitions of probability, multiplication law, addition law, independence, mutually exclusive events, Bayes' Theorem
12. Significance tests	Components of significance tests, tests on means and proportions, interpretation of tests, power
13. Qualitative variables: goodness of fit and association	X^2 goodness-of-fit test, X^2 test for two-way tables

There are a number of areas which are not covered in the textbook, and material for some of these is available on the website. The principal resources here are the Excel work sheets with data for most of the examples. Also there is a chapter on using and testing probability models. This has a number of case study applications of the probability distributions discussed in the book. There is also a report of the analysis of a large market research survey. This covers most of the important topics of the book and draws together many of the ideas into an analysis with appropriate use of graphs and summaries, together with their interpretation within the context of marketing a product. This chapter deals with the use of the statistical methods for the analysis, and not with the interpretation of the concepts.

Note on calculations

Throughout this book there are lots of numerical calculations. Generally I have tried to use a large number of decimal places to minimise the effects of rounding error. If readers work through the numerical calculations by themselves they may find that they obtain slightly different numerical results, particularly if they use a computer program and particularly when there is a sequence of calculations with intermediate steps. I have adopted the system of having each calculation depending upon the immediately preceding result. This is not the most accurate way of doing the calculations but does mean that there is a logical consistency from one line to the next. For example, on page 203 there is a calculation of $(85-142)/52$ which is equal to -1.10 to 2 decimal places followed by a probability calculation based upon -1.10 to give an answer of 0.1375. If you use a computer program for this calculation then you do not need the intermediate step of working out the value of $(85-142)/52$ and you will get an answer of 0.1365. Using the computer program is the more accurate method but the differences are slight.

Acknowledgements

The idea for this book grew out of a Teaching and Learning Technology Program project in the Department of Statistics and Modelling Science at Strathclyde University to develop software for teaching elementary statistics to bioscience students. Our colleagues on this project, which resulted in the Quercus software, were Steve Blythe, Bill Gurney, Jim Kay, Athol Korabinski and Colin Aitken. Steve Blythe was involved in the initial stages of this project and wrote the first draft of chapter 7 on Time Series Models.

We are very grateful for the considerable support we have received from Liz Gooster at Arnold publishers and her predecessors, Kirsty Stroud and Nikki Dennis.

All of the textbook was written while I was working as a research statistician in the Division of Biostatistics and Epidemiology at the European Institute of Oncology in Milan. I would like to thank the head of this Division, Peter Boyle, for his support and encouragement during this project.

1

Introduction

- Statistics has a role to play in business and society
- Economic and business decision-making can be assisted by a statistical analysis of relevant data
- Understanding the key components of a statistical study
- Illustration of statistical techniques useful in business and management.

Statistics and statistical ideas have a great importance in modern life. It is virtually impossible to read a newspaper without finding references to some aspect of statistics. Typical examples are government figures such as the retail price index or the crime count; the results of political opinion polls; a medical study which concludes that a particular foodstuff is associated with an increased risk of a disease or that a particular treatment is better for treating a disease; and market surveys indicating what are the most frequently watched television programmes. All of these are related through their use of statistical ideas and methods.

A common feature of these studies is that some quantity is measured, and that this measurement is carried out on a sample of items or individuals. The results are analysed with a view to summarizing the state of the system or to looking at relationships. Statistics is the science which deals with the collection, presentation and interpretation of data. Most often this is quantitative numerical data, but qualitative non-numerical data are also important. There are statistical implications whenever data are collected in a survey or experiment, whenever data are summarized, whenever the results are presented and interpreted. The most important aspect is the interpretation of the results. The correct interpretation depends upon a sound methodology for the collection of the appropriate data, suitable techniques for summarizing the data and presenting the results in an honest and enlightening fashion.

In this chapter we will begin with a series of topical areas in society and business where statistical ideas come to the fore.

1.1 Opinion polls

In the run-up to any election a large number of opinion polls are published in the newspapers and on television. Every month a number of opinion polls are published, even if there is no election. In the United Kingdom these are normally carried out by a relatively small number of well-established market research companies. Their purpose is to estimate public opinion at a point in time. Generally this is concerned with political issues, such as the popularity of political parties, policies or individuals.

Data from a sequence of opinion polls published during the run-up to the elections for the Welsh Assembly in 1999 are shown in Table 1.1. These show the voting intentions of the electorate, together with the sample sizes. There is some variation in the voting percentages and also in the numbers of individuals questioned.

Table 1.1 Opinion polls in the run-up to the 1999 Welsh Assembly elections

| Date | Sample size | Voting intentions (%) | | | | Polling company | Poll type |
		Lab	PC	Con	LD		
5 May	1501	47	26	14	10	NOP	Phone
30 Mar	971	54	21	13	12	Beaufort	Face to face
11 Mar	1015	51	23	16	8	Beaufort	Face to face
12 Feb	1501	54	20	16	9	NOP	Phone
23 Dec	962	59	18	14	8	Beaufort	Face to face
25 Nov	1002	62	18	12	7	Beaufort	Face to face
1 Oct	992	57	19	12	8	Beaufort	Face to face
18 Sept	1500	50	24	16	8	NOP	Phone
16 Jul	1008	56	20	14	9	Beaufort	Face to face

Source: http://www.bbc.co.uk, 5 May 1999.
The percentages quoted are those for the 'first past the post' section of the election, not the 'top-up vote' for the proportional representation part. Lab, Labour; PC, Plaid Cymru; Con, Conservative; LD, Liberal Democrat.

All of these opinion polls are supposed to be samples from the electorate. This is a population of around 2.2 million people over 18, and yet the sample is minuscule by comparison. Because a sample is used as opposed to the whole population, some error is bound to arise as a result of the sampling procedure. Statistical methods will show you that if the sample is randomly selected then the error on a quoted percentage is about plus or minus 3% based on a sample size of around 1500. When we look at samples and estimating percentages from samples then we will look in detail at the meaning of this error, its implications and how we can make comparisons allowing for it.

A key word in the above paragraph was 'random'. Many statistical ideas are based upon the notion of randomness. Probability models are used to describe random behaviour, and some probability models will be looked at in this book. The important aspect of selecting a sample at random is that we should induce no bias in the estimation of whatever it is we are interested in. If a random sample for a public opinion poll were selected solely by one interviewer walking around the streets of Aberystwyth, say, then this would not be a random sample from the whole of Wales. It would not even be a random sample from all people in Aberystwyth, as only those who were out and about at the same time as the interviewer could be in the sample. The selection of a random sample is a difficult task, and there are various ways to go about it. Without knowing the advantages and disadvantages of samples, it is not easy to use the good information in them properly.

Chapters: *business surveys, sampling distributions, large surveys, probability, probability models, significance testing.*

1.2 Supermarket relocation

This example is based upon a statistical consultancy problem which was brought to me by one of the students who had been in my statistics class a few years previously. She was working as a management accountant for a chain of supermarkets, and her department was grappling with the problem of how to decide upon the most suitable location for new stores.

About nine months previously the chain had opened up a supermarket in a new location, and this had turned out to be such a financial disaster that the company was forced to close the store with the loss of a huge amount of money and prestige. Previously they had used the judgement of the management team in reaching a decision about opening new stores. In the light of the previous disaster they were looking to develop a model which they could use to predict if a new store in a new location would be successful. They hoped that this would lead to more objective decision-making and reduce their likelihood of making a huge error.

The financial department had detailed records about all the stores. This included, among others, average weekly turnover, store size, store location, availability of car parking, and presence of competitors' supermarkets in the vicinity. The company also suspected that the number of people living within a 'ten minute drive to shop' area around the supermarket and their social class demographics would be important. This information had come from checkout surveys of shoppers in the stores which had shown that only a minority of customers had travelled far to shop in the stores.

The supermarket company had been in contact with a market research institute which had used the small-area statistics based upon the 1991 Census to develop a marketing package for retailers. Given any location in the United Kingdom, the market research company sold information about the number of people living within a 10-mile radius and demographic information such as the proportion of households in the area with no cars or the proportion of people in the area in social classes A–C. All of this information is anonymous and is based upon the postcode. This makes it a very powerful tool for analyses based upon small areas.

The market research company had charged the supermarket company in the region of £20 000 for the demographic information on each of their current stores. For any new location a prediction of the viability of a new store would cost £200. The accountants in the supermarket company had no real idea of the details of the prediction model. The reason why I was consulted about this problem was to assess the feasibility of the prediction model and to see if there was any way the company could do the predictions without resorting to the market research company every time.

Because this was a consultancy contract I had to sign a confidentiality statement saying that I would not disclose any results or anything else which might be detrimental to the company. Consequently, all the numbers in this example are fictitious. The setting is real, however, and quite typical. I was recently speaking to a statistical colleague who had been asked to give a lecture entitled 'When do statistics lie?' His lecture was to be one word: 'Always'. While we might dispute this generally, in this example it is true.

Some graphs showing the relationships between the weekly turnover of all the supermarkets in the chain and the size of the shop, the number of people living within a 10-minute drive, the percentage of families in social classes A–C within a 10-minute drive and whether or not there is a competing supermarket close by are presented in Figure 1.1. All of these graphs and the lines drawn on them will be discussed later in this book.

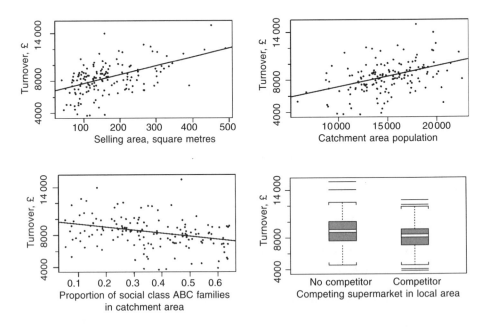

Figure 1.1 Relationship between supermarket turnover and selected characteristics.

Three of the graphs are known as scatter plots, and they purport to display the association between the two variables drawn. From them we can see that the turnover tends to increase as the size of the store increases, as the population increases and as the proportion of families in social classes A–C decreases. The lines on these graphs are known as regression lines and they can be used to predict the turnover. For example, in a shop with a surface area of 400 square metres we would predict a turnover of around £11 000 per week. The last plot is known as a boxplot and shows the association between turnover and whether or not there is a competing supermarket in the neighbourhood. As the box is lower down when there is a competitor, this suggests that the turnover is lower when a competitor is close by.

All of these observations make economic sense, and by combining all the information the company hoped to build a good prediction model which would help them in their decision-making. While the analysis showed the company the important features of their supermarkets in relation to the turnover, the prediction of where to open new supermarkets wasn't overwhelmingly successful. You can see from the graphs one reason for this. There is a great deal of variation in weekly turnover even for shops with similar characteristics. For shops with about 100 square metres of selling area the weekly turnovers range from just under £4000 to over £10 000.

This example illustrates some of the important aspects of statistics and its application in business. Information is gathered, some of it routine business information and some readily available demographic information. Relationships between quantities are investigated. Statistical models are developed and quantities relevant to these models are estimated from the data. Predictions are made and compared to the data. Overall there is much variation which cannot be accounted for by the model, and a statistical analysis is concerned with making inferences and decisions in the presence of variability. Without the variation there would be little need for statistical ideas, but as this variation is pervasive there is much need for good statistical analysis.

Chapters: *graphs, relationships, linear regression, correlation, estimation, probability models, significance testing.*

1.3 Financial indices

Every day, the Financial Times index for the London Stock Exchange, the Dow Jones index for the New York Stock Exchange, the Hang Seng index for Hong Kong and the Nikkei index for Tokyo are reported on UK television and radio news programmes and in the newspapers. Every month, the change in the UK retail price index over the previous year is reported in order to give a measure of the rate of inflation. These are all examples of index numbers which are used to compare changes in the average prices of a number of different commodities.

The retail price index, shown in Figure 1.2 for the period from 1961 until 1997, measures the change from one month to the next in the cost of a 'representative' basket of goods and services typically bought by an 'average' household. Many wage negotiations, social security and pension rises are based upon the rate of inflation, so this index affects virtually everyone living in the UK. The graph shows huge cyclical changes in the UK

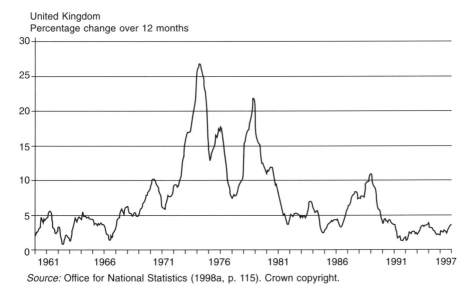

Source: Office for National Statistics (1998a, p. 115). Crown copyright.

Figure 1.2 The retail price index, 1961–1997.

rate of inflation over this time. The percentage changes are always positive, which means that at no time has there ever been a fall in the cost of the goods.

The prices of goods and services in the UK compared to those in France, Italy, United States and Japan are shown in Figure 1.3. This is a graph of what are known as price relatives, and the value of 100 throughout the period refers to prices in the UK. If a country has a value of more than 100 then goods and services in that country would appear to be more expensive than in the UK, while if the index was less than 100 then they would appear to be cheaper. From the graph you can see that in the 1970s a UK visitor to France, Japan and the USA would find these countries dearer – in 1977 about 40% dearer. A UK visitor to Italy in the same period would find prices there about the same as in the UK, though in the late 1970s Italy became relatively cheaper. At the end of the series, you can see that to a UK resident in 1995, Japan would seem about 80% more expensive than the United Kingdom, France about 25% more expensive, while Italy and the United States would have seemed about 10% cheaper.

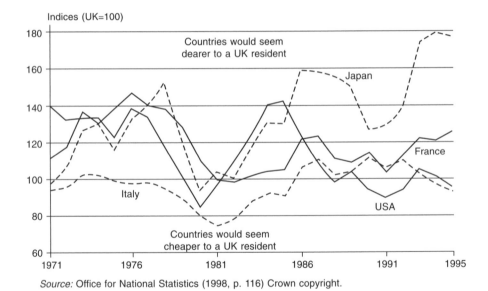

Source: Office for National Statistics (1998, p. 116) Crown copyright.

Figure 1.3 Relative price levels, 1971–1995.

The analyses of these graphs can be linked to economic trends and political and social changes. For example, in 1973 there was a huge increase in the price of oil from the Gulf states, and this would have been a contributory effect on the huge increase in the rate of inflation in the UK which is evident from Figure 1.2. Other rises and falls can also be linked to the effects of external events and government policies. The statistical aspects of these graphs lie in the presentation of the information, the collection of the data, the construction of the indices and the exact interpretation of the indices and the changes in the indices.

Chapters: *index numbers, time series, large surveys.*

1.4 Official statistics

A huge volume of statistics is produced by the statistical services of governments all over the world. An example, taken from an Italian newspaper, the *Corriere della Sera* of 22 July 1999, is shown in Figure 1.4. This graph shows the continuing loss of service and industrial jobs in Italy over a period of one year. What we have is called a time series plot of the percentage change in employment relative to one year ago. In April 1999, the number of people employed in industry was 2.9% below the value in April 1998, which in turn was 2.2% below the figure in April 1997. The changes in the service industries have been much less dramatic. These figures are taken from the Italian government statistical service, ISTAT, and are comparable to government statistics everywhere.

I dati dell'Istat di aprile: solo nelle aziende alimentari e nei servizi le nuove opportunità

Grandi imprese, occupazione in calo

Persi altri 25 mila posti, ma la tendenza sembra rallentare

ROMA — Continuano a diminuire i posti di lavoro nella grande industria: 25 mila occupati in meno in aprile rispetto a un anno fa. Secondo le rilevazioni Istat i settori dove di sono registrate le maggiori flessioni sono stati il comparto della produzione di mezzi di trasporto (-5,6%), della produzione di apparecchi elettrici e di precisione (-4,5%), dell'industria della carta, stampa, editoria (-3,5%) e della produzione di articoli in gomma e materie plastiche (-3%). Per contro, hanno presentato un risultato in controtendenza i dati relativi agli occupati nelle industrie alimentari, delle bevande e del tabacco, con un +5%. Complessivamente, nei primi quattro mesi dell'anno, l'occupazio-

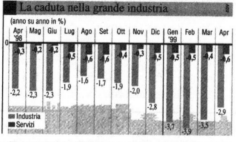

nell'intermediazione finanziaria (-0,9%); mentre sono risultati in rialzo i posti di lavoro nel com-

nel raffronto con lo stesso mese del '98, un calo più pesante, pari a 30 mila posti di lavoro in me-

giorni lavorativi rispetto all'aprile '98, le ore effettivamente lavorate per dipendente, al netto dei cassintegrati, hanno mostrato una diminuzione tendenziale dell'1,4%. È diminuita anche l'incidenza delle ore straordinarie, passando in un anno dal 4,5% al 3,8%. Nei primi quattro mesi dell'anno le ore lavorate per dipendente sono diminuite dell'1,9%, rispetto allo stesso periodo '98 e sempre a parità di giorni lavorativi (l'incidenza delle ore di straordinario è scesa di un punto percentuale, passando dal 4,9% al 3,9%).

La rilevazione Istat ha preso in esame anche l'andamento degli stipendi: la retribuzione lorda media per dipendente, calcolata per gli occupati al netto del

Figure 1.4 Unemployment in Italy.

Official statistics such as these are used by governments to make policy decisions about the economy. They may be used by trade unions in connection with measures designed to stimulate new jobs in certain regions. They may be used by political and social commentators. They will be used by academics in the pursuit of research into the performance of governments. They will be used to compare countries and regions, and to compare the current situation with the past.

There are many different types of official statistics, and virtually all countries have a government statistical service which is responsible for collecting and presenting and interpreting this information in a fair and unbiased way. The most important aspect of these figures is making sure that they are correct, unbiased and free from any government manipulation. It is also important that they measure appropriate economic and social quantities. The interpretation of official statistics is always accompanied with many political overtones and, depending on your standpoint, there is always the chance that the figures will be interpreted to suit a particular political style. The interpretation of official statistics can be influenced severely by the manner in which they are presented and in the manner in which they are collected.

In the UK, official statistics are published by the Office for National Statistics. This is a government agency responsible for the compilation, analysis and dissemination of a great deal of economic, social and demographic information. It works in partnership with others in the Government Statistical Service to provide the parliament, government and the wider community with the statistical information, analysis and advice needed to improve decision-making, stimulate research and inform debate. It aims to provide an authoritative and impartial picture of society and a window on the work and performance of government, allowing the impact of government policies and actions to be assessed (Office for National Statistics 1998b, p. ii). Further information about the Office for National Statistics is available at http://www.statistics.gov.uk.

The information comes from a number of sources including the population census and register of births and deaths, large government-sponsored surveys such as the Labour Force Survey, the General Household Survey and the Family Expenditure Survey, local authority data, social security data, tax returns and crime figures.

Chapters: *business surveys, large surveys, index numbers, time series.*

1.5 School and hospital league tables

In the UK school league tables are published on an annual basis. These give the numbers of pupils in the school and the percentage of pupils passing the national examinations at various levels. These include the percentage of pupils obtaining five or more GCSE O levels at grades A–C (in England, Wales and Northern Ireland) or Standard grades 1–3 (in Scotland). Data for schools in Surrey and Sussex in 1998 are presented in Table 1.2.

This type of table is required as part of the 1991 Parent's Charter from the Department of Education and Science and was part of an initiative by the Conservative governments of the late 1980s and early 1990s to promote the use of indicators by which institutions can be compared. The aim of this process is ultimately to evaluate the performance of the institutions, but it could also be used as a means of trying to improve performance overall. In Table 1.2 you can see that the percentage of pupils who obtain five or more GCSEs at levels A*–C ranges from 22% to 100%. This is an enormous range, and if there are factors which differ systematically among the schools and which are directly associated with the performance of the pupils then it may be possible to modify these factors and hence bring about an overall improvement in performance.

A brief glance at Table 1.2 will show you that the 13 schools at the top of the list are independent schools and seven of them are single-sex, while all the schools at the bottom are mixed. If you were completely uncritical in your analysis you might conclude that independent schools are better. This is much too simple an explanation, and you can also see that all but one of the 13 independent schools at the top are selective, and the one that is not is very small. This means that pupils are selected into the independent schools, most likely on the basis of some measure of achievement at an earlier age. Consequently, the raw comparison of the schools on the basis of the figures presented in Table 1.2 is flawed as like is not being compared with like. It is clear that the performance of a school must be related to the prior ability of the pupils, and schools which select pupils for entry on the basis of this prior ability would be expected to have higher results at the end.

A relationship between final outcome and prior ability is clearly seen in the plot in Figure 1.5 where the average A-level points scores of A-level candidates in the school is

Table 1.2 School league table for Surrey and Sussex

Name	Type	Entry policy	Gender	Total % pupils	GCSE	A-level points
Dunottar Day for Girls, Reigate	IND	S	G	410	100	20.9
The Cornerstone, Epsom	IND	N	M	49	100	
City of London Freemen's School, Ashtead	IND	S	M	753	98	24.3
Greenacre School for Girls, Banstead	IND	S	G	388	97	21.5
Manor House, Leatherhead	IND	S	G	329	95	
Woldingham School, Caterham	IND	S	G	554	95	22.8
Reigate Grammar	IND	S	M	810	93	23.9
Notre Dame, Lingfield	IND	S	M	520	90	22.5
St John's, Leatherhead	IND	S	B	413	90	20.7
Caterham School	IND	S	M	692	87	22.3
Epsom College	IND	S	M	662	85	23.8
Worth School	IND	S	B	118	78	18.8
Ewell Castle, Epsom	IND	S	B	445	73	14.4
The Ashcombe, Dorking	C	C	M	1371	73	20.2
St Bede's, Redhill	VA	C	M	1460	71	20.2
Oxted County	C	C	M	1938	70	17.2
St Teresa's, Dorking	IND	S	G	325	68	21.2
Rosebery, Epsom	GM	C	G	1233	67	18.9
Imberhorne	C	C	M	269	67	15.2
Howard of Effingham, Leatherhead	C	C	M	1539	66	22.3
Sackville	C	C	M	280	66	17.0
Therfield, Leatherhead	C	C	M	1344	64	18.0
Glyn ADT Technology, Epsom	GM	C	B	1245	63	14.3
The Priory CofE (Aided), Dorking	VA	C	M	554	58	14.2
Warlingham School	C	C	M	1456	58	14.5
Reigate School	C	C	M	751	55	
Hazelwick	C	C	M	316	55	16.4
The Holy Trinity C of E	VA	C	M	163	54	18.3
Oakwood School, Horley	C	C	M	1294	53	
St Wilfrid's RC	VA	C	M	110	50	14.9
The Warwick, Redhill	C	C	M	784	47	
The Beacon, Banstead	GM	C	M	1226	45	12.5
St Andrew's RC, Leatherhead	VA	C	M	615	44	17.5
Epsom and Ewell High	GM	C	M	1129	36	13.3
Box Hill, Dorking	IND	S	M	249	33	9.4
de Stafford College, Caterham	GM	C	M	643	31	8.6
The Royal Alexandra and Albert, Reigate	VA	C	M	504	31	
Thomas Bennett Community College	C	C	M	189	25	8.5
Ifield Community College	C	C	M	146	24	12.1
Hurtwood House, Dorking	IND	N	M	289	22	20.3
Reigate College	FE		M	1109		15.3
North East Surrey College	FE		M	1391		11.4
East Surrey College, Redhill	FE		M	1516		9.4
Crawley College	FE		M	1768		6.6

Type: IND, independent; C, county; VA, voluntary aided; GM, grant maintained; FE, sixth form college (further education). Entry policy: S, selective; C, comprehensive; N, non-selective independent. Gender: B, boys only; G, girls only; M, mixed. % GCSE: percentage of 15-year-olds who achieved at least five grades A* to C. A-level points: average points score per candidate (A = 10, B = 8, C = 6, D = 4, E = 2).

plotted against the percentage of 15-year-olds who obtain five or more GSCEs at grades A*–C. As the GCSE percentage increases the A-level points score also increases. The straight line on the graph represents fitting a statistical model, known as a linear regression model, to these data. This model describes the relationship between the average A-level points score and the GCSE percentage. Using the graph we can read off that a school with a GCSE percentage of 50 is predicted to have an average A-level points score of 15.4 points per candidate; for a school with a GCSE percentage of 70 the prediction is 18.5 points. The two schools highlighted are two which have a greater A-level points score per candidate than is predicted by the statistical model. You could argue that these schools are worth studying with a view to seeing if there is anything related to the practices in the school which would lead to their having a much better score than predicted. It could turn out to be something simple which is unimportant such as able pupils bypassing GCSE exams, or it could be something important to do with the way the A-level syllabus is taught.

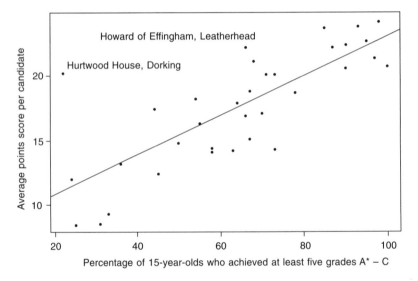

Figure 1.5 Relationship between A-level score and GCSE.

The comparison of schools is seen to be a very complex affair and one that requires much statistical analysis. The comparison of the raw figures is misleading as there are many features of schools and pupils which are not taken into account. The statistical modelling analysis tries to take into account the most important effects and relationships with a view to making sure that the comparisons are as equitable as possible. These are often known as adjusted comparisons, and the purpose of statistical modelling here is to make comparisons of like with like so that there is no bias involved in the comparison.

Chapter: *graphs, relationships, correlation, linear regression, estimation.*

1.6 Common features of statistical studies: what exactly is statistics?

Each of these examples has certain common features and these will be discussed throughout the book. One of the key features is that there is a **population** under study and that we have some information about this population. Generally the information comes from only a small subset of the population, known as a **sample** from the population. This is certainly true in the example of the opinion polls in Section 1.1, but was not the case in Section 1.2 where we had information on the whole population of supermarkets at that time. Although it is not obvious, there is sampling involved in the retail price index (Section 1.3) and other official statistics (Section 1.4), as most of these indices are based upon large government-sponsored surveys.

A key feature that we will come to is that we want to be able to say something about the population based upon the sample. This is the process of making an **inference** about the larger population based upon knowing something about the much smaller sample. In order for this to be valid you need the sample to be the same as the population, only smaller. Clearly this is impossible, and what we need is a sample that is **representative**. This is quite a hard concept to describe very rigorously, but what we need is that the sample should not be biased relative to the population.

Although it may not be obvious, inferences are based upon **statistical models**. These are mathematical descriptions of some properties of the data in the sample. In Section 1.5 there was a model which could be described as a straight-line model. The equation of a straight line involves two quantities known as the intercept and the slope. In statistical terminology these are known as **parameters** of the model as they are quantities whose values are unknown. From looking at Figure 1.5 you can see that all of the points in the graph lie roughly in a straight line, but the exact intercept and slope are not known. In a statistical analysis these unknown parameters are then estimated from the data, in this case from the GCSE percentages and the A-level points. So **estimation** of parameters of models from sample data is another key feature of a statistical analysis.

Once we have a model and we have estimated values for the unknown parameters of the model, then we want to use the model. There are two main uses. The first is in the **interpretation** of the model by describing the relationships, possibly in numerical terms. For example, in the boxplot in Figure 1.1 we can estimate that the turnover is £1000 per week less if there is a competitor to the supermarket in the neighbourhood. This will provide information to the managers about the likely effect of a competitor opening a supermarket close to one of their own.

The second main use is in making **predictions** from the model. We saw one type of prediction in connection with Figure 1.5 where we were looking at the predicted A-level score, based upon a statistical model, and comparing the observed score from schools to see which schools did better than they were predicted to do. In Section 1.2 we had the predicted turnover of a new supermarket in a new location to help in the decision-making process of the managers. What we see here is the use of statistical models as an aid to management decision-making processes. Thus the correct interpretation of the model is vitally important.

In order for a model's predictions and estimated parameters to be valid we need to be sure that the model is a good description of the data. Thus we have to test the fit of the model and to **test hypotheses** about the model. This has not been illustrated here, but in Table 1.1 we might want to test if there really has been a significant decline in the support for Labour over the period of the studies. This testing is also accomplished within the framework of statistical models. Hypotheses testing is an important aspect of statistical work, but it is not nearly as important as its use of textbooks and academic publications would suggest. Estimation of parameters and correct interpretation are much more important that these tests of effects.

Although statistics is inherently numerical and based on mathematical models, it is very firmly located within the process of scientific research. The conclusions of a statistical investigation are much more important than the numerical calculations. Without the conclusions the numerical summaries are of limited use. Also, if the calculations have errors, or are based on false assumptions, the conclusions are worthless. Part of the statistical process involves a critical consideration of the assumptions that you are making as part of the analysis and, if possible, assessing the impact of the assumptions on your conclusions.

Statistics is the science which deals with the collection, summary, interpretation and presentation of data. The data are often numerical but not always, though there is generally some numerical component such as counting the frequency of responses. As far as data collection is concerned, we are primarily interested in collecting appropriate information and making sure that the sample we have used is representative of the population that we wish to study. When summarizing the data, we are taking a large amount of information and summarizing the most relevant aspects of it in a much smaller quantity of summary statistics. Appropriate summaries lead to concise interpretation, which can be understood and interpreted by all. A good presentation of the results, often in a clear graph or table, will always assist the interpretation and utility of the statistical analysis.

Statistical models pervade the whole process, and they are in the background of all the work in this book. The correct interpretation of the data will depend upon the correct model being used. The models are normally probability models which provide information on the chances that certain observations are made or on the values which are expected to occur. With this framework you can see that statistics is concerned with making decisions in the face of uncertainty. In the exact sciences such as physics, within the limits of measurement error, exactly the same observations are made each time an experiment is carried out in exactly the same way. In the social sciences and business world there is not such good control of the experimental or survey process to ensure that the observations are highly reproducible. Statistical techniques are required in precisely the situations which are common in the social sciences and business – where surveys are carried out, where there is a great deal of variation from one individual in the population to another, where it is not possible to carry out a direct experiment and where it is necessary to make observations on disparate individuals.

1.7 Statistical techniques in business

Stray *et al.* (1994) carried out two surveys of students who had undertaken studies for a Master of Business Administration (MBA) degree. There were 420 respondents in all, students who had taken an MBA course and who were now working in business and management. The main reasons given for the benefits of using statistical techniques were that they

- improve problem understanding (60% of past students);
- reduce subjective judgement (60%);
- permit a deeper analysis of the problem (59%);
- enable justification of decisions (56%);
- give more confidence in the final decision (52%).

This survey paints a very rosy picture of the reasons why a knowledge of statistical techniques and analysis is vital for students of business and management. The most common techniques are

- data collection and presentation (98% of students aware, and 98% used the technique)
- probability analysis (95%, 67%)
- forecasting and prediction (92%, 70%)
- regression and correlation (92%, 61%)
- confidence intervals (81%, 45%)
- hypothesis tests (78%, 34%)
- survey design (74%, 47%)
- analysis of tables (60%, 32%)
- index numbers (56%, 34%)

To give such a high profile to statistical education for business managers on the basis of this one survey is good from my point of view as a statistician, but in the interests of fairness and of presenting an honest interpretation you should be aware of the limitations of the survey. It would be a good idea to try and read Stray *et al.*'s paper. Their survey was carried out with a self-completed questionnaire, and not all past students responded. This means that you have to worry about possible bias associated with the study. One likely bias is that those past students who were more interested in statistics would have completed the questionnaire and returned it, while those who were not interested in statistics and didn't use it during their work stuffed the questionnaire in the bin. Consequently, all of the percentages probably overestimate the use of statistics.

In the earlier sections of this chapter virtually all of these techniques have been illustrated and you can see from the percentages that their use by MBA students working in business and management is not inconsiderable, even allowing for some overestimation. Obviously managers are not going to be using these techniques all the time, but it is clear that in order to be an effective manager in business it is necessary to be aware of statistical ideas and to know when they are appropriate and when they are not. This, then, is the goal of the book.

2

Summarizing business surveys: populations and samples, variables and variability

- Populations and samples
- Representative samples and bias
- Different types of variables that can be recorded and their effect on subsequent analysis
- The differences between qualitative and quantitative measurements
- Summarizing a set of observations through measures of location and variation
- Understanding the differences between the sample mean and median
- Understanding the differences between the sample standard deviation and semi-inter quartile range
- Knowing when to use the different summary statistics and the effect of outliers on them
- Knowing how to interpret the summary measures and to compare them.

2.1 Populations and samples

Most statistical information comes from a sample of a population. The **population** is the name given to the collection of items about which we require information. Most business surveys are surveys of existing customers, potential customers, firms or individuals within firms. Generally the population will be a collection of individuals, but it need not be so. In some instances it may be a collection of households or a collection of firms. If a business survey is concerned with the quality of the manufactured items then the population will be the items produced at various stages of production.

A **sample** is just a subgroup or a subset of the population. A sample is selected because it is much quicker to analyse the data from a small group than from the whole population. Sometimes in quality control surveys the sampling and testing procedure can result in the destruction of the item and so a sample is necessary.

One aspect of statistical work is the process of summarizing the information in a sample and using this information to come to conclusions about the population. The key aspect of this work is that the sample is **representative** of the population. This idea is embodied in the notion that the sample mimics all the characteristics of the population so that the information you get from the sample is the same as the information that you would get from the whole population if it were studied.

2.1.1 Selecting samples

Achieving a representative sample is not a simple matter and great care is required in the selection of the sample from the population. Some aspects of this process are discussed in Chapter 5. Market research firms go to great lengths to try to obtain representative samples of the target population and they tend to use a method of sampling known as quota sampling. Essentially this entails giving the interviewer a list of the different types of people to interview – for example, two unmarried men aged 20–30, one unmarried man over 65, four married women aged 20–40 with children, one divorced man. The interviewer then selects individuals corresponding to these characteristics to fill up the quota. Contacting the individuals is generally left to the discretion of the interviewer and may include contacting people in a shopping centre or railway station, or knocking on doors in residential areas.

This is not the only way to achieve a representative sample, and statisticians tend to prefer the use of a random sampling mechanism. The key feature of this is that the items selected from the population are selected at random without any subjective choice. In many cases this ensures that any item in the population has a known chance of being selected. With a quota sample method some individuals, particularly those in minority groups, however you define them, are not going to be selected if they do not fall into one of the predefined quotas. Also the chance of any one individual being selected can depend on some preferences of the interviewer.

The selection of a random sample will not guarantee a representative sample but, provided non-response or refusal to participate in the survey is minimal, it should provide a sample which is free of any subjective bias. A random sample is selected by a mechanism such as drawing a number out of a hat or rolling a die, or any mechanism which depends upon chance. This means that the decision as to which items are included in the sample is controlled by a random mechanism and thus any subjective bias is eliminated. Quota samples are generally not free of this bias, though in good surveys by reliable market research firms attempts are made to eliminate it or, at least, minimize its effects.

The use of random samples does not completely eliminate the possibility that any particular sample has some unintended or unnoticed bias. It is always possible that a random sampling procedure will result in a biased sample. In large samples this is a remote possibility. The likelihood can also be reduced by considering carefully the experimental design and by using more sophisticated random sampling procedures. These are discussed in more detail in Chapters 5 and 10.

2.1.2 Business survey

We finish off this section with an illustration of a small business study planned by a firm of estate agents in a large town which also serves an area of countryside and villages adjacent to the town. The estate agents are undertaking a customer satisfaction survey. From a business perspective, they are looking for information from their existing and potential customers about the factors which are important in choosing estate agents and the factors which are important for the quality of service. This information should enable them to improve their service, obtain more customers and increase their profitability.

Figure 2.1 shows a plan of part of the area in which the estate agents operate, and the sampled households are marked with a black spot. This illustrates the process of statistical inference. The estate agents want to have reliable information about all of the households in their cachement area. Contacting everyone would be an expensive and time-consuming task. A sample is quicker and cheaper. The process of statistical inference uses the information in the selected sample to describe the whole population, even although only a part of the population is sampled. The basis for this is the belief that the sample is representative of the whole population. Consequently, the information obtained from the sample should be the same as the information which would be obtained from the whole population. It will not be exactly the same, as there is some error introduced by taking a sample. This is known as **sampling error**.

Sampling error is not the same as bias. Bias will arise if the sample is not representative. For example, a biased sample is one which includes too many households living in detached houses or large semi-detached ones and not enough living in terraced houses, see sample marked with crosses. By improving the design and conduct of the survey you would hope to eliminate bias. Sampling error will always be present, even if there is no bias in the sample whatsoever. Sampling error arises because of the nature of sampling. A second random sample is illustrated by open circles in Figure 2.1. The responses from

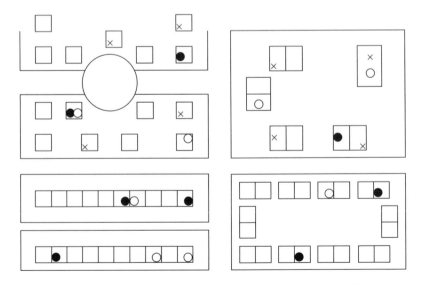

Figure 2.1 Map of part of the area of the estate agents' survey showing households sampled.

the different households are not going to be exactly the same and this is the source of sampling error. This error can be minimized by taking larger samples, but is always going to be present to some extent. **Statistical inference** is the process of making statements about a population based upon information in a sample, taking into account sampling error.

2.2 Types of variable

A possible questionnaire for use in the estate agents' survey is presented in Figure 2.2. The questions in this questionnaire are very similar to questions in many business surveys and opinion polls. We use them here to illustrate the different types of variables.

A **variable** is the general name given to a quantity measured or recorded in a study. This name is used to emphasize that these quantities may vary from one unit in the study to another. There are four main types of variables, in two groups, summarized in Table 2.1. The two groups are **qualitative** and **quantitative** variables. Within the qualitative group there are **nominal** and **ranked** (also known as **ordinal**) variables, and within the quantitative group there are **discrete** and **continuous** variables.

2.2.1 Qualitative variables

Questions 3, 4 and 4a are all examples of nominal variables. In fact questions 3 and 4 are special cases of this general term as they have only two possible values and so are sometimes referred to as **binary** variables. The key element here is that only the name or label of the answer is important. Question 4a will have the name of an estate agent as the response and this is clearly a nominal variable. Question 4b is also going to be a nominal variable but is not going to be an easy variable to code for a statistical analysis as the reasons given may be varied and overlapping, with many different values.

Most of the questions in the questionnaire are examples of ranked variables where the respondent is asked to give a grade to the response. In question 2 the responses are on a four-point graded scale of unlikeliness. Question 3 is recorded on a four-point scale of importance. Beside each response to question 3 is a number giving the grade of 1, 2, 3, 4. This is the way in which the data will be stored in the computer database before the statistical analysis, but note that the number only holds ordinal information in that 2 has a greater importance than 1, 3 greater than 1 and 2, and 4 greater than 1, 2 and 3. Although numbers are used, there is no more meaning to the numbers for this question beyond their order. Questions 6, 8 and 9 are also examples of ordinal variables.

2.2.2 Quantitative variables

The two parts to question 7 give rise to discrete counted responses of 0, 1, 2, . . . individuals. Questions 8 and 9 can also be thought of as discrete variables as they convey some quantitative information in the limits of the classes. In one sense all of the nominal

Northwest Dallas Realty Survey

1. How long have you lived in your current home? ____Years

2. How likely would you say you are to sell your home within the next one, two, and three-year time frames?

   Within one year. . . . . ☐ Not at all likely ☐ Somewhat unlikely ☐ Somewhat likely ☐ Very likely

   Within two years . . . ☐ Not at all likely ☐ Somewhat unlikely ☐ Somewhat likely ☐ Very likely

   Within three years. . . ☐ Not at all likely ☐ Somewhat unlikely ☐ Somewhat likely ☐ Very likely

3. How important would each of the following be to you in deciding which Realty firm to choose to sell your home? Please use any number from 1 (Not at all Important) to 4 (Extremely Important).

	Not at all important	Somewhat important	Very important	Extremely important
Realty firm has many of signs in the neighbourhood	☐1	☐2	☐3	☐4
Realtor's sales agent visited me at my home	☐1	☐2	☐3	☐4
Realty firm sold a neighbour's home	☐1	☐2	☐3	☐4
Positive recommendation of a friend, relative, or business associate	☐1	☐2	☐3	☐4
Realty firm has a national presence	☐1	☐2	☐3	☐4
Sales agent is knowledgeable about the neighbourhood	☐1	☐2	☐3	☐4
Sales agent has been in the business for a long time	☐1	☐2	☐3	☐4
Positive recommendation of a friend or relative	☐1	☐2	☐3	☐4
Realty firm has large ad in the yellow pages	☐1	☐2	☐3	☐4
Realty firm has friendly sales agents	☐1	☐2	☐3	☐4
Firm charges lower sales commissions than most firms	☐1	☐2	☐3	☐4

3a. Would you rather deal with a large Realty firm (one that has a lot of sales agents), or a small Realty firm (one where the owner is the only sales agent)?

   ☐ Prefer large firm   ☐ Prefer small firm

3b. Why?

_____

_____

_____

4. If you were going to sell your home, do you know which Realty firm you most likely would use?

   ☐ Yes (Continue with question 4a)   ☐ No (Skip to question 5)

4a. Which firm would you most likely use?—————————————————————————

4b. Why?—————————————————————————————————————

5. Listed below are a few Realtors who do business in your neighbourhood. Please indicate how familiar you are with each of them.

	Never heard of them	Heard of them, but never met anyone from the firm	Heard of them, and met at least one of their agents
Garden Realty	☐	☐	☐
Brown & Co.	☐	☐	☐
Prudential	☐	☐	☐
Better Homes	☐	☐	☐

6. Please rate each Realty firm indicated on the items below using the following scale:

   1 = Excellent
   2 = Good
   3 = Fair
   4 = Poor

If you have heard of a particular Realty firm but know nothing about it, rate it based on whatever impressions you have of it. If you have never heard of a Realty firm before, please leave its column blank.

	Garden realty	Brown & Co.	Prudential	Better Homes
Experience of sales agent(s)	―――	―――	―――	―――
Ability to sell homes quickly	―――	―――	―――	―――
Ability to obtain the best price for a seller	―――	―――	―――	―――
Knowledge of the neighbourhood	―――	―――	―――	―――
Courteousness of sales agent(s)	―――	―――	―――	―――
Commission charged is reasonable	―――	―――	―――	―――

7. Including yourself, how many people live in your household? _____ Adults _____ Children

Questions 8 through 10 are optional questions to help us better understand the Realty needs of people in your neighborhood. Please feel free to leave these items blank if they make you feel uncomfortable

8. What is your age?

☐ Under 21 ☐ 35–44 ☐ 55–64

☐ 21–34 ☐ 45–54 ☐ 65 or older

9. Into which of the following groups does your household's total annual income fall, before taxes?

☐ Under $25,000 ☐ $35,000–$49,999 ☐ $75,000–$99,999 ☐ $200,000 or more

☐ $25,000–$34,999 ☐ $50,000–$74,999 ☐ $100,000–$199,999

10. What is the approximate market value of your current home? $_____

Please return this questionnaire to The Business Research Lab using the enclosed postage-paid envelope. Thank you very much for taking the time to participate in this survey.

Figure 2.2 Estate agent survey questionnaire. (Taken from http://busreslab.com/survey-htm)

and ranked variables can be thought of as discrete (they can only take discrete values) but they are not quantitative and so are not considered to be discrete variables.

Questions 1 and 10 are continuous variables as they can take any values in a given range. In question 1 the range is restricted to be greater than 0 years, while in question 10 it is again going to be a positive value. Although the responses to question 1 are likely to be 1, 2, 3, 4, . . . years, the distinction between questions 1 and 7 is that responses of 1.5

Table 2.1 Types of variable

Type of variable		Definition	Examples
Qualitative	Nominal	A variable taking a set of distinct values with no natural order	Gender
	Ranked or ordinal	A variable taking a set of distinct values where there is a natural comparative order to the values	Grade of satisfaction, degree of preference
Quantitative	Discrete	A count of the number of times something occurs	The number of children in a family, the number of employees in a factory
	Continuous	A variable which can take any value within a given range	Height, weight, time

years and 0.5 years, for example, are possible whereas such responses are not possible for question 7. Although you may find that the responses to many questions are discrete in practice, when you are classifying the variables it is important to take into account the possible values, not just the ones recorded.

2.2.3 Classifying variables and measurement scales

It is important to be able to classify the variables in the studies. This is not always an easy task to accomplish and in the example there are occasions where the same variable can be thought of in more than one way. The statistical analysis that you perform depends upon the type of variable. The presentation of the data is also influenced by the type of variable. You will see in Chapter 3 that, for example, bar charts are appropriate displays for qualitative data but not for continuous data. Also, means and standard deviations, which are summary statistics of data, may be used with quantitative variables but are not appropriate for qualitative variables. Such situations are present in all statistical analysis, so it is important to be able to distinguish among the four types of variables.

If you consult other textbooks you may find that they use a slightly different classification of the variables. Some just make the distinction between qualitative, counted and measured variables, while others go into detail about the scales of measurement of the variables.

In terms of the scales of measurement on which the variables are recorded there is a clear order. Nominal variables are measured on a **nominal** scale where only the name is important and there is no concept of order or size or distance. Ordinal variables are recorded on an **ordinal** scale where there is a concept of size but not of distance. Thus with a variable recording the degree of satisfaction at three levels – unsatisfied, satisfied, very satisfied – we know that satisfied is better than unsatisfied but not how much better. Although the responses might be coded as 1, 2, 3, the fact that satisfied is recorded as 2 and unsatisfied as 1 does not mean that satisfied is twice as good as unsatisfied and very satisfied three times as good. Such a conclusion requires the observations to be measured on a **ratio** scale. Similarly, the fact that the difference between the codes for satisfied and unsatisfied is the same as the difference between the codes for very satisfied and satisfied does not mean that there is any information about the relative positions of unsatisfied, satisfied and very satisfied other than their order. If the distance between the codes meant something then this would imply that the variable was recorded on an **interval** scale.

Continuous variables can be measured on either a ratio scale or an interval scale. These scales all support the concepts of size and distance. A ratio scale has a fixed zero, for example, temperature in kelvin, height in centimetres, weight in kilograms, house prices in £, wages in £. With such scales it is correct to say that someone is twice as tall or twice as heavy. Discrete counts are also recorded on a ratio scale as there is a fixed zero. With an interval scale there is no fixed zero but distances can be measured. Examples include temperature in degrees Celsius or Fahrenheit, time of the day in minutes.

There is a nice discussion in Wright (1997) about the differences between ratio and interval scales. This paper shows that it is often not clear what scale a variable is recorded on. For many statistical procedures the choice of technique is not really influenced by the difference between a ratio scale and an interval scale. However, the interpretation of the data are. If a ratio scale is used then the comparison of £1 with £2 is exactly the same as the comparison of £100 with £200. If the cost is treated as an interval scale then the difference between £201 and £200 is exactly the same as that between £2 and £1.

In the next two sections we discuss how to summarize the information provided in a survey. We begin with quantitative variables, and in Section 2.4 we turn to qualitative variables. As quantitative variables hold numerical information we summarize them with numerical summaries of the main features of the data. Qualitative data are best summarized with tables.

2.3 Location and variability in quantitative variables

The two most important features of a set of responses to a quantitative variable are the average value and the variability in the responses. There are a number of measures of the average value, and the most common are the mean and the median. The average of a set of responses is also known as the **location** of the responses. There are also two common measures of **variability** in a set of responses, and they are the semi-interquartile range and the standard deviation. While these two sets of summary measures are measuring the same quantities, they do so in different ways and so are not interchangeable.

2.3.1 Location: mean and median

The **mean** of a sample of observations is the arithmetic average, and is obtained by adding up all the numbers and dividing by the number of observations. It is a relatively straightforward calculation and is done automatically on many calculators, spreadsheets and statistical programs.

The **median** is slightly simpler to obtain than the mean, but is not so readily available on calculators. In order to obtain the median you need to sort the observations into increasing order of magnitude. The median is the middle observation in this ordered list of numbers. If there are two numbers in the middle, which will happen if there are an even number of observations, then the median is the average of these two numbers.

The calculations for both measures are illustrated with a small example.

Example 2.1: Staying-on rates in Glasgow
The following data give the voluntary staying-on rates for S4 pupils in a sample of 30 schools in Glasgow. These data are based on the Scottish School Leavers Survey 1985. This is part of a series of surveys carried out by the Centre for Educational Sociology at Edinburgh University and funded by the Scottish Office Education Department in the 1980s (see http://www.ed.ac.uk/ces/projects/projectindex.htm#ssls; http://www.scotland.gov.uk/library3/education/ssls-01.asp; http://www.scre.ac.uk/rie/nl55/NL55StatsSurvey.html).

The staying-on rate is simply the ratio, expressed as a percentage, of the number of pupils who were old enough to leave school at the end of S4 but who decided to stay on into S5 divided by the total number of pupils in S4 who were old enough to leave school at the end of S4. In Scotland, there has to be a distinction between voluntary staying-on rates and staying-on rates in general, as a substantial portion of pupils in S4 are not old enough to leave school at the end of S4 as they are not yet 16, which is the minimum age for leaving school.

35	39	46	22	27	51	19	26	29	34	31	40	23	49	38
12	15	27	31	32	42	35	37	29	33	44	27	19	33	31

The total of the 30 staying-on rates is 956 and so the mean is

$$\text{mean} = \frac{956}{30} = 31.867\%.$$

Sorting the observations into increasing order of magnitude results in the following ordered list:

12	15	19	19	22	23	26	27	27	27	29	29	31	31	31
32	33	33	34	35	35	37	38	39	40	42	44	46	49	51

There are 30 numbers so there are two in the middle. The 15th largest number is 31 and the 15th smallest is 32. The median is the average of these two and so is equal to

$$\text{Median} = \frac{1}{2}(31 + 32) = 31.5\%.$$

If the set of observations are denoted $x_1, x_2, \ldots, x_n$, where there are n observations in the sample, then the sum of the observations is often represented by the following notation:

$$x_1 + x_2 + \ldots + x_n = \sum_{i=1}^{n} x_i.$$

This is just a concise way of writing the sum of the observations and is expressed as 'the sum of x_i from i equal to 1 until i equal to n'. There are other equivalent ways of writing this sum. These are used when there can be no confusion over the index of the sum (i) and the total (n) is also clear. These ways are to write the sum of the observations as

$$\sum x_i \quad \text{or} \quad \sum_{i}^{n} x_i;$$

sometimes you will even see the index omitted, as in $\sum x$. Using this notation gives a more concise formula for the sample mean, which is denoted $\bar{x}$:

$$\bar{x} = \frac{1}{n} \sum_{1}^{n} x_i.$$

The mean can be calculated to any number of decimal places, and in the example the calculation was presented to three decimal places, which is excessive as the data are only recorded to the nearest unit. Generally you should not quote the mean to an excessive amount of decimal places in any report. A good rule is to quote the mean to one more decimal place that the original data. In the example the data are rounded to the nearest unit so the mean should be presented to the nearest tenth of a unit, i.e. 31.9%. You do not have to take the same trouble with the median as it will be one of the original data points or an average of two of them. The median is often denoted by M, though this is by no means a universal notation.

2.3.2 Comparison of the mean and median

In Example 2.1 the numerical values of the mean and median are very similar to each other. However, they are not to be used interchangeably depending upon your whim. The two measures summarize the location of a set of observations in distinct ways. The median can be thought of as the typical value in the sense that 50% of the observations are above the median and 50% are below. Thus the median is the observation in the middle of the distribution.

The only information which goes into the calculation of the median is the information in the middle of the distribution. This is just one value or the average of two values. The mean, on the other hand, uses each observation equally and measures the location of a set of observations in the sense that it is the arithmetic average of all of them, treating all the observations equally.

There are situations when it is more appropriate to use the median than the mean and other situations in which the reverse is the case. There are also cases in which the numerical values of the two measures are much the same. In Table 2.2 we investigate the value of the median and mean using some small sets of data so that their main properties can be illustrated. The table illustrates some important points:

1. When the distribution of the observations is exactly symmetric then the mean and median are identical. If the distribution is approximately symmetric then the mean and median will have roughly the same value. This means that as the mean is based upon

Table 2.2 Comparison of mean and median

Data	Mean	Median	
2,3,3,4,4,4,5,5,6	4	4	This distribution is symmetric and the two measures are identical.
2,3,3,4,4,4,5,5,8	4.2	4	Changing an extreme observation by a little amount has no effect on the median, but there is a small change to the mean.
4,4,4,6	4.5	4	The median is normally one of the observations on the data set (except possibly when there are an even number of observations) while the mean need not be so.
4,4,4,8	5	4	One large observation has no influence on
4,4,4,12	6	4	the median but a huge influence on the mean.
4,4,4,24	9	4	
2,2,2,3,3,5,6,9	4	3	When the distribution is clustered and has a
2,2,2,3,5,6,19	5.25	3	long tail then the median takes a value in the cluster while the mean is outside it.

all of the observations and the median only on one or two then the mean is a better measure of location with symmetric distributions.

2. The median is not sensitive to any changes in the extremes, while the mean is more sensitive to individual values.

3. If there are extremely large (or extremely small) observations then they have an undue influence on the mean but not on the median. If there are extreme observations then the median is a more reliable measure of location.

4. If the distribution is skew such that there is a group of observations clustered together with a long tail of stragglers towards high (or low) values then the median better reflects the group of observations clustered together and the mean is influenced by the extreme values. In cases of skew distributions then the median is to be preferred.

The choice between the mean and the median is dictated by the shape of the distribution of the observations in the sample. If the distribution is symmetric then use the mean; if it is skew or has extreme observations distant from the rest of the data then use the median. The best way to see the shape of the distribution is by plotting it in a graph, and appropriate graphs are discussed in Chapter 3.

2.3.3 Variability and range

Variability is one of the fundamental concepts in statistics. Without variability there would be no statistics. Many people find it difficult to deal with the presence of variability. In a marketing survey you may find that in two families with similar educational background, incomes, housing expenditures and children's ages one spends £150 pounds per week on food and the other only £80. Obviously how much a family spends per week on food is the result of a fairly complex set of decisions, and it is no surprise that there is a great deal of variability when dealing with measurements on individuals or households.

Essentially the presence of variability means that you cannot make predictions which are 100% accurate. In a physical science experiment you may be able to make predictions which are highly accurate. If you know how much power a kettle element has and the initial temperature then you can predict accurately how long it will take to boil 1 litre of water starting at 15°C. Even then, the actual measurements will not be exactly the same as the prediction as a result of random variation which arises from factors outside the control of the experimenter.

Part of the scientific process involves the search for the causes of the variability in experimental or survey results, so you should not think of variation in your surveys as a nuisance. Rather variation is a natural phenomenon and one which, if explored, can lead to new discoveries. For example, investigation of the random variation in the colour of sweet peas lead Mendel to the discovery of genetics. Genetics is based on chance, which is measured by probabilities, and chance leads to variation in experimental results. In a marketing study an investigation into the factors which are associated with the decision of households to spend a large or a small amount of money on commodities can lead to better advertising targeting and marketing of the products.

In this sub section we will focus on the measurement of variability by a single figure. This is important because we can then compare variability in different groups. The essence of this comparison is to see if the variability is the same or different in the groups. If the marketing study shows that the variation in weekly expenditure of food for owner-occupier households with two children is £80 per week while the variability for households with two children living in local authority rented accommodation is £50, then this means that predictions about the level of expenditure will be more precise in the latter case as the variability is lower. It also means there is more scope for looking for factors to explain the variability between households among the owner-occupier group as the variability is greater.

We will discover later how to control the effects of variability in surveys and experiments (see Chapters 5 and 11). One way is to have a good expimental procedure where as much as possible is standardized. If variability is low then you can have more confidence in your experimental results. Even then you will still require statistical techniques to help you to interpret the results of your experiments. There are corresponding techniques for use in surveys.

The simplest measure of variation is the sample **range**, which is

$$\text{range} = \text{maximum} - \text{minimum}.$$

In Example 2.1 the range is $51 - 12 = 39\%$. This is an obvious measure, but it is not used very often as it depends on the two most extreme values.

The two common measures of variability are the **standard deviation** and the **semi-interquartile range**. These are used in conjunction with the mean and median, respectively. The calculation of the semi-interquartile range is much easier than the calculation of the standard deviation. We will illustrate their calculation with an example.

2.3.4 Sample standard deviation

The sample standard deviation is the 'average distance of each observation in the sample from the sample mean'. This is a relatively straightforward concept but the formula for the calculation looks quite daunting. The n observations in the sample are denoted $x_1, x_2, \ldots, x_n$. The distance of the ith observation from the mean is $x_i - \bar{x}$, and we need to take the 'average' of these. The slight trouble is that some of these deviations will be positive, when x_i is greater than $\bar{x}$, and some will be negative, when x_i is less than $\bar{x}$, with the result that the sum, $\sum(x_i - \bar{x})$, will always be zero.

In order to overcome this and derive a suitable measure of variability it is necessary to make the deviations all positive. There are two ways of doing this; one is obvious and not used, while the other is more obscure and is used in the calculation of the standard deviation. The obvious way is to take the absolute values of the deviations, $|x_i - \bar{x}|$, and take the average of them. This results in a measure of variability known as the mean absolute deviation, which is not used very much as its statistical properties are difficult to evaluate.

The more obscure way is to square the deviations, $(x_i - \bar{x})^2$, which makes them all positive, calculate the 'average' of the squared deviations and then take the square root.

This is the process of the calculation of the standard deviation. The 'average' of the squared deviations is known as the **sample variance** and is given by

$$s^2 = \frac{1}{n-1} \sum (x_i - \bar{x})^2.$$

The sample standard deviation is the square root of this and is

$$s = \sqrt{\frac{1}{n-1} \sum (x_i - \bar{x})^2} \,.$$

Example 2.2: Regional variations in average weekly earnings

Table 2.3 was obtained from the Office of National Statistics website (http://www.statistics.gov.uk). The data refer to the average gross weekly earnings as at April 1996 of full-time workers in 13 different regions of the United Kingdom. The overall figure for the UK is also shown.

Table 2.3 Average gross weekly earnings by region, April 1996 (£)

Region	Females	Males		
United Kingdom	282.3	389.9		
		x_i	$x_i - \bar{x}$	$(x_i - \bar{x})^2$
North East	252.4	347.7	−26.354	694.5
North West GOR*	262.4	369.0	−5.054	25.5
Merseyside	271.3	361.7	−12.354	152.6
Yorkshire and the Humber	252.5	350.7	−23.354	545.4
East Midlands	248.7	352.9	−21.154	447.5
West Midlands	256.9	360.1	−13.954	194.7
Eastern	279.9	382.3	8.246	68.0
London	364.9	514.3	140.246	19 668.9
South East GOR*	292.7	412.7	38.646	1 493.5
South West	261.1	364.8	−9.254	85.6
Wales	250.5	345.5	−28.554	815.3
Scotland	262.0	363.6	−10.454	109.3
Northern Ireland	256.9	337.4	−36.654	1 343.5
Sum		4 862.7		25 644.5

*Government office region. For statistical comparisons and administrative purposes the United Kingdom was divided up into 13 regions. Before 1998 they were known as standard statistical regions, but with the reorganization of local authorities in 1998 some changes were introduced. As a result there are some slight differences between the standard statistical regions and the government office regions.

The mean gross weekly earnings among full-time male workers over the 13 regions are

$$\bar{x} = \frac{\sum x_i}{n} = \frac{4862.7}{13} = £374.054.$$

Calculation of the sample variance by the direct way of forming the differences, $x_i - \bar{x}$, squaring them, $(x_i - \bar{x})^2$, and then summing gives

$$s^2 = \frac{1}{n-1} \sum (x_i - \bar{x})^2 = \frac{25644.5}{12} = 2137.04,$$

$$s = \sqrt{2137.04} = £46.2282.$$

The information would be summarized by saying that the mean gross average weekly earnings among males in full-time employment over the 13 regions is £374 with a standard deviation of £46.

If you compare the mean of £374 to the figure in Table 2.3 for the United Kingdom of £390 you will notice a large discrepancy. In the calculation of the mean and standard deviation we have treated each of the 13 regions equally. This is fine when we want to look at regional differences on a regional level, but you can argue that the regions are not all of the same size and this information should be taken into account. There is a much greater population in the London and South East regions than there is in the North East and Wales, and as the former regions have the highest weekly wages the UK average figure is greater than the average over the regions calculated assuming each region is treated equally. The way to take into account the different populations in the different regions when calculating the mean is to calculate a weighted mean, and this will be illustrated in Chapter 4.

The reason for the use of 'average' in quotes is that the divisor used to get the average is $n - 1$, and not n. Consequently it is not a real average, although in all but the smallest samples there will be little numerical effect. The reason for the divisor of $n - 1$ is to ensure that the sample variance, s^2, is unbiased relative to the population variance. The exact meaning of unbiased in this context will be discussed in Chapter 10. Essentially it means that there is no tendency for the sample variance to be consistently too large or too small relative to the population variance.

A second, related reason, is that there are only $n - 1$ bits of free information left in the sample once the sample mean has been estimated. A sample of size n contains n pieces of information and the estimation of the mean uses up one of them, leaving $n - 1$. This is related to a term, the degrees of freedom, which will appear at numerous points throughout the book (see Chapters 11–13 and 15).

2.3.5 Semi-interquartile range

The semi-interquartile range is a based upon the **quartiles** of the distribution. The median is the observation which splits the data into two equal parts, and the quartiles split the data into four equal parts. The median is thus the second quartile. The semi-interquartile range is half of the distance between the first and third quartiles.

The median, denoted by M, is calculated by sorting the observations in order of increasing magnitude and taking the observation in the middle. The other two quartiles are known as the lower quartile (Q_L) and the upper quartile (Q_U). The simplest way to calculate them is as the median of the lower and upper halves of the data, respectively. The lower quartile is the median of the lower half of the data and the upper quartile is the median of the upper half of the data.

If there is an odd number of observations then the median is the single observation in the middle. There is then the question as to which half of the data this observation goes into: the lower half or the upper half? As it is not clear, the simplest solution is not to include it in either half. This is the convention adopted here.

Example 2.3: Regional variation in average gross weekly earnings

Taking the data for males in Table 2.3 and rearranging in order of magnitude, we calculate the quartiles as shown in Figure 2.3. There are 13 observations so the median is the seventh largest, which is £361.7. As there are an odd number of observations the middle value is not included in either the top half or the bottom half of the data. There are six observations in the bottom half and the lower quartile is the median of them, i.e. the average of the third and fourth largest. The upper quartile is the average of the two observations in the middle of the top half.

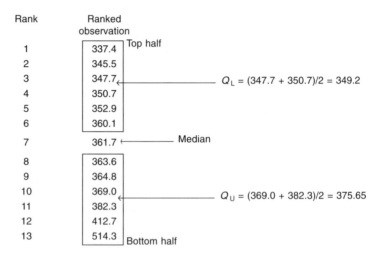

Figure 2.3 Quartiles of male average gross weekly earnings.

The convention for the calculation of the quartiles used here is simple and commonly used. It is not, however, universally used nor is it completely correct. It is, however, simple and unambiguous. You will find that there are slight differences in the ways that other textbooks describe the calculation of the upper and lower quartiles. Different computer programs also adopt different procedures. If the sample is a reasonable size then there is little numerical difference among the different procedures.

The most correct method of calculating the quartiles is to use the following interpolation. If there are n observations then the median is observation $(n + 1)/2$ in order. If there are 9 observations then the median is the 5th largest, if there are 10 observations then it is the 5.5th largest, i.e. the average of the 5th and 6th largest observations. Following this process through to the lower and upper quartiles gives us the lower quartile as observation $(n + 1)/4$ in order and the upper quartile as observation $3(n + 1)/4$ in order. If there are 10 observations then the lower quartile is observation 2.75 and is between the 2nd and 3rd observation three-quarters of the way up, i.e.

$$Q_L = x_{(2)} + \frac{3}{4}(x_{(3)} - x_{(2)}).$$

The upper quartile is at position 8.25 and is a quarter of the way between observations 8 and 9, i.e.

$$Q_U = x_{(8)} + \frac{1}{4}(x_{(9)} - x_{(8)}).$$

This is a slightly more complex process but is the one used in statistical programs such as Minitab.

The semi-interquartile range (*SIR*) is half of the distance from the lower quartile to the upper quartile:

$$SIR = \frac{1}{2}(Q_U - Q_L).$$

This can also be written as

$$SIR = \frac{1}{2}[(Q_U - M) + (M - Q_L)].$$

This shows that *SIR* can be interpreted as the average distance of the quartiles to the median. In fact, M could be replaced by any observation in the range (Q_L, Q_U) in the above and so *SIR* is a more general average distance between the quartiles and any point between them.

Example 2.4: Calculation of the semi-interquartile range

From Example 2.3, we have

$$M = 361.7,\ Q_L = 349.2,\ Q_U = 375.65,$$

$$SIR = \frac{1}{2}(Q_U - Q_L) = \frac{1}{2}(375.65 - 349.2) = 13.225.$$

Thus the median average gross weekly earnings for males over regions in the United Kingdom is £361.7 with a semi-interquartile range of £13.2.

2.3.6 Comparing the standard deviation and the semi-interquartile range

In situations where the mean is used to measure the location the standard deviation should be used to measure the variability. When the median is the most appropriate measure of location then you should use the semi-interquartile range to measure the variability. For symmetric distributions with no outliers use the mean and standard deviation; for skew distributions or for distributions with outliers use the median and semi-interquartile range.

If there is an outlier among the observations then the effect of this will be to inflate the standard deviation but leave the semi-interquartile range unchanged, as in the following examples:

Row	Data	Mean	Median	Standard deviation	Semi-interquartile range
1	3,4,4,5,5,5,6,6,7	5	5	1.225	1
2	3,4,4,5,5,5,6,6,8	5.11	5	1.41	1
3	3,4,4,5,5,5,6,6,10	5.33	5	2.00	1
4	2,4,4,5,5,5,6,6,8	5	5	1.69	1
5	0,4,4,5,5,5,6,6,10	5	5	2.60	1

In rows 1–3 we see that as the largest observation is made progressively larger, and so tends to become an outlier with respect to the rest of the observations, the mean and the standard deviation increase but the median and semi-interquartile range remain unchanged. Looking at rows 1, 4 and 5, we see that as the largest and smallest observations are moved away from the centre in a symmetric fashion the mean and median are equal and unchanged, but the standard deviation increases while the semi-interquartile range remains constant.

The semi-interquartile range only depends upon the middle 50% of the observations, while the standard deviation depends upon all of the observations in the data. The standard deviation is based upon all observations treated equally, while the semi-interquartile range places no weight on the top 25% or on the bottom 25%. Thus the standard deviation is sensitive to outliers or skew distributions while the semi-interquartile range is not. The standard deviation is better when the distribution is symmetric with no outliers as all observations play a part in its calculation.

As a rule, the standard deviation should not be quoted to more than two significant figures more than the original observations, i.e. one more than the mean. If the original observations are in units then it is acceptable to quote the standard deviation to two decimal places. You may also quote it to no or 1 decimal place, but to quote three or four or more would be excessive.

2.3.7 Interpretation of measures of variability

The interpretation of a standard deviation is one of the most difficult things in elementary statistics. Perhaps the best way to begin is to take measurements of the same quantity in two different groups, A and B. If the standard deviation in one group is 10 and in the other it is 20 then we can say that the second group is twice as variable as the first. (Standard deviations are bounded below by zero, i.e. they have a fixed zero and so are measured on a ratio scale.) So standard deviations are used to compare variabilities.

The harder task is in the interpretation of a single standard deviation. Roughly speaking, the standard deviation is the average distance of each observation from the mean, but that is a rather abstract idea to use.

Others try to interpret the standard deviation in terms of the range of observations. If we have a set of data which is symmetric, with no outliers and most of the observations in the middle of the distribution of the data (see Figure 3.2a), then about 95% of all observations are expected to lie within the range mean plus or minus 2 standard deviations, 68% within plus or minus one standard deviation and 99.7% within plus or minus 3 standard deviations. This actually is related to the normal distribution, which will be discussed in Section 9.4.

The semi-interquartile range is the average distance of the two quartiles from the median and so is related to the middle 50% of the distribution. The quartile range from Q_L to Q_U is the range of the middle 50% of the observations. Again two semi-interquartile ranges can easily be compared by looking at their ratio. With a symmetric distribution with a single peak we would expect that the semi-interquartile range to be approximately equal to two-thirds of the standard deviation.

When summarizing a set of observations the measure of location should always be accompanied by its appropriate measure of variability. It is common to summarize a set of observations by giving the mean and standard deviation. This convention can also be used for the median and semi-interquartile range, but this is less common.

Example 2.5: Comparison of male and female earnings

The data in Example 2.2 can be summarized as follows.

	Mean	Standard deviation	Median	Semi-interquartile range
Males	374.1	46.2	361.7	13.2
Females	270.2	31.1	261.1	11.6

It is not very surprising that the wages for males are on average higher than for females. The means are larger than the medians, and this tells us that the data are skewed towards high values. Secondly, if the data were symmetric then we would expect the semi-interquartile range to be about two-thirds of the standard deviation. This is not the case, both semi-interquartile ranges being much smaller than two-thirds of the standard deviations. This also suggests that the distributions are not symmetric. Going back to the original data in Example 2.2, we can see that the average gross weekly earnings in London are much larger than in the other regions. London is an outlier and we know the effect of such values on the mean and standard deviation. Thus the median and semi-interquartile range are the most suitable measures here.

Our main interest with these data was an investigation of regional variation. Comparing the standard deviations, we see that the male–female ratio is 46.2/31.1 = 1.49, while for the semi-interquartile ranges it is 13.2/11.6 = 1.14. As the semi-interquartile range is the most relevant, we see that there is slightly more regional variation in average weekly earnings among male full-time workers than among females, about 14% more.

Comparison of the variations in groups of observations tells you about the variability. In groups where the variability is higher you are less sure about the averages quoted as there is greater variability. With reported results of a mean of 20 and a standard deviation of 2 then you would conclude that 95% of the observations are expected to lie within the range 16 to 24. If the mean were 20 and the standard deviation 5 then the range would be 10 to 30 for 95% of observations. Similarly with the median. The coefficient of variation is a dimensionless measure of variability which is usually expressed as a percentage. It is given by

$$CV = \frac{s}{\bar{x}} \times 100.$$

The higher the coefficient of variation the greater the relative variability. If the mean is 20 and the standard deviation 2 then $CV = 10\%$. If the mean is 50 and the standard deviation 5 then $CV = 10\%$ and the relative variability is identical. If, however, the mean is 50 but the standard deviation is 3 then $CV = 6\%$ and so the relative variability is less. The coefficient of variation is not used a great deal, but it is a useful summary for comparing relative variabilities when the means are quite different in different subgroups of observations. It is possible to calculate a coefficient of variation based upon the median and semi-interquartile range. This is

$$CV = \frac{SIR}{M} \times 100,$$

and would be used in similar circumstances to the coefficient of variation based upon the mean and standard deviation.

In Example 2.5 the coefficients of variation for males and females based upon the median and semi-interquartile range are 3.6% and 4.4%, respectively. This means that regional variation is small relative to the level of average earnings. However, taking into account the median wage over the regions, there is more relative regional variation in wages among female full-time workers than among males. The reversal compared to the comparison of the semi-interquartile ranges occurs because there is a relatively modest difference in semi-interquartile ranges, that over regions for males being only 14% greater than that for females, but there is a big difference in medians, the median weekly wage among females being 28% smaller than that for males.

This small example brings out a number of important points about a statistical analysis:

1. It is very easy to manipulate data in such a way that it can appear to give contradictory information. Often this occurs because the presentations are not completely equivalent. Using the standard deviations, we would claim that there was 43% more regional variation among males than females; using the semi-interquartile range, it is only 14%; and using the coefficients of variation based on medians, there is 22% more variation among females. We have already discussed that the standard deviations are not appropriate here because of the outliers. The comparison of the semi-interquartile ranges is a comparison of absolute variability, whereas the comparison of the coefficients of variation is a comparison of relative variability. These are not the same thing, so it is no surprise that they give different results. Both are valid provided you are clear in using the correct interpretation.
2. You should always be very careful about what you have calculated and the implications of these calculations. Here we have treated each region in the same way as any other region. This can be criticized as this places as much weight on smaller regions as it does on larger regions.
3. You should state very clearly what calculations you have done and your reasons for doing them.

The final point is that the decisions about the most appropriate summary statistics depend upon the shape of the distribution. This means that, in practice, you should always look at the shape of the distribution by using appropriate graphical techniques (Chapter 3) before going on to summarize the distribution using single-number summaries.

2.4 Qualitative variables – proportions

Qualitative variables are summarized by a table of proportions or percentages. While there are the concepts of location and variability for such variables, there are no single-number summaries for the location and variability. Generally the location is summarized by giving the percentages in the most common groups. The single most common value of the variable is known as the **mode**, and this is occasionally used. It is impossible to measure the variability adequately and all that can be done is to present the values of the variable and the percentages of observations. This is known as the **frequency distribution**.

Example 2.6: Employment categories over time

The data in Table 2.4 were published by the Office for National Statistics (http://www.statistics.gov.uk). Most of the data were collected from the Labour Force Survey (http://www.statistics.gov.uk/nsbase/themes/labour_market/surveys/labour_force_text.asp). The numbers are in millions and are unadjusted totals. With many government surveys there is some adjustment of the raw figures from the surveys to counteract the effects of non-response. At other times it is necessary to adjust the figures for seasonal effects, as employment is traditionally better in the summer than in the winter.

Table 2.4 Male and female employment

	Millions		Percentages	
	1984	1997	1984	1997
Men				
in employment	14.1	14.7	66.8	65.9
ILO unemployed	1.9	1.3	9.0	5.8
economically inactive	5.1	6.3	24.2	28.3
all aged 16 and over	21.1	22.3	100.0	100.0
Women				
in employment	9.9	12.0	43.4	50.8
ILO unemployed	1.3	0.7	5.7	3.0
economically inactive	11.5	10.9	50.4	46.2
all aged 16 and over	22.8	23.6	100.0	100.0

The International Labour Office (ILO) measure of unemployment refers to people without a job who were available to start work in the two weeks following their interview and had either looked for work in the four weeks before interview or were waiting to start a job they had already obtained. Employment status is the qualitative variable here and it has three possible values: in employment, ILO unemployed, and economically inactive. The economically inactive category will include people who are at home looking after children as well as sick people, retired people and students.

We shall consider the distribution over the three categories in men and women separately to compare 1984 with 1997. The data are in millions of individuals, but we need to convert them into percentages to make a valid comparison as the total number of individuals aged 16 or over changes over these two periods.

We see that 66.8% of men aged 16 and over were economically active in 1984, and this dropped slightly to 65.9% in 1997; the percentage who were unemployed dropped from 9.0% to 5.8%; and the percentage who were economically inactive rose from 24.2% to 28.3%. There were more dramatic changes among women. The percentage of women aged 16 or over in employment rose from 43.4% to 50.8%. Unemployment fell from 5.7% to 3.0% and the percentage economically inactive fell from 50.4% to 46.2%.

With these percentages you can see the changing pattern of employment in the United Kingdom over the 13-year period from 1984 to 1997. This can be characterized by an increase in the employment of women.

There are two ways to compare percentages. We can say that there has been an increase of 7.4 percentage points in the percentage of women aged 16 or over in employment, from 43.4% to 50.8%. An alternative way is to state that there has been a 17.1% increase in the percentage of women in employment. This is calculated as 7.4/43.4, expressed as a percentage. Both are correct, but they represent different comparisons, and it is vital that you are clear about which one is calculated and presented. The percentage change in the percentages will always appear greater than the absolute points change in the percentages.

Changing the base is a common way of manipulating percentages. On the surface the unemployment rates as a percentage of all men over 16 are 5.8% in 1997. However, if you believe that economically inactive people should not be considered then an unemployment rate can be calculated as $1.3/(22.3 - 6.3) = 8.1\%$. This is the percentage of economically active men over 16 who are not in employment. The corresponding figure for women is $0.7/(23.6 - 10.9) = 5.5\%$.

Thus you can see how the unemployment figures can be influenced by definitions. A way of reducing the figure is by classifying as many people who are unemployed as economically inactive. To increase the figure you want as many economically inactive people to be considered unemployed. Over the period 1980 to 1995 there were a large number of changes in the definition of who was considered to be unemployed, and this is one of the reasons for the use of the ILO definition in the table as it has remained fairly static. A good source of information on government statistics, particularly with regard to health is Kerrison and Macfarlane (2000).

3

Graphs for investigating distributions and relationships

- Knowing how to construct and interpret the various graphs
- Understanding the differences between the different types of graph and when each is appropriate
- Developing experience in providing an interpretation of the distribution through the use of appropriate graphs
- Understanding the importance of distributional shape
- Knowing how to find outliers in a distribution
- Beginning to appreciate the association between two variables and to know the difference between associations involving two quantitative variables, two qualitative variables, and a quantitative and a qualitative variable
- Understanding that you can be manipulative when constructing a graph and consequently that there are good and bad ways of presenting information.

3.1 Introduction

The most effective way to present information is by means of a visual display. Graphs are frequently used in statistical analyses both as a means of uncovering patterns in a set of data and as a means of conveying the important information from a survey in a concise and accurate fashion.

Although the raw summaries from market research surveys are generally presented in tabular form, it is extremely common to present the main summary information graphically in order to heighten its visual impact. Many publications from the Office of National Statistics make extensive use of graphs and tables; for example, the *Social Trends* series is a fascinating collection of information about people, lifestyles and society in Britain which make heavy use of good informative graphical presentations.

In this chapter we will look at a number of techniques which can be used for presenting data in a graph. For the most part such graphs will be used to summarize numerical data measured on an interval or ratio scale. They will also be appropriate for counted data where there are a large number of different values. Some of the graphs will be appropriate for nominal or ordinal data. One of the key points made throughout this chapter is that not all graphs are appropriate for all types of data.

In practise all of these graphs are readily drawn using computer programs. Statistical programs tend to produce graphs which are more likely to be correct than those output by non-statistical programs. Spreadsheets are one of the most useful types of program for constructing graphs, but the user needs to be particularly careful that the type of graph and the data used in the graph are correct. There are also special purpose graphics packages for drawing graphs. When using computer programs it is necessary to be very careful that the graph accurately reflects the data, otherwise it will not be possible to interpret the graph correctly.

3.2 Histogram

3.2.1 Frequency distribution

The most important graph for looking at the distribution of a numerical variable is the **histogram**. This is a pictorial representation of the frequency distribution of a variable; an example is presented in Figure 3.1. This graph is a histogram of the A-level points scores for the data on Schools in Surrey and Sussex which are presented in Chapter 1. You can see at a glance that the range of scores is from about 6 to 26 points and that there are two classes with a frequency of 7 schools, 14–16 points and 20–22 points. It is easy to construct histograms like this with a computer program without paying too much attention to the details of the construction. However, in order to understand the good and not so good properties of histograms it is necessary to look in a little detail at this construction. As the histogram is based upon the frequency distribution we will investigate this first.

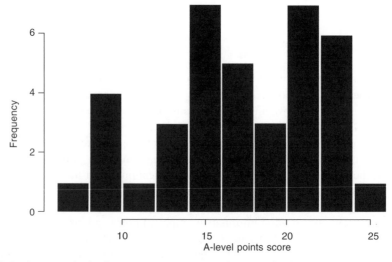

Figure 3.1 Histogram of A-level points scores in Surrey and Sussex schools.

Numerical data, such as those in Table 1.2 may be summarized by grouping them into class intervals. This is a step in the data reduction process as we take the 38 average A-level points scores for the individual schools and group them into a smaller number of distinct classes. A **class interval** is a range of values of the variable and a **frequency distribution** is the set of all class intervals spanning the range of the variable together with their associated frequencies. The **frequency** of the class is simply the number of observations in the class.

A frequency distribution for the A-level points score is shown in Table 3.1. This has nine class intervals each of width 2 points. The **relative frequency** is simply the frequency in each group divided by the total number of observations. The histogram in Figure 3.1 is based upon this frequency distribution, though the relative frequency could have been used equally well. Relative frequencies are normally used when constructing histograms to compare the distribution of a variable over two or more groups where the total numbers of observations are not the same in the different groups. This histogram is not the only one that could have been constructed, and we could have used different class widths or different starting points. All of these would have been valid frequency distributions, so you should note that the frequency distribution is not unique and how you summarize the data depends upon choices that you make. We will investigate the effect of these choices on the histogram a little later.

Table 3.1 Frequency distribution of the A-level points scores

Class interval	Frequency	Relative frequency
6 < Points ≤ 8	1	0.026
8 < Points ≤ 10	4	0.105
10 < Points ≤ 12	1	0.026
12 < Points ≤ 14	3	0.079
14 < Points ≤ 16	7	0.184
16 < Points ≤ 18	5	0.132
18 < Points ≤ 20	3	0.079
20 < Points ≤ 22	7	0.184
22 < Points ≤ 24	6	0.158
24 < Points ≤ 26	1	0.026
Total	38	1.000

The main feature of a frequency distribution is that it is made up of a set of non-overlapping class intervals, which span the range of the observed data. Generally these intervals all have the same width as in Table 3.1, though this is not necessary. There are a number of quantities to be defined about class intervals in a common-sense fashion. In Table 3.1 the first class contains all values with a points score greater than 6 but less than or equal to 8 and the class boundaries are 6.0 for the lower one and 8.0 for the upper boundary. The **class width** is the difference between the upper boundary and the lower one, while the **class mid-point** is the average of the two boundaries. Thus the class mid-point is 7 points and the width is 2 points.

It is possible to be quite pedantic about the definition of these quantities and take into account the rounding of the digits. This would mean that a score of 6 points actually lies somewhere between 5.95 and 6.05, and a score of 8.0 represents a score between 7.95 and 8.05. Then the class boundaries would be 6.05 and 8.05, again with a width of 2.0 points but with a mid-point of 7.05. These differences, while they may be important in some circumstances, are not likely to bother us greatly.

3.2.2 Interpretation of a histogram

The class widths are important in drawing a histogram as the **frequency of a class is represented by the area** in the histogram. Thus, in Figure 3.1, the area of the box representing the class interval 6–8 is proportional to the frequency. As the classes in Figure 3.1 are all of the same width, both the height and the area are proportional to the frequency. If the classes are all of exactly the same width then you do not need to bother about the difference between area and height and can concentrate on height alone. You only need to bother about the difference between area representing frequency and height representing frequency when some of the classes are of a different width to the rest.

There are a number of features of a distribution which you should look for as an aid to interpreting the distribution:

- The location of the middle of the data. If the distribution has one peak and is symmetric then the location is given by the peak.
- The range of the data, defined by the smallest and largest observations.
- The shape of the distribution. There are two basic shapes which are common and important. The first is what is called symmetrical with a single peak, and is illustrated in the graph in Figure 3.2a. The second are known as skew distributions, and the other three plots in Figure 3.2 are all examples of this shape. The skew can be positive if the long tail is towards high values (Figure 3.2b), or negative if the long tail is towards low values. If the skew is extreme such that the peak of the distribution is at one of the ends, then the distribution may be described as being 'J-shaped' (Figure 3.2c) or 'reverse J-shaped', depending on the direction of the tail.
- The number of obvious peaks in the data. Figures 3.2a–c have just the one peak. In Figure 3.2d there is evidence of two peaks, one of which is more pronounced than the other. A distribution with two peaks is known as bimodal, and is often indicative of two subgroups of individuals within the same population.
- The presence of outlying points, separate from the rest of the data, at the extremes. This is not illustrated in Figure 3.2 but is present in Figure 3.5 and other plots in this chapter.

In Figure 3.1 we would report that the middle of the distribution was about 16–18 points and that the range was from 6 to 26 points. There are two peaks, one at 14–16 points and the other at 20–22 points. There are no obvious outlying points at the extremes. The shape of the distribution is harder to describe because it has two peaks and does not correspond to one of the more common shapes. We might describe it as slightly skew with a longer tail towards lower values.

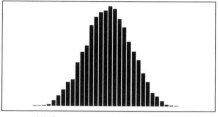

(a) Symmetric with a single peak

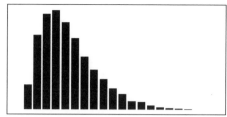

(b) Skew with a long tail to high values
(positive skew)

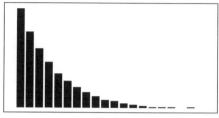

(c) Skew with a long tail to high values
(J. shaped)

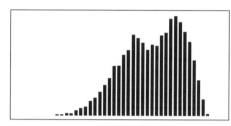

(d) Two peaks and skew with a long tail to
low values

Figure 3.2 Examples of distributional shapes.

3.2.3 Effect of class limits on histograms of small sets of data

In the illustration there are only 38 schools, and this is really quite a small number for the effective use of a histogram. With a small number of observations the interpretation of the histogram can be influenced by the choices we make for the class intervals and their widths. In Figure 3.3 the effect of using classes of differing width is illustrated. When the class interval is small, the histogram is very jagged with much detail; when the width is large, there are few intervals and there is no detail. Clearly the most useful class width is going to lie between these two extremes, and the usual choice is to have somewhere between 10 and 20 intervals. This obviously depends upon the number of observations you have, and the more observations you have the more intervals you can usefully use. Most computer programs have an option for computing the number of classes automatically.

While it is obvious that the class interval width and the number of classes will have an influence on the interpretation of the histogram, it is not immediately apparent that the decision about where to start the intervals may also have an impact. In Figure 3.4 all the classes have the same width of 5 points and the only difference is in the starting points of the intervals. You can see that quite different shapes are produced as a result of this choice. The peak changes and the general shape is affected. When the limits are 5, 10, 15, 20 the shape is clearly skew with a long tail towards low values and a peak at 20–25, while with 3, 8, 13, 18, 23, 28 the shape is more symmetrical and the peak is at 18–23. Part of the problem here is that there are only 38 observations, but the example illustrates that in small samples the choice of the class widths and intervals makes a difference to your interpretation.

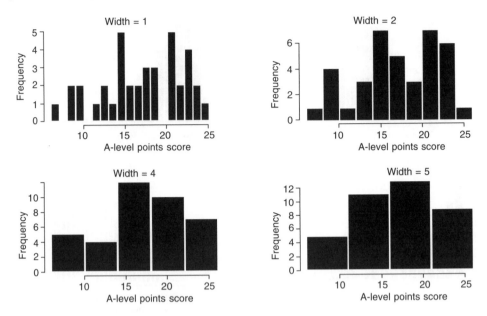

Figure 3.3 Effect of different class widths on the histogram.

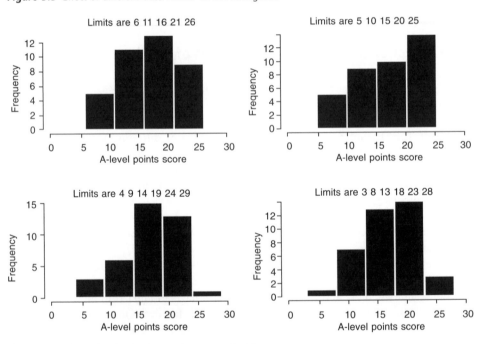

Figure 3.4 Effect of different class midpoints on the histogram.

3.2.4 Histograms with classes of unequal width

When the class intervals are of different widths then it is vital that you take this into account when constructing histograms. This is accomplished through the calculation of the density, which is the frequency per unit width. The correct calculations and the problems associated with incorrect drawing of the histogram are illustrated with data on pre-tax income for all taxpayers in the United Kingdom in 1995/96 (Table 3.2).

Table 3.2 Distribution of pre-tax income among taxpayers in the UK, 1995/96

Class interval, £	Frequency, 1000s	Lower boundary, £	Class width, £	Density per £1000
3 525–3 999	547	3 525	475	1 152
4 000–4 499	582	4 000	500	1 164
4 500–4 999	714	4 500	500	1 428
5 000–5 499	885	5 000	500	1 770
5 500–5 999	861	5 500	500	1 722
6 000–6 999	1 610	6 000	1 000	1 610
7 000–7 999	1 730	7 000	1 000	1 730
8 000–9 999	3 040	8 000	2 000	1 520
10 000–11 999	2 720	10 000	2 000	1 360
12 000–14 999	3 470	12 000	3 000	1 157
15 000–19 999	3 930	15 000	5 000	786
20 000–29 999	3 790	20 000	10 000	379
30 000–49 999	1 360	30 000	20 000	68
50 000–99 999	455	50 000	50 000	9.1
100 000 or over	126	100 000	900 000	0.14

Source: Office for National Statistics (1998b, Table 15.2).

There are 15 class intervals, but the last one has no upper limit. It is known as an 'open-ended interval' and poses some difficulties for drawing the histogram. One way to cope with it is by assuming that there is an upper limit, and I have chosen £1 000 000. Another way is to omit it from the graph; this is a reasonable thing to do here as this group contains a very small percentage of individuals, less than 0.5% of all taxpayers. Whatever you do will involve some choice, and the key thing is to try to make this choice as realistic as possible.

Among the 14 intervals with a known boundary the class widths range from £3525 to £50 000, and if we do not take this information into account when forming the histogram we get the plot Figure 3.5a. Here each class is plotted as having the same width and the height of the box is proportional to the frequency. This histogram erroneously suggests that the distribution of income is skew with a long tail towards low values and that most people have incomes at the upper end of the range with a mode between £15 000 and £20 000. In Figure 3.5b we take into account the class widths but again have the frequency proportional to the height of the box and not the area. This graph is now skew with a long tail towards high incomes but still with an appreciable number of individuals earning more than £30 000.

The correct histogram is Figure 3.5c, where the area of the box represents the frequency. This is a very skew distribution with most individuals at the lower income levels and a long tail towards the very high incomes. The peak of the distribution is down at the £5000–£8000 level and not at the £15 000–£20 000 as in the two incorrect histograms.

In the correct histogram the value plotted on the vertical axis, and from which the area is derived, is the **density**, which is the frequency per unit width. It is calculated as

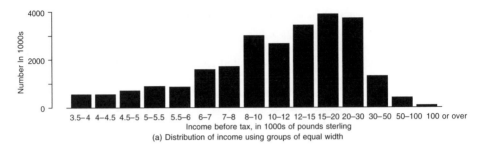

(a) Distribution of income using groups of equal width

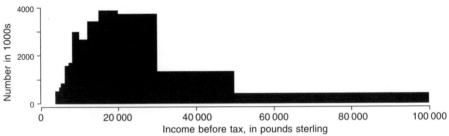

(b) Distribution of income using groups of unequal width but not correcting for the density

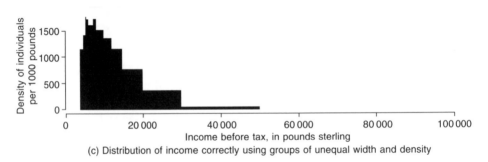

(c) Distribution of income correctly using groups of unequal width and density

Figure 3.5 Distribution of incomes – unequal classes.

$$\text{density} = \frac{\text{frequency}}{\text{class width}}.$$

This means that when you calculate the area of the box representing a class interval it is given by the product of the width and the density, which is the frequency, and so we have the area of the box representing the frequency which is a necessary property of the histogram. In Table 3.2 there are 547 000 taxpayers in the lowest income category of incomes, greater than or equal to £3525 but less than £4000. This has a width of £475, and the density is given by

$$\text{density} = \frac{547}{0.475} = 1152 \text{ thousand taxpayers per £1000}.$$

Thus the coordinates of the box representing the lowest class interval are (£3225, 0), (£3225, 1152), (£4000, 1152), and (£4000, 0).

This example shows that you have to be very careful when representing a frequency distribution with unequal class intervals as a histogram. This is an area where computer programs run into most difficulties when drawing histograms. You will find that you have to check carefully that the graph is correct. If you are starting off with the raw data, as we did in Figure 3.1, then it is normally a good idea to use classes of equal width. If you are using data which are already published and have unequal class widths, as in Table 3.2, then you have to be more careful when constructing the histogram.

3.3 Stem and leaf plot

3.3.1 Parts of a stem and leaf plot

In small samples a stem and leaf plot is a useful way of organizing the data into a graphical presentation. This is a relatively recent invention dating back to the mid-1970s with a book on exploratory data analysis by the American statistician, John Tukey (1977). This is a philosophy in statistics of looking closely at your data for interesting patterns which may help in your analysis. The stem and leaf plot serves much the same purpose as the histogram, and has some advantages and some disadvantages.

The capital–asset ratios of the world's 30 largest banks in 1980 are presented in Table 3.3, together with the country in which each bank is based. This ratio is a rough measure, used in management accounting, of the cash value of the bank compared to the value of the property it owns. The purpose of the analysis is to investigate the variability in this ratio and to see if there is any association between the ratio and the country in which the bank is based.

Table 3.3 Capital–asset ratios of the world's 30 largest banks, 1980

3.6	US	3.7	US	5.8	Fr	1.4	Fr	1.1	Fr	1.6	Fr
5.4	UK	3.1	WG	3.5	Ja	3.6	US	3.8	Ja	3.6	Ja
3.6	Ja	2.8	WG	4.0	Ja	5.3	UK	3.0	US	0.4	Ja
3.2	WG	5.8	SA	6.9	SA	3.3	Ca	3.5	Ne	2.8	WG
2.7	Ja	4.4	US	3.2	It	3.2	Ja	7.0	UK	3.6	Ja

US, United States; Fr, France; UK, United Kingdom; Ja, Japan; WG, West Germany; Ca, Canada; SA, South America; It, Italy; Ne, Netherlands

Sources: Moody's (1983); Lackman (1987).

Each number in Table 3.3 has two significant digits, and a stem and leaf plot simply separates these out into the individual digits. The leading digit, before the decimal point, is put into the stem of the plot, and the trailing digit, after the decimal point, goes into what are known as the leaves of the plot. A stem and leaf plot where each class is of width 0.5 units is presented in Figure 3.6.

There are three distinct parts to the stem and leaf plot. The first column, beginning 1, 1, 3, gives what are known as the **depths**. The second column, beginning 0, 0, 1, 1, 2, contains the **stems** of the plot, while the rest of the plot comprises the **leaves**. From the stems and leaves you can reconstruct the original raw data in Table 3.3. There is an observation with a stem of 0 and a leaf of 4, corresponding to 0.4; the next are at a stem

Leaf Unit = 0.10

1	0	4
1	0	
3	1	14
4	1	6
4	2	
7	2	788
13	3	012223
(9)	3	556666678
8	4	04
6	4	
6	5	34
4	5	88
2	6	
2	6	9
1	7	0

Figure 3.6 Stem and leaf plot of capital–asset ratios.

of 1 with leaves of 1 and 4, corresponding to 1.1 and 1.4 respectively; the next again at a stem of 1 with a leaf of 6, corresponding to 1.6. We can carry on in this fashion, reconstructing the original data from the plot right up until the largest observation of 7.0.

The column of depths gives the cumulative numbers of observations from the two extremes into the centre of the distribution. Thus in Figure 3.6 we have one observation less than or equal to 0.4, three observations of 1.4 or less, four less than or equal to 1.6, and seven less than or equal to 2.8. Starting from the maximum, we have one observation with a value of 7.0 or more, two with ratios of 6.9 or more, four with ratios of 5.8 and above, six of 6.3 and over and eight greater than 4.0. There are 30 observations in total, and this means that there are 15 in each half of the data. The remaining group, 3.5–3.9, is not assigned to the top half or the lower half of the data and so is left on its own. The number of observations in this group is written down in brackets to signify that it is a frequency, and not a cumulative frequency, and that the middle of the distribution is in this class interval.

The points to look for in the interpretation of a stem and leaf plot are the same points as for the histogram. In fact, you can view a stem and leaf plot as a histogram with classes of equal width where frequency is measured from left to right rather than from bottom to top. The same points concerning range, location, shape, subgroups and outliers are important.

The interpretation that we would give to Figure 3.6 is that the range of capital–asset ratios is from 0.4 to 7.0. The location of the main peak is at 3.5–3.9 and the middle value (median) of the distribution is $(3.5 + 3.6)/2 = 3.55$. The distribution is reasonably symmetric about the middle, but there is evidence of outliers at a low value of 0.4 and at high values of 6.9 and 7.0.

Relative to the interpretation of the histogram, the stem and leaf plot provides extra information. Firstly, we know the exact range, rather than the limits of the class boundaries. Secondly, we can find the middle value of the distribution with ease. Finally, the exact data values are preserved and so you can see if there is any pattern to them.

3.3.2 Class width and stem and leaf plots

Stem and leaf plots will always have a limited choice for the width of the class intervals compared to the possibilities that are open in histograms. In stem and leaf plots you are only allowed to have widths of 1, 2 or 5 (or 10, 20, 50; or 0.1, 0.2, 0.5; etc.). Stem and leaf plots where the intervals are of width 1 unit, 2 units and 0.2 units are presented in Figure 3.7. As for histograms, it is best to have about 10–20 class intervals, depending on

```
Leaf Unit = 0.10
   1        0     4
   4        1     146
   7        2     788
  (15)      3     012223556666678
   8        4     04
   6        5     3488
   2        6     9
   1        7     0

Leaf Unit = 1.0
   4        0     0111
  (18)      0     222333333333333333
   8        0     445555
   2        0     67

Leaf Unit = 0.10
   1        0     4
   1        0
   1        0
   2        1     1
   2        1
   3        1     4
   4        1     6
   4        1
   4        2
   4        2
   4        2
   5        2     7
   7        2     88
   9        3     01
  13        3     2223
  15        3     55
  15        3     666667
   9        3     8
   8        4     0
   7        4
   7        4     4
   6        4
   6        4
   6        5
   6        5     3
   5        5     4
   4        5
   4        5     88
   2        6
   2        6
   2        6
   2        6
   2        6     9
   1        7     0
```

Figure 3.7 Stem and leaf plots with different class widths.

the number of observations. The best two stem and leaf plots for the data in Table 3.3 are the one in Figure 3.6 or the first one in Figure 3.7, where the class width is 1 unit. The outliers are not so apparent in the latter, but otherwise the interpretation is much the same. The other two stem and leaf plots in Figure 3.7 are not so useful. One, where the class width is 2 units, has too few intervals (4), which makes it difficult to retain the exact data and to see the outliers, while the other has too many intervals (34), each of width 0.2 units. There is not much summary and all that you can see is a set of straggling points, apparently with subgroups everywhere.

3.3.3 Stem and leaf plot of a skew distribution

As a final example of the use of stem and leaf plots, we interpret the distribution of the distances that students lived from the University of Strathclyde in 1993. These data came from a sample of first-year students, and the data were collected in a questionnaire in response to a question: 'How far do you travel each day to get to the university?'. A stem and leaf plot of these data is presented in Figure 3.8. Their distribution is evidently skew. The stem and leaf plot is slightly more informative than a histogram would be, as the original data are retained.

```
Leaf Unit = 1.0
  2      0     01
  9      0     2233333
 23      0     44444445555555
 38      0     666666677777777
(10)     0     8888889999
 32      1     000000011
 23      1     2233
 19      1     4455555
 12      1     67
 10      1
 10      2     00
  8      2     22
  6      2
  6      2     6
  5      2     8
  4      3     0
  3      3     3
  2      3
  2      3     6
  1      3     9
```

Figure 3.8 Distances from university.

From the stem and leaf plot it is easy to see that

- the range is from 0 to 39 miles;
- the peak is at 6–7 miles, which was the distance away for 15/80 = 18% of the students;
- the middle value (median) is (8+8)/2 = 8 miles;
- the distribution is skew to the left, with a long tail to the right (high values);
- 57/80 = 71% of students live less than 12 miles away, that is, within Glasgow.

You will also notice that, relative to the number of observations at 10 and 15 miles, there are fewer observations at 9, 11, 14 and 16 miles. This is in accord with what you might expect, in that distances are rounded to the nearest 5 units once they start getting larger.

3.4 Boxplot

3.4.1 Illustration of a boxplot

Stem and leaf plots and histograms are good displays for investigating the distribution of a variable and for looking at important features of it. They are less good for comparing the distribution for different subgroups of a population. There are techniques for doing so by drawing two stem and leaf plots, back to back, but you need to have roughly the same number of observations in each sample for a fair comparison. With larger samples you can draw histograms of the relative frequency distributions and align their scales so that the distributions can be compared.

A stem and leaf plot is a rearrangement of the original data into a graphical display which retains as much of the information in the original data as possible. The histogram is a graphical display based upon a summary of the original data into class intervals and their frequency. The boxplot is a graphical display based upon an even greater summary of the original data. It is based upon only five values, known as the 'five-number summary' of a distribution. These five values are the median, the upper and lower quartiles and the maximum and minimum observations, and all have already been described in Chapter 2.

The boxplot for the capital–asset ratio data of Table 3.3 is presented in Figure 3.9. Within the box there is a line representing the median value. The box is drawn from the lower quartile, Q_L, to the upper quartile, Q_U, representing the distribution of the middle 50% of the data. The lines from the box towards the maximum and minimum observations

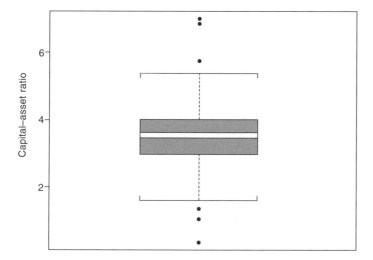

Figure 3.9 Boxplot of capital–asset ratios.

represent the spread of the observations in the tails of the distribution. These lines do not reach the maximum and minimum values but stop short, and three outliers are drawn as separate points with high ratios of 5.8, 6.9 and 7.0, while three points are also noted as outliers with low ratios of 1.4, 1.1 and 0.4.

3.4.2 Outliers and boxplots

One good feature of the boxplot is that the outliers are noted (see Figure 3.10). A point is labelled as an **outlier** if it is far away from the rest of the data. The criterion for deciding if a point is an outlier is to calculate the following limits, known as **fences**. First calculate the interquartile range $(Q_U - Q_L)$. Then the **inner fences** are located at

$$Q_L - \frac{3}{2}(Q_U - Q_L) \quad \text{and} \quad Q_U + \frac{3}{2}(Q_U - Q_L),$$

while the **outer fences** are calculated as

$$Q_L - 3(Q_U - Q_L) \quad \text{and} \quad Q_U + 3(Q_U - Q_L).$$

Points which are more extreme than the outer fences are labelled as **probable outliers**, while points lying between the inner and outer fences are known as **possible outliers**. In the capital–asset ratio data the median is $M = 3.55$, the lower quartile $Q_L = 3.0$ and the upper quartile $Q_U = 4.0$. The interquartile range is thus $4.0 - 3.0 = 1.0$ and the inner fences are at

$$3.0 - \frac{3}{2} \times 1.0 = 1.5 \quad \text{and} \quad 4.0 + \frac{3}{2} \times 1.0 = 5.5,$$

with the outer fences at

$$3.0 - 3 \times 1.0 = 0.0 \quad \text{and} \quad 4.0 + 3 \times 1.0 = 7.0.$$

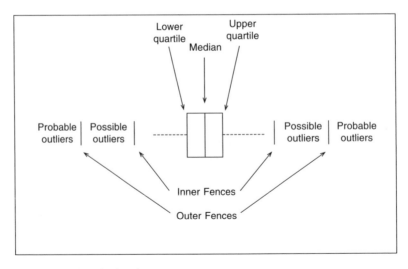

Figure 3.10 Components of a boxplot.

There are no points more extreme than the outer fences, though one has the same value (7.0), and there are outliers at six points between the inner and outer fences. In fact there are seven observations at these six points as there are two observations at 5.8.

As with many graphical procedures, one will normally use a computer to draw the graph. These are not normally available in spreadsheets and are usually only to be found in specialized statistical programs. While all computer programs drawing boxplots will give one which is of the same general shape as that in Figure 3.9 there may be some minor differences in the exact definition of the quartiles, the exact formula used to determine the outliers, the shape of the box and the extension of the dotted lines from the box towards the extremes. You may also find that the boxplot is drawn horizontally as in Figure 3.10, rather than vertically as in Figure 3.9. Generally the lines will extend as far as the largest observation less than or equal to the upper inner fence and the smallest observation greater than or equal to the lower inner fence.

3.4.3 Interpretation of boxplots

In the interpretation of boxplots the main features that you try to interpret are:

- the location of the middle value;
- variability, through the total range or the range from the lower to the upper quartile;
- the shape of the distribution, in terms of how symmetric or skew it is;
- outliers.

The interpretation that would be given to Figure 3.9 is that the capital–asset ratio has a median of 3.55, with a range from 0.4 to 7.0. The middle 50% of the data range from 3.0 to 4.0. The distribution is reasonably symmetric, with outliers at both ends. Going back to the original data in Table 3.3, we can see that the outliers with small ratios are a US bank and two of the four French ones, with a third French bank having a ratio of 1.6. The outliers with large ratios correspond to two South American banks, a UK one and the fourth French one.

The main use of boxplots is in the comparison of a number of groups of data. They are particularly useful for comparing locations and variability of distributions in a number of subgroups. They will also help to differentiate skew distributions from symmetric ones. Boxplots of the distributions in Figure 3.2 are shown in Figure 3.11. Figure 3.11a illustrates a symmetric distribution, Figure 3.11b a skew distribution with an outlier at the maximum, and Figure 3.11c an extremely skew distribution (J-shaped), again with an outlier at the maximum. From Figure 3.11d you would conclude that the distribution was slightly skew with a long tail towards low values; however, from the corresponding histogram in Figure 3.2d, you know that the distribution has two peaks. The peaks of a distribution cannot be represented in a boxplot, and this is one of the problems with boxplots.

3.4.4 Comparing boxplots

Among all the graphical techniques, boxplots are the most useful for comparing distributions. Among the schools in Surrey and Sussex there are 13 county schools, 4 further education

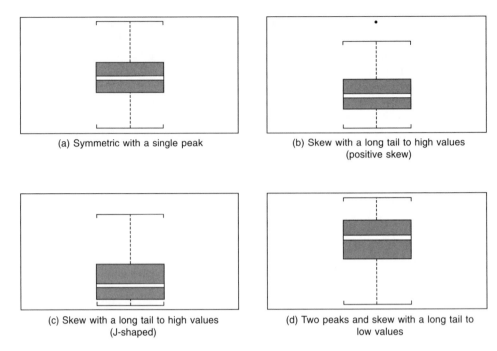

(a) Symmetric with a single peak   (b) Skew with a long tail to high values
(positive skew)

(c) Skew with a long tail to high values   (d) Two peaks and skew with a long tail to
(J-shaped)   low values

Figure 3.11 Boxplots of different shapes.

(sixth-form) colleges, 5 grant maintained, 16 independent and 6 voluntary aided. Even
with such small numbers of schools per group, the boxplots of the A-level points scores
in Figure 3.12 show some interesting differences. Firstly, all the distributions are reasonably
symmetric. Secondly, the independent schools tend to have the higher scores and the

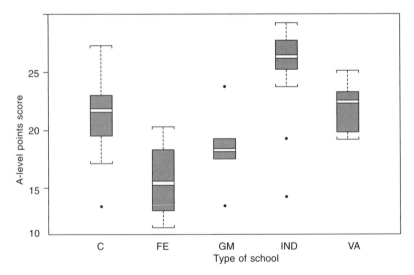

Figure 3.12 Boxplots of the A-level points scores. County (C), Further Education College (FE), Grant Maintained
(GM), Independent (IND), Voluntary Aided (VA).

sixth-form colleges the lower. There are a couple of outliers with relatively low scores among the independent schools and one low outlier among the county schools.

3.5 Cumulative distribution function plot

The median and quartiles which are used in the formation of the boxplot split the distribution of a numerical variable into four equal parts. The quantities which split the data into three equal parts are known as the tertiles, and those which split the data into 5 equal parts are known as the quintiles. The general names for these quantities are the **quantiles** or **percentiles**. One way to obtain the quantiles or percentiles is through the use of the **cumulative distribution function** plot. There are two versions of this plot, one based upon the original data and the other based upon a grouped frequency distribution.

3.5.1 Original data plot

The cumulative distribution function plot of the A-level points scores for the Surrey and Sussex schools in 1998 is shown in Figure 3.13. This graph is calculated by sorting the observations into order of increasing magnitude and calculating the cumulative relative frequencies (Table 3.4). There are 38 schools with an A-level points score, and the smallest value is 6.6. Thus there is one observation less than or equal to 6.6 and the cumulative relative frequency is $1/38 = 0.0263$. The next smallest observation is 8.5, and there are two observations less than or equal to 8.5, giving a cumulative relative frequency of $2/38 = 0.0526$.

The cumulative distribution function plot is just a plot of the sorted observations against the cumulative relative frequencies. Thus 6.6 is plotted against 0.0263, 8.5 against 0.0526, 8.6 against 0.0790, and so on. In order to emphasize the discrete nature of the

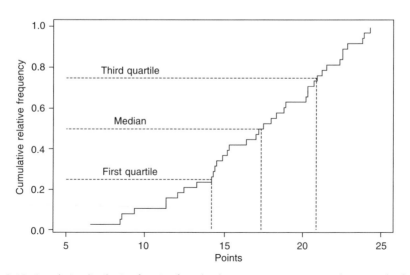

Figure 3.13 Cumulative distribution function for A-level points scores in Surrey and Sussex schools.

Table 3.4 Cumulative relative frequencies for A-level points scores

Observation	Frequency	Cumulative relative frequency	Observation	Frequency	Cumulative relative frequency
6.6	1	0.02632	17.5	1	0.52632
8.5	1	0.05263	18.0	1	0.55263
8.6	1	0.07895	18.3	1	0.57895
9.4	2	0.13158	18.8	1	0.60526
11.4	1	0.15789	18.9	1	0.63158
12.1	1	0.18421	20.2	2	0.68421
12.5	1	0.21053	20.3	1	0.71053
13.3	1	0.23684	20.7	1	0.73684
14.2	1	0.26316	20.9	1	0.76316
14.3	1	0.28947	21.2	1	0.78947
14.4	1	0.31579	21.5	1	0.81579
14.5	1	0.34211	22.3	2	0.86842
14.9	1	0.36842	22.5	1	0.89474
15.2	1	0.39474	22.8	1	0.92105
15.3	1	0.42105	23.8	1	0.94737
16.4	1	0.44737	23.9	1	0.97368
17.0	1	0.47368	24.3	1	1.00000
17.2	1	0.50000			

calculations the points are plotted as a step function. This means that the graph changes in discrete steps: the value of the vertical axis on the plot stays at 0.0263, its value for the first observation, right up until the second observation (8.5), whereupon it jumps up to 0.0526. It remains at this value until the third observation (8.6) where it jumps up to 0.0790, and so on.

You can use the cumulative distribution function graph to read off the percentiles of the distribution, and the median and quartiles are drawn on Figure 3.13. Reading these off gives a median of 17.35, a lower quartile of 14.2 and an upper quartile of 20.9.

3.5.2 Grouped data plot or ogive

If the data are grouped into a frequency distribution then a slightly different method is used to draw the cumulative frequency distribution. A grouped frequency distribution of the A-level scores is presented in Table 3.5, along with the upper class boundaries and the cumulative relative frequencies. The plot of the cumulative frequency distribution is shown in Figure 3.14. This is a plot of the cumulative relative frequency against the upper class boundary, but note that this time the points are joined together with straight lines rather than forming a step function. In this context the cumulative frequency plot is sometimes called an **ogive**. If all we have is the grouped frequency distribution we do not know where in the class intervals the points lie so we have to assume them to be equally spread throughout the class, and thus straight lines are used to form a continuous curve rather than a step function.

Using Figure 3.14 to read off values for the median and quartiles gives 14.2 for the lower quartile, 17.25 for the median and 21.3 for the upper quartile. These are not exactly

Table 3.5 Grouped frequency distribution of the A-level points scores

Class interval	Frequency	Cumulative relative frequency	Upper class boundary
6 < Points ≤ 8	1	0.02632	8.05
8 < Points ≤ 10	4	0.13158	10.05
10 < Points ≤ 12	1	0.15789	12.05
12 < Points ≤ 14	3	0.23684	14.05
14 < Points ≤ 16	7	0.42105	16.05
16 < Points ≤ 18	5	0.55263	18.05
18 < Points ≤ 20	3	0.63158	20.05
20 < Points ≤ 22	7	0.81579	22.05
22 < Points ≤ 24	6	0.97368	24.05
24 < Points ≤ 26	1	1.00000	26.05

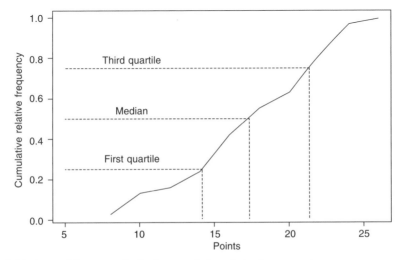

Figure 3.14 Grouped frequency distribution cumulative relative frequency plot.

the same as obtained from Figure 3.13, and you should not expect them to be identical as different representations of the same data are being used. This illustrates one of the features of estimation which is a key idea in statistics. If you have the same data but summarize it in different ways then you will generally obtain different values for the estimates. The differences may not be very great but they can still be quite confusing. In this instance the estimates based upon Figure 3.13 are probably the better ones to use as they are based on a set of data which has not previously been summarized into class intervals as has Figure 3.14. If the sample were much larger, with in excess of 100 observations, then you would expect the differences between the two approaches to be much smaller.

3.6 Bar chart and pie chart

All of the graphical techniques discussed so far have been of use for quantitative data. We now turn our attention to graphical displays for qualitative data such as arise in tables. The two common displays are the **pie chart** and the **bar chart**. Figure 3.15 presents different ways of representing graphically the proportions of different types of school appearing in the league table shown in Table 1.2.

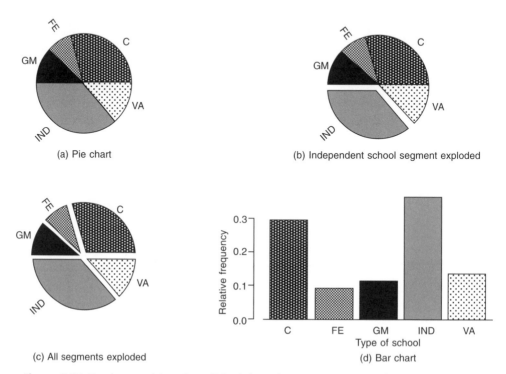

Figure 3.15 Pie charts and bar chart. (IND: independent; C: county; VA: voluntary aid; GM: grant maintained; FE: sixth form college (further education))

3.6.1 Pie chart

In a pie chart the various categories of interest together make up a complete circle or pie. Each individual category is represented by a slice of the pie in accordance with its frequency or relative frequency. Each slice forms an angle at the centre of the circle, and the size of each angle is in proportion to the frequency or relative frequency. Thus, for example, of the 44 schools in Table 1.2, 13 are county schools, so the angle corresponding to county schools is $(13/44) \times 360° = 106.4°$.

Pie charts and their variants seem to be the most frequently used graphical displays and three versions are shown in Figure 3.15. Figure 3.15a is the standard pie chart with no fancy bits. Figure 3.15b is an exploded pie chart, where one segment is separated from the rest in order to draw attention to it, while in Figure 3.15c all the segments are exploded. There are many versions and most of them seek to highlight part of the chart at the expense of the rest. This is a misuse of the pie chart as its prime purpose is to show each area in relation to the total. Thus in the standard pie chart it is easy to see that the independent schools are the biggest single sector.

3.6.2 Bar chart

The bar chart is quite similar to the histogram, but because the variable is qualitative there is no numerical information in the division into categories and so all the bars are of the same width. This means that the height of the bar is proportional to the frequency. In many respects the bar chart is easier to read than the pie charts in Figure 3.15 because you are always reading off the same scale. Thus the bar chart makes it easier to see that the county schools are 30% of the total. A bar chart is better if individual frequencies are of interest. It is also more flexible, as we will see, in that it can be used with more than just frequency data. Also comparisons of categories are easier with a bar chart as it is much easier to compare the lengths of two or more bars than it is to visually compare angles at different orientations.

The data in Table 3.6 illustrate the redistribution of household income through taxation in the United Kingdom in 1995/96. Incomes and other quantitative variables are often grouped into classes of equal frequency and the quintile groups are shown in the table, where 20% of the households are in each group. The table gives the average total original income of each household in the quintile group, then the average cash benefits that are added to give the average gross income for all households in the quintile group. The last row gives the average final income after tax for households in the quintile group. The values recorded in the table are not frequencies, but average amounts of money. A bar chart can be used to give an illustration of the data in this table, but a pie chart could not be used here as there are no frequencies.

Table 3.6 Redistribution of income through taxes and benefits, 1995/96

	Bottom fifth	Next fifth	Middle fifth	Next fifth	Top fifth
Total original income[1]	2 430	6 090	13 790	22 450	41 260
Benefits in cash[2]	4 910	4 660	3 360	2 130	1 190
Gross income[3]	7 340	10 750	17 150	24 580	42 450
Final income[4]	8 230	10 200	13 990	17 980	29 200

[1] Total household income from wages, salaries, occupational pensions, investments.
[2] Retirement pensions and income support.
[3] Sum of total original income and benefits in cash.
[4] After deduction of taxes (direct, indirect, national insurance and local) from and addition of benefits in kind (education, health service, housing and travel subsidies and school meals) to gross income.

Source: Office for National Statistics (1998a, Table 5.20).

In the bar charts in Figure 3.16 you can see the average original income rising over the household income quintile groups and the benefits falling. Gross income rises at a slower rate than total original income as a result of the cash benefits which are a major source of income for the lowest two quintile groups. The redistribution of income through tax and benefits should be evident from the smaller differences between the quintile groups on final income than on gross income. However, one thing which hampers the interpretation of these bar charts is that the scales for the incomes and benefits are not the same. This

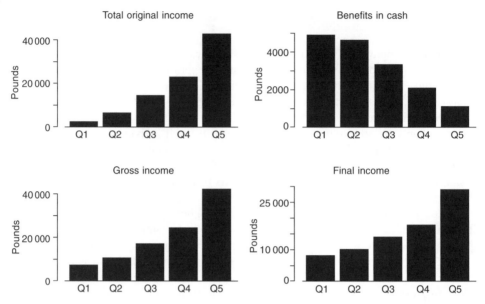

Figure 3.16 Bar charts of income.

is corrected in Figure 3.17, where a multiple bar chart is plotted on a single scale. Here you can more easily see that the differences among the quintiles are smaller for average final income than for total original income firstly as a result of cash benefits and secondly as a result of taxation and benefits in kind.

Multiple bar charts are an extremely useful way of presenting data where the main reason for the presentation is the comparison of the response in a number of groups. Here

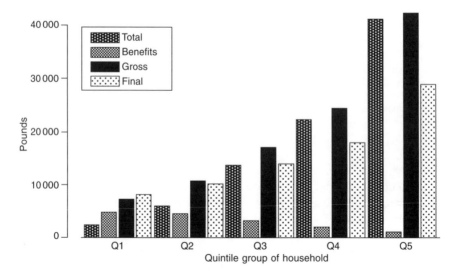

Figure 3.17 Multiple bar charts of income.

we are comparing incomes in the quintile groups. Stacked bar charts where the groups are stacked on top of each other, are useful graphs when the main interest is in showing the components of the total. In Figure 3.18 the gross income is plotted as a function of the quintile groups and each bar has two subsections, one representing the total original income and the other the cash benefits. In Figure 3.18a the vertical axis is the average income in the group, in £, while in Figure 3.18b it is the percentage of the gross income from the two averages. In the latter you can see that in the lowest quintile the average cash benefits make up about 70% of the average gross income, while in the highest quintile they are less than 5%. Figure 3.18a concentrates on the absolute incomes, while Figure 3.18b focuses on the relative contribution of the two sources.

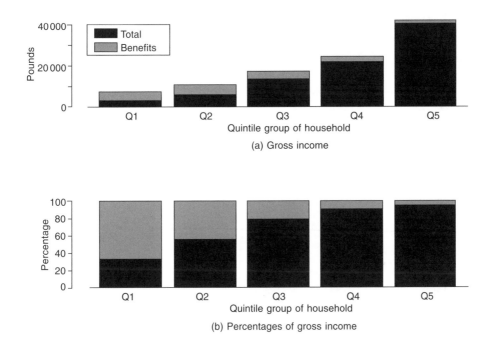

Figure 3.18 Stacked bar charts of income.

3.7 Scatter plot

3.7.1 Two quantitative variables

In Section 3.4 we saw that boxplots were very useful displays for investigating distributions in a number of different subgroups (see Figure 3.12). This type of plot is investigating the relationship between one quantitative variable and one qualitative variable. Relationships between two qualitative variables are often presented in tables, and bar charts can then be used for graphical display purposes (see Figure 3.17). In this section we will look at the most important graphical display used to investigate relationships between two quantitative variables. We have already seen examples of the scatter plot (Figure 1.5). A scatter plot or scatter diagram is a two-dimensional plot with one of the variables plotted along the horizontal axis and the other along the vertical axis.

The data in Table 3.7 were taken from the Fortune 500 website (http://www. fortune.com) on 26 November 1999. For the top 30 companies in the list, ranked according to their value, we have the profits expressed as a percentage of revenue and the earnings per share. There was no information for the three insurance companies on earnings per share. We wish to see if there is any relationship between the earnings per share and the percentage profits.

Table 3.7 Top 30 companies from Fortune 500, November 1999

Company	Profits/revenue (%)	Profits/assets (%)	Profits/equity (%)	Earnings per share	Percentage return
General Motors	1.8	1.1	19.7	4.18	21.4
Ford	15.3	9.3	94.3	17.76	86.8
WalMart	3.2	9	21	1.98	107.7
Exxon	6.3	6.9	14.6	2.58	22.4
General Electric	9.2	2.6	23.9	2.8	41
IBM	7.8	7.3	32.6	6.57	77.5
Citigroup	7.6	0.9	13.6	2.43	−6.8
Phillip Morris	9.3	9	33.2	2.2	22.8
Boeing	2	3	9.1	1.15	−32.4
AT&T	11.9	10.7	25.1	3.55	26.2
Bank of America	10.2	0.8	11.2	2.9	1.3
Mobil	3.6	4	9.3	2.1	24.3
Hewlett Packard	6.3	8.8	17.4	2.77	18.7
State Farm Ins Cos	3	1.2	3.2	na	na
Sears Roebuck	2.5	2.8	17.3	2.68	−4.4
EI du Pont de Nemours	11.4	11.3	31.5	3.9	−9.7
Proctor & Gambol	10.2	12.2	30.9	2.56	15.9
TIAA-CREF	2.3	0.3	13.3	na	na
Merrill Lynch	3.5	0.4	12.4	3	−7.5
Prudential	3.2	0.4	5.4	na	na
Kmart	1.7	4	9.5	1.01	33.1
American International	11.3	1.9	13.9	3.57	33.6
Chase Manhattan Corp	11.7	1	15.9	4.24	32.5
Texaco	1.8	2	4.9	0.99	0.5
Bell Atlantic	9.4	5.4	22.7	1.86	22.5
Fannie Mae	10.8	0.7	22.1	3.23	31.7
Enron	2.2	2.4	10	2.02	39.9
Compaq	−8.8	−11.9	−24.2	−1.71	49
Morgan Stanley Dean Witter	10.5	1	23.2	5.33	21.4
Dayton Hudson	3	6	17.6	1.98	62.1

Source: http://www.fortune.com, 26 November 1999
na: Not available.

A scatter plot of the earnings per share and profit as a percentage of revenue is shown in Figure 3.19. The first thing that stands out from the plot for all companies with a value recorded for earnings per share is the presence of two points which are far away from the rest (Figure 3.19a). These might be referred to as outliers and these have been labelled.

In 1998 Compaq made a loss and its earnings per share were negative. This was associated with the purchase of Digital in January 1998. Ford reported huge earnings per share, nearly three times as great as the next highest value, and also had the highest percentage profit. Among all companies it appears as if there is a tendency for the companies with high earnings per share also to have a large value for the percentage profit, while those with a low profits tend to have low earnings per share.

The presence of the two extreme points distorts the graph slightly and it is not easy to see the relationship among the majority of the companies. Figure 3.19b is a plot of the same variables but with the two extreme points for Ford and Compaq excluded. Now it is clear that there is considerable variability in the association, but it does appear as if there is a slight positive association in that high profits are associated with high earnings and low earnings with low profits. The plot also gives the impression that there are two groups of companies on the basis of profit as a percentage of revenue: companies with a profit of 2–4% and an earnings per share between $1 and $4, and companies with higher profits.

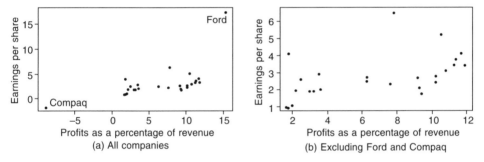

Figure 3.19 Earnings per share and profits as a per cent of revenue.

In a scatter plot to investigate any association between two quantitative variables it is often immaterial which variable is plotted on the horizontal axis and which is plotted on the vertical axis. Later on, when we come to think about describing any possible relationship by means of a statistical model, we will see that it is a good idea to put any variable which might be thought of as coming first, or being causal, on the horizontal axis, and secondary variables whose values might be affected by the first one on the vertical axis. In this sense we may think of earnings per share as being affected by profits, so we choose to put the profits on the horizontal axis and the earnings on the vertical axis. If one of the variables is time, it is usual to put it on the horizontal axis.

3.7.2 Scatter plot matrix

In Table 3.7 there are five variables in all and we can investigate, pictorially, the associations among them in a scatter plot matrix. This is a plot of all combinations of scatter plots and is shown in Figure 3.20. The plots in the first row all have profit as a percentage of revenue on the vertical axis and the other four variables along the horizontal axes. The plots in the first column all have profit as a percentage of revenue on the horizontal axis and the other four variables along the vertical axes. The plot that we have just been discussing in Figure 3.19a is the graph in the fourth row and first column of Figure 3.20. The plot in the first row and the fourth column of Figure 3.20 contains exactly the same data but has profits on the vertical axis and earnings per share on the horizontal axis.

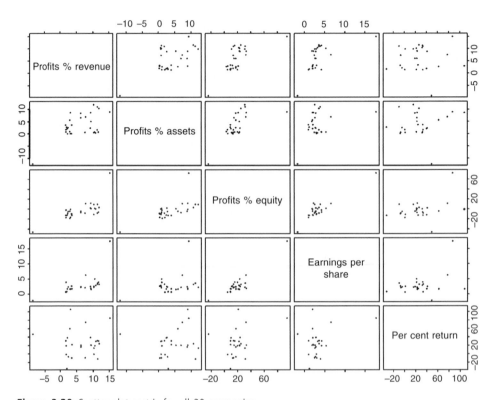

Figure 3.20 Scatter plot matrix for all 30 companies.

The scatter plot matrix is used to investigate associations between pairs of variables in a set of variables. It is clear that in Figure 3.20 the two companies, Ford and Compaq, are quite isolated in virtually all plots. If you look at the plots of profit as a percentage of equity against earnings per share (row 4, column 3 and row 3, column 4) then you can see that there appears to be a strong association between these two variables as most of the points lie on or near the diagonal line. This would mean that we could attempt to describe the relationship by a straight line and use this line for prediction purposes (see Chapter 6). Among the other plots, there appear to be associations between all three profit variables in that they all tend to be high or all low together. The percentage return for the investor does not appear to have strong associations with any other variable as there is no apparent pattern in the plots along the last row or along the last column.

In Figure 3.21 the same scatter plot matrix is drawn but now with the two extreme companies, Ford and Compaq, excluded. If you look at the same plots of profit as a percentage of equity against earnings per share in the same two positions as in Figure 3.20 then you will see that the association between these two variables does not appear to be anywhere near as strong. There is still a tendency for both to be high or both to be low in the same company, but the graph does not show most of the points lying in a straight line along the diagonal. The plot of profit as a percentage of assets against profit as a percentage of equity (row 3, column 2 and row 2, column 3) shows the points arranged in a sort of band going from the bottom left in the graph to the top right. If you look hard enough you might see a similar band shape in the plots of profit as a percentage of revenue against profit as a percentage of equity (row 3, column 1 and row 1, column 3).

Clearly the visual impression of the associations among the variables has been seriously influenced by the two companies which were omitted. The purpose behind showing these two examples is to introduce the idea that the interpretation of association in graphs can be influenced by just one or two influential points. In the example there were only 30 companies and so the potential for one or two observations to be seriously influential is great. In larger samples you would not expect just one or two companies to have so great an effect. However, they may do and this is one of the important reasons why you should always plot data first to look at the patterns among the variables and the possible influence of subgroups.

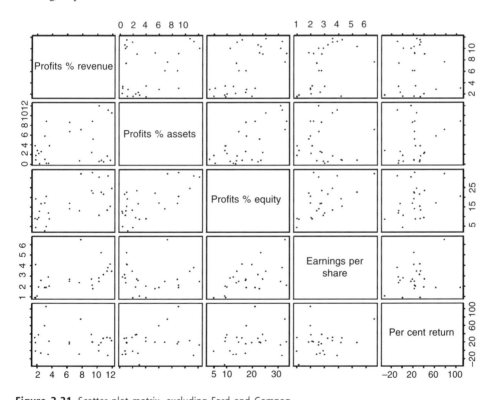

Figure 3.21 Scatter plot matrix, excluding Ford and Compaq.

3.8 Time series plot

A time series is a series of observations equally spaced in time. Usually we are interested in investigating how the values of the variable change with time. A time series plot is simply a type of scatter plot with time on the horizontal axis. The vertical axis has the units of the variable of interest, and may or may not have a suppressed zero – that is, the vertical scale may run from zero to a value at least that of the maximum observation, or between a convenient range encompassing both the minimum and maximum values. Usually, no symbols are plotted at each point, but rather the sequence of points on the plot are joined with straight lines; this does not imply that we believe that the variable of interest changed linearly between observations, and is simply a guide to the eye. If there are many points it is difficult to perceive the temporal pattern readily without the connecting lines. This is particularly true if there is marked seasonal fluctuation, as seen in Figure 3.22.

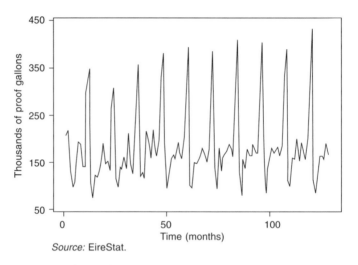

Source: EireStat.

Figure 3.22 Monthly sales of spirits for home consumption, Republic of Ireland, from January 1982 to May 1992.

In Figure 3.22 we plot the monthly volume of spirits sold for domestic use in the Republic of Ireland between 1982 and 1992. Along the vertical axis we have the monthly sales volumes, and along the horizontal axis we have time measured as the number of months from January 1981. Thus 1 corresponds to January 1981, 13 to January 1982, and the last observation of 125 corresponds to May 1992.

This series shows a pronounced seasonality with vastly increased sales over the Christmas/ New Year period. There is some evidence of an increasing trend in the time series, and as in each year the sales for any month tend to fluctuate, we can see that random (unattributable) effects are also present. Even closer scrutiny suggests that there is some difference in character between the early part of the series (up to about 1986) and the remainder, with greater variability (spread) latterly. This we might attribute to some economic or social change which has tended to decrease the summer sales of spirits while allowing the winter sales to continue growing – in other words, an external factor – although we have no direct evidence of this and it is only a suggested interpretation, unconfirmed by any other sources.

Most time series will display some or all of these characteristics to some degree. Many time series plots appear throughout Chapters 4 and 7, some showing a trend (a sustained general increase or decrease in the values of the series over the period of observation), some seasonality, and others cyclical patterns. Seasonality is present in virtually all time series which have data recorded every quarter, every month, or even every day in the week. Weekly sales figures may even have two sets of seasonal behaviour – a (roughly) four-week cycle representing monthly fluctuations and a 12-month cycle for annual ones. The time series plot of the percentage changes in the retail price index in Figure 1.2 illustrates irregular cyclical behaviour which is not related to seasonal behaviour though is often alluded to as a business cycle, with peaks in 1962, 1965, 1969, 1971, 1975, etc.

3.9 Misleading displays

When presenting graphs in a report it is very easy to make use of a misleading display which will misrepresent the true situation. One illustration has already been given in the histograms of Figure 3.5, where the area of the class is not used to represent frequency. A second common way to misrepresent data is to manipulate the vertical axis of the plot. In histograms with equal class widths and bar charts the vertical axis represents frequency and so should have a minimum value of 0. If you change the minimum then you can find yourself with graphs which lend themselves to different visual interpretations, even though they are based upon the same data.

An illustration is given in Figure 3.23 which gives two poor representations of the bar chart of cash benefits in the quintile groups of income presented in Figure 3.16. In Figure 3.23a, where the vertical axis in the plot ranges from 0 to £15 000, the scale is too small and this serves to minimize the visual differences between the groups. In Figure 3.23b the scale is enlarged by omitting the 0 and only plotting the values from £1000 to £5000. This serves to emphasize the differences between the five groups. If you pay careful attention to the vertical scales then you can reconstruct the data values but if you just pay attention to the overall display then both of these graphs are misleading.

A third major way of presenting a misleading display is to introduce an extra dimension. An example of a three-dimensional bar graph is presented in Figure 3.24. This is taken from a report by the market research company NOP on the Information Technology

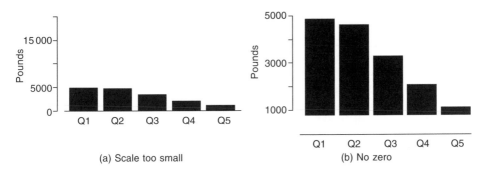

(a) Scale too small   (b) No zero

Figure 3.23 Manipulating bar charts.

Survey published in January 1998 and available on the NOP website (http://www.nop.co.uk). This chart is based upon the responses by 1250 users and suppliers of IT personnel.

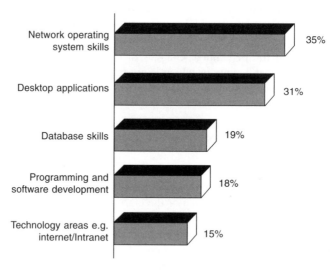

Figure 3.24 Bar chart from Information Technology Survey.

This graph is not particularly misleading as the widths and depths of the bars are the same so that the frequency is proportional to the height. The major difficulty with the graph is that there is no horizontal axis scale – though the percentages are inserted to compensate, which is a good point. However, the colour which stands out in the plot is the black representing the sides of the bars, not the grey face to the bars. This is unfortunate as it is the length of the grey bars which is proportional to the frequency not the length of the diagonal on the black side. One could make this graph much more misleading by making the height and depth *both* proportional to the frequency – then the *volume* of the larger bars will be greater and this would distort the presentation.

The final examples are taken from a graphical presentation in the *Guardian* of 27 April 1997 (Figure 3.25). There are four types of chart here. There is a bar chart of weekly household spending on petrol by decile income group which is a good display. The scatter plot of economic growth by energy use is also informative. However, the other two graphs both have misleading aspects to them.

In the time plots of index numbers of pollutants the time scale has been squashed such that the 10-year gap between 1980 and 1990 occupies the same distance on the scale as the three-year gap between 1990 and 1993. This means that the rise in the pollutants from 1980 to 1990 is visually much greater than it should be. From the graphs in Figure 3.25 you get the impression that there has been a sharp rise in carbon monoxide and nitrogen dioxide emissions until 1990 followed by a sharp decline, a sharp rise in carbon dioxide until 1990 followed by a levelling off, and a sharp rise in black smoke until 1990 followed by a slower rise. A graph with the correct scale for the time axis is presented in Figure 3.26, and you can see that the interpretation is much changed. There is now no apparent change of slope for black smoke and carbon dioxide and the reductions in carbon monoxide

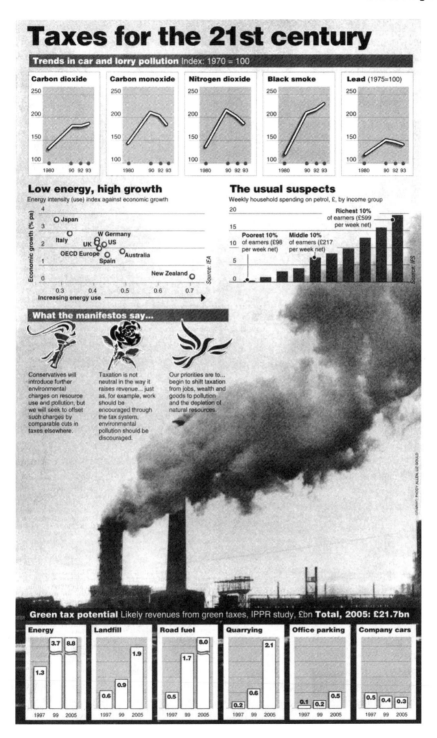

Figure 3.25 Graphs from the *Guardian*, 27 April 1997.

and nitrogen dioxide are not as steep as they appeared in Figure 3.25. Even Figure 3.26 is not perfect, as it would be much better if there were data for the whole period from 1980 until 1990 rather than just the two end years.

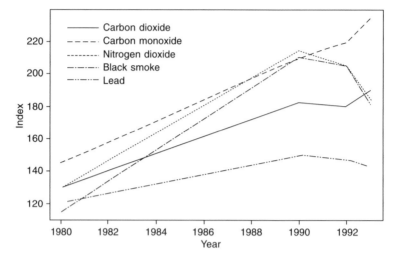

Figure 3.26 Trends in car and lorry pollution (1970=100).

The bar charts of likely revenues from green taxes for energy and road fuel have a gap in them to indicate that the values are off the scale. Also the vertical axis scale is not stated, though you can work it out from the values as ranging from 0.0 to 2.2. Any value in excess of 2.2 is represented by a break in the bar. Thus for energy the values for 1999 and 2005 are both represented by bars of the same height and are only differentiated by the values. This means that there is no information in the display and you may as well not have had it.

3.10 Summary

In this chapter we have looked at a number of different graphical procedures which can be used to investigate the distribution of a qualitative or quantitative variable and relationships between two variables. They can also be used to present information in a report so that you can draw attention to the most important features of the distribution and compare distributions over a number of subgroups. Graphs have two purposes: one is to aid the interpretation by discovering properties of a set of data, and the other is to aid the presentation of the most important information in the data set.

When you are in the process of trying to discover the features of a set of data you will generally try out a number of different graphical displays through a method of analysis which has come to be known as exploratory data analysis. The goal here is to look at the data through a number of displays to try to uncover interesting features concerning the structure of the data. Each set of data is likely to have its own peculiarities, but in general the structure focuses on the shape of the distribution, the location and variation, and the presence of outliers or multiple peaks. These points are the ones to pay attention to in analysing graphical displays.

Graphs are a very useful way of presenting information. You must ensure that the graphs you create are fair and are not misleading. Each graph should be accompanied by a title, together with labels for the axes, including units of measurement, and a scale for the axes. The axes should not have breaks in them, or different scales at different points of the axis. One key point about any display is that it should be clear and convey the important information unambiguously. In many publications you will find displays of information which are misleading to a greater or lesser degree. In most cases this has probably just arisen from carelessness and is not part of a deliberate attempt to mislead. You will find that many computer programs give a wide variety of choice for graphical displays and you will have to be careful that you choose the correct display for the purpose. The displays illustrated in this chapter are the most important ones in statistical analysis and interpretation and cover most situations.

4

Index numbers

- Understanding the concept of a weighted average and its relationship to index numbers
- Knowing how to calculate simple index numbers and change the base period
- Knowing how to calculate the Paasche and Laspeyres weighted index numbers, to interpret them and to compare them
- Appreciating how the retail price index is calculated and how to interpret changes in this and other indices.

4.1 Introduction

The summary statistics of a distribution (Section 2.3) are usually used to describe the state of a population at a particular instant using a sample from that population. Often two or more groups need to be compared, and this can be achieved by comparing the summary statistics. The key condition is that the quantities to be compared are indeed comparable. In most situations this will be obviously true, as when comparing the average family size in different geographic areas, or the proportion of households with two or more cars over different household compositions (households with a single adult, households with two adults and no children, households with at least one child over 16, etc.). The meaning of the variable of interest is the same over the groups.

If you are interested in the comparison of incomes in different regions then the median income in each region will obviously be a useful quantity to compare. Distributions of income are generally very skew, with most individuals at a low level but a few with a great deal of income, which means that the median is usually the best measure of location. If you then go on to consider that income is a measure of affluence then you need to think not just about the income but also about the costs involved in living in the different areas.

In the regional income example in Section 2.3 the average income is much greater in London and the South East than in the rest of the United Kingdom. Does this mean that people in these areas are very much better off than people in the North East? In order to answer this question, we need to take into account the costs of living in such areas and these may be very different. House prices are generally much greater in London and the South East than in the rest of the UK; the costs of food, travel and entertainment may also be much greater.

In order to measure the 'cost of living' a great deal of information has to be taken into account. You need to consider not only the cost of the important commodities that you need to live but also how often they are bought. Items which are very expensive contribute to the cost of living if they are bought infrequently, while relatively inexpensive items can contribute equally to the cost of living if they are bought in great numbers.

The cost of living is an average which is weighted by the amounts purchased. In the Section 4.2 we will look at the calculation of a weighted average. Then in Section 4.3 we will look at simple index numbers used to standardize a series, before moving on to more specific calculations of index numbers in general, including a discussion of the cost of living index in the UK, which is more correctly known as the retail price index. This index is derived from a number of government-sponsored surveys including the General Household Survey and monthly surveys of consumer prices in different areas. A case study in the use of index numbers and standardization of data is presented in the book's website.

If a set of quantities are to be compared over time then there is a second dimension to the comparisons, namely the constancy of the units of measurement. This is particularly the case when dealing with money. In 1980 the starting salary for a university lecturer was about £8 000 per year. In 1998 it was about £16 000. Does this mean that new lecturers in 1998 were paid twice as well as new lecturers in 1980? In numbers of pounds, yes – but the real question is the value of these pounds. One of the effects of inflation over this period is that the value of the pound has decreased, which means that the two salaries cannot be compared as they refer to pounds which had different values.

The retail price index is a measure of inflation, and we can use its value to adjust prices to a common basis to make valid comparisons over time. In 1980 the retail price index was 66.8, but by 1998 it had risen to 162.9 (January 1997 = 100). This means that £244 = £162.9/66.8 in 1998 had the same value as £100 in 1980. This implies that the starting salary of £16 000 is equivalent to £6561 in terms of the value of the pound in 1980. Thus it would appear as if the 'real' value of the starting salary is less than it was in 1980. (Data on the retail price index were taken from http://www.statistics.gov.uk/statbase).

Figure 4.1 was posted on the MORI International website (http://www.mori.com). It shows the relationship between time and an index describing the mood of the nation. This index is standardized to have a value of 100 in April 1993. The lower the index the more pessimistic the nation, and the higher the index the more optimistic it is. The graph shows

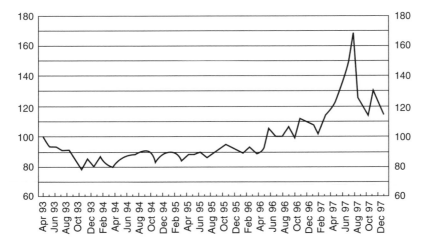

Figure 4.1 Mood of the nation index.

that in 1994 and 1995 the mood of the nation was pessimistic relative to the starting point of April 1993. From 1996 the mood became more optimistic soaring to a peak value of 169 in June 1997, just after the general election. Since then the index has decreased but is still higher than the base time point of April 1993.

The index is calculated from surveys carried out by MORI. Every month a representative quota sample of around 2000 adults aged 15+ are interviewed across Great Britain. Interviews are conducted face-to-face, at home, as part of the monthly MORI Financial Services Omnibus. Data are weighted to match the profile of the population.

The index is calculated from questions on the economic optimism of the public ('Do you think that the general economic condition of the country will improve, stay the same, or get worse over the next 12 months?'), the fear of redundancy of those in work ('How concerned would you say you are about the possibility of being made redundant or becoming unemployed over the next 12 months?') and the number who say they are unemployed. Thus the index is a summary of three components.

This example illustrates a number of features of index numbers. Firstly, they are often used for comparisons over time relative to a fixed point in time, known as the base time. Secondly, they are measured on a dimensionless scale – that is to say, there are no units to attach to an index number. Thirdly, a number of different components go into the calculation of the index number (in the mood of the nation there were three). Fourthly, the trends in the index are interpreted in relation to external events which are anticipated to have an effect on the index.

4.2 Weighted averages

With a sample of observations denoted $x_1, x_2, \ldots, x_n$, where there are n observations, the sample mean is

$$\bar{x} = \frac{1}{n}\sum_{1}^{n} x_i.$$

In order to calculate a weighted average, each observation is assigned a weight w_i which must be greater than or equal to zero. Negative weights are not allowed. The weight reflects the importance of each observation, with more important observations having greater weight. The weighted average is calculated by multiplying each observation by its weight, adding up the total, and dividing by the total of the weights:

$$\bar{x}_{\text{weight}} = \frac{\sum w_i x_i}{\sum w_i}.$$

In the calculation of the sample mean all observations are treated equally with the same weight, $w_i = 1/n$, so that the sum of the weights $\sum w_i = 1$. Thus the ordinary sample mean is a special case of a weighted mean, with all observations given the same weight. In fact, you can also think of the median as being a weighted mean, with all observations except the one in the middle having a weight of zero.

In Example 2.2 the average weekly earnings over the 13 regions of the United Kingdom was calculated. The raw average for males treating all regions equally, was £374, which was low compared to the figure for males in the United Kingdom as a whole of £390. We explained the discrepancy by citing the fact that not all regions are of the same size, and we noted, in particular, that London and the South East are the two regions with the highest average weekly earnings and the largest populations. In Example 4.1 we take the population of each region into account.

Example 4.1: Average weekly earnings
The second column of Table 4.1 gives the total resident population of each region in 1995, and these will be used as the weights. In fact, these weights are probably not completely correct. For males we should be using the total number of males in the regions who are in full-time work, while for females we should be using the total number of females in the regions who are in full-time work. These figures are not available, and we use the total population as an approximately correct weight.

Table 4.1 Gross weekly earnings and total resident population by region, 1995

Region	Total resident population, 1995 (Thousands)	Gross weekly earnings, April 1995 (£)		
		Females	Males	
United Kingdom	58 605.8	282.3	389.9	
	w_i		x_i	$x_i w_i$
North East	2 605.1	252.4	347.7	905 793.27
North West GOR	5 472.7	262.4	369.0	2 019 426.30
Merseyside	1 427.2	271.3	361.7	516 218.24
Yorkshire and the Humber	5 029.5	252.5	350.7	1 763 845.65
East Midlands	4 123.9	248.7	352.9	1 455 324.31
West Midlands	5 306.4	256.9	360.1	1 910 834.64
Eastern	5 257.4	279.9	382.3	2 009 904.02
London	7 007.1	364.4	514.3	3 603 751.53
South East GOR	7 847.2	292.7	412.7	3 238 539.44
South West	4 826.9	261.1	364.8	1 760 853.12
Wales	2 916.8	250.5	345.5	1 007 754.40
Scotland	5 136.6	262.0	363.6	1 867 667.76
Northern Ireland	1 649.0	256.9	337.4	556 372.60
Sum	58 605.8			22 616 285.28

We illustrate the calculations using the data for males. The weighted sum is $\sum w_i x_i$ = 22 616 285.28 and the sum of the weights is $\sum w_i$ = 58 605.8 giving the weighted mean over regions of the average weekly earnings for males as

$$\bar{x}_{\text{weight}} = \frac{\sum w_i x_i}{\sum w_i} = \frac{22\,616\,285.28}{58\,605.8} = 385.9.$$

This figure is much closer to the published figure in the table of £390. The slight difference of £4 is likely to be a result of using the total population as weights rather than the male working population. The effect is very slight, as you would expect the total population in each region to be roughly proportional to the total number of males in full-time employment. The weighted mean of £386 is a good bit larger than the (unweighted) mean of £374. The weighted mean takes into account the differing population sizes and the influence of the larger regions.

From the point of view of the calculations it makes no difference what scale the weights are in, as all that is important is the ratio of the weights. In Table 4.2 the weights are expressed as the raw populations, the proportions, summing to 1, and the percentages, summing to 100. Calculating the weighted averages in Example 4.1 with the weights as the proportions, the percentages of the raw populations will give exactly the same answer. Using weights as proportions or percentages is useful, however, as you can see how much the regions contribute to the weighted average. In Table 4.2 you can see that London contributes just under 12% and the South East over 13%. Thus two regions contribute over 25% of the total weight. Other regions such as the North East and Northern Ireland have a low weight.

Table 4.2 Comparison of weights

Region	Population	Proportion	Percentage
North East	2605.1	0.0445	4.45
North West GOR	5472.7	0.0934	9.34
Merseyside	1427.2	0.0244	2.44
Yorkshire and the Humber	5029.5	0.0858	8.58
East Midlands	4123.9	0.0704	7.04
West Midlands	5306.4	0.0905	9.05
Eastern	5257.4	0.0897	8.97
London	7007.1	0.1196	11.96
South East (GOR)	7847.2	0.1339	13.39
South West	4826.9	0.0824	8.24
Wales	2916.8	0.0498	4.98
Scotland	5136.6	0.0876	8.76
Northern Ireland	1649.0	0.0281	2.81
Totals	58 605.8	1.0000	100.00

From the point of view of the weighted average, London is 11.96/4.45 = 2.7 times as important as the North East because the populations is 2.7 times as large. It does not matter whether you use the raw populations or the proportions you will get exactly the same answer for the ratio.

$$\frac{7007.1}{2605.1} = \frac{0.1196}{0.0045} = \frac{11.96}{4.45} = 2.7.$$

This illustrates that it is not the numerical value of the weights which is important but their relative values through the ratios. The relative values of the populations, the percentages and the proportions are exactly the same.

4.3 Simple index numbers

4.3.1 Calculation and interpretation of simple index numbers

Simple index numbers are used to assess the relative change in a quantity over a period of time.

Example 4.2: Viewing figures

The data in Table 4.3 were published in the *Guardian* on 27 April 1997. They accompanied a small article about the effect of the decision to extend the *Nine O'Clock News* to an hour during the run-up to the general election which was held on 1 May 1997. The programme is only transmitted from Monday to Friday, which explains the gaps in the dates.

Table 4.3 Viewing figures and index numbers with different base periods

Day of the week	Date of transmission in April 1997	Number of viewers (millions)	Index base 1 April	Index base 14 April	Index base 17 April
Tuesday	1	4.6	100.0	139.4	86.8
Wednesday	2	4.1	89.1	124.2	77.4
Thursday	3	5.1	110.9	154.5	96.2
Friday	4	3.8	82.6	115.2	71.7
Monday	7	3.9	84.8	118.2	73.6
Tuesday	8	4.2	91.3	127.3	79.2
Wednesday	9	3.5	76.1	106.1	66.0
Thursday	10	5.0	108.7	151.5	94.3
Friday	11	4.2	91.3	127.3	79.2
Monday	14	3.3	71.7	100.0	62.3
Tuesday	15	3.7	80.4	112.1	69.8
Wednesday	16	3.5	76.1	106.1	66.0
Thursday	17	5.3	115.2	160.6	100.0
Friday	18	3.9	84.8	118.2	73.6
Monday	21	3.7	80.4	112.1	69.8
Tuesday	22	3.9	84.8	118.2	73.6
Wednesday	23	4.2	91.3	127.3	79.2

A better interpretation of the patterns and trends in these figures can be seen when plotting the data as in Figure 4.2. Here it can be seen that the daily viewing figures are around 4 million, with three separate peaks over 5 million on 3, 10 and 17 April. These were all Thursdays, so you might ask why the peaks occurred here. For example, the peaks may be related to events which happened on these days or to the programmes just before or just after the news.

Simple index numbers are useful in a series of this sort as it is easy to see the relative change from a particular point, namely the base period. In order to calculate a simple index number it is necessary to choose a base period, or base date in this case. Often the beginning of the series is chosen but that is not necessary. The simple index number is given by

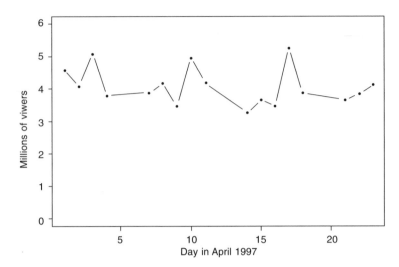

Figure 4.2 *Nine O'Clock News* viewing figures.

$$\text{Index} = \frac{\text{Value}}{\text{Base Value}} \times 100,$$

and usually it is sufficient to quote it to one decimal place.

We take 1 April to be the Base Period and the value there is 4.6 million; its index number is therefore 100 by definition. For the second day the index is

$$\frac{4.1}{4.6} \times 100 = 89.1,$$

and for the third it is

$$\frac{5.1}{4.6} \times 100 = 110.9.$$

The others can be calculated in a similar fashion and are shown in the fourth column of Table 4.3. The interpretation of these is that relative to the viewing figures for 1 April there was a reduction of $100 - 89.1 = 11.9\%$ in the viewing figures for 2 April. On 3 April, there was a $110.9 - 100 = 10.9\%$ increase in the viewing figures relative to the baseline date of 1 April. Everything in the interpretation of the index number is concerned with the fixed base relative to which the index is always measured.

In the final two columns of Table 4.3, the indices are presented with different base dates. In the first it is 14 April, and in the second 17 April is used. Numerically the values are different as they refer to different bases. Perhaps the easiest way to look at the trends in an index series is to plot it, and this is done in Figure 4.3. The graph shows that although the three series have numerically different values the pattern in the indices is exactly the same. In the series with 14 April as the base all the indices are greater than or equal to 100 as 14 April was the day with the lowest viewing figures. The indices for the series with 17 April as base are all less than or equal to 100 as this date was the day with the highest viewing figures.

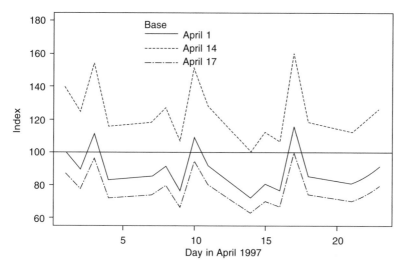

Figure 4.3 Viewing figures index series.

4.3.2 Changing base period

It is a straightforward matter to change the base period of a series using only the information in one index series. To change an index with base period X into an index with base period Y, all you need to do is to divide each value of the index with base period X by the value of the index with base period X at period Y. This can be expressed in a formula as

$$\text{Index}_Y(t) = \frac{\text{Index}_X(t)}{\text{Index}_X(Y)} \times 100,$$

where $\text{Index}_X(t)$ is the index value of the series at period t using base X. Using the series with 1 April as base, it can be converted into a series with 14 April as base by dividing through by the value of the index (with base period 1 April) on 14 April, which is equal to 71.7, and multiplying by 100. The index (based on 14 April) for 1 April is thus

$$\text{Index}_{14}(1) = \frac{\text{Index}_1(1)}{\text{Index}_1(14)} \times 100 = \frac{100.0}{71.7} \times 100 = 139.5.$$

This is the same figure as that in Table 4.3 calculated from the original data, to within rounding error. Similarly, the index (based on 14 April) for 17 April is

$$\text{Index}_{14}(17) = \frac{\text{Index}_1(17)}{\text{Index}_1(14)} \times 100 = \frac{115.2}{71.7} \times 100 = 160.7.$$

Again this is the same as that given in Table 4.3.

The purpose of these calculations is to show you that you can go from an index series with one base to an index series with another base very easily. The relevant figures for the calculations are summarized in Table 4.4. These show quite different numbers, with the values of the series influenced markedly by the choice of base. This has also been illustrated in Figure 4.3. If we compare 1 April to 17 April we see that the change in the

Table 4.4 Effect of changing base period

Date of transmission in April 1997	Number of viewers (millions)	Index base 1 April	Index base 14 April	Index base 17 April
1	4.6	100.0	139.4	86.8
14	3.3	71.7	100.0	62.3
17	5.3	115.2	160.6	100.0
Value increase from 1 April to 17 April		15.2	21.2	13.2
Percentage increase from 1 April to 17 April		15.2%	15.2%	15.2%
Percentage decrease from 17 April to 1 April		13.2%	13.2%	13.2%

index value depends upon the base. However, index numbers should always be interpreted in terms of relative change – that is, percentage change – and these are exactly the same irrespective of the base. The clear message is that you should only talk about relative changes with index numbers, and that the absolute change in the index depends upon the choice of base year.

The distinction between absolute change and relative change is an important one in any comparison. A good strategy is to talk about a points difference in connection with an absolute change in an index and a percentage difference in connection with a relative difference in an index.

A single series of an index number will enable you to assess the changes in the value of the index over time. A more interesting use of index numbers is in the comparison of a number of different series over time. The series plotted in Figure 4.4 give the average

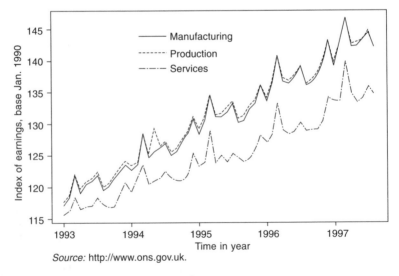

Source: http://www.ons.gov.uk.

Figure 4.4 Average earnings index (base = 1990)

earnings in three sectors of industry in the UK from 1993, at monthly intervals, until mid-way through 1997. The index numbers were calculated relative to a base year of 1990 and they are the actual figures, not the seasonally adjusted figures.

The graph shows that average earnings have increased, and that the indices for manufacturing and production are very similar to each other in numerical value. The series for services is increasing at a much slower rate. The seasonal patterns to the index numbers can also be seen, with peaks occurring in March and July of each year. Generally, the seasonal pattern is the same in all three sectors. Reading off from the graph, it is easy to see that, compared to 1990, average earnings in manufacturing and production had risen by 39% and 40% respectively by the beginning of 1997, while in the services sector the average earnings had only risen by 34%.

The same three series of average earnings are plotted in Figure 4.5 relative to a base year of 1993. There is no change to the shape of the lines compared to Figure 4.4 except that we can now see the increase in the average earnings relative to a common base at the beginning of the series. Relative to 1993, average earnings rose by 19% in manufacturing, 19% in production, and 16% in services. These are smaller differences compared to the comparison of 1997 with 1990, as the latter is over a longer time period.

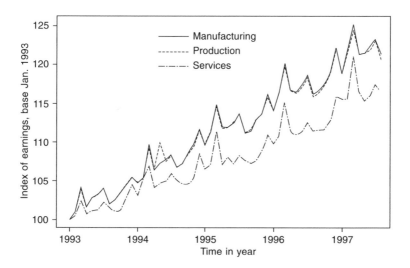

Figure 4.5 Average earnings index (base = 1993)

Over the period from 1993 to 1997, the differences between the average earnings indices for manufacturing and production compared to services appear to be greater in Figure 4.4 than Figure 4.5. This occurs because the series are not starting off from the same point in 1993. Part of the difference over the period from 1993 to 1997 in Figure 4.4 is made up of the discrepancies between 1990 and 1993. For the comparison among the sectors over the period from 1993 to 1997 only, Figure 4.5 is a more appropriate plot as the series have been standardized to be at the same point at the beginning of the time period.

4.4 Weighted index numbers

4.4.1 Comparing prices and quantities

In the preceding sections index numbers were used to investigate changes in a price or quantity over time. The comparisons were first carried out within a series – to give the relative change in a variable from one time to another time. The relative changes in two or more indices were also compared. The data presented in Table 4.5 illustrate the comparisons of a range of items, each with different quantities and prices, which are to be compared over two time periods and two countries. The data represent the weekly purchases and prices for a selection of three foodstuffs. The aim is to provide an assessment of the overall price changes between 1995 and 1998 and then to compare the changes over the two countries. This type of calculation is similar to the calculation which is carried out for the retail price index or any other composite index number.

Table 4.5 Food basket in UK and Italy. (Prices are in pounds (£) in UK and line in Italy

		1995		1998	
	Unit	Quantity	Price	Quantity	Price
UK					
Beer	Bottle	10	1.5	15	1.8
Cheese	kg	1	3.76	0.8	4.4
Bread	400 g	3	0.8	4	1.2
Italy					
Beer	Bottle	3	1 300	5	1 500
Cheese	kg	2	17 000	2	18 500
Bread	400 g	5	3 000	4	3 500

Two simple approaches are illustrated below, both of which give misleading answers. The first is to calculate the price indices for the three foodstuffs separately and then average them. For the UK data, this gives an index of 129.0. The second is to take the average price in each year and then to calculate the price index for the averages, which for the UK would be $100 \times 2.47/2.02 = 122.1$.

		1995		1998		Price
	Unit	Quantity	Price	Quantity	Price	index
UK						
Beer	Bottle	10	1.5	15	1.8	120.00
Cheese	kg	1	3.76	0.8	4.4	117.02
Bread	400 g	3	0.8	4	1.2	150.00
Average			2.02		2.47	129.01

Both of these indices are incorrect as they do not take into account the number of items which are bought each week. In this sense they treat a bottle of beer the same as 1 kg of cheese and 400 g of bread, despite more units of beer and bread being bought each week than of cheese. Also the two indices are different, as the first (129.0) is an average of three ratios and the second (122.1) is the ratio of two averages. These two quantities are seldom the same. They would be the same if all the 1995 prices were equal for all products or if all the 1998 prices were zero!

In the following table the notation for the prices and quantities is laid out. The prices are denoted P and the quantities consumed are denoted Q. The first subscript refers to either the base year, b, or the current year, c. The second subscript stands for the commodities: 1 for beer, 2 for cheese and 3 for bread.

	Unit	Base year		Current year	
		Quantity	Price	Quantity	Price
Beer	Bottle	Q_{b1}	P_{b1}	Q_{c1}	P_{c1}
Cheese	kg	Q_{b2}	P_{b2}	Q_{c2}	P_{c2}
Bread	400 g	Q_{b3}	P_{b3}	Q_{c3}	P_{c3}

One solution is to calculate the value of the basket of goods in the two years and then to calculate the index from the two values. The value of a purchase is the price times the quantity (PQ). This represents the amount spent on the commodity. The total values of the baskets of goods in the two time periods are as.

$$V_b = \sum_i P_{bi} Q_{bi},$$

$$V_c = \sum_i P_{ci} Q_{ci},$$

so the value index is

$$\text{Value index} = \frac{V_c}{V_b} \times 100 = \frac{\sum P_{ci} Q_{ci}}{\sum P_{bi} Q_{bi}} \times 100.$$

For our basket of UK goods the calculations are as follows:

	1995			1998			
	Quantity	Price	Value	Quantity	Price	Value	
UK							
Beer	10	1.5	15	15	1.8	27	
Cheese	1	3.76	3.76	0.8	4.4	3.52	
Bread	3	0.8	2.4	4	1.2	4.8	
			21.16			35.32	166.92

The value of the basket in 1995 is £21.16 and in 1998 it is £35.32, giving a value index of 166.9. From 1995 to 1998 both the prices and the quantities have changed so this index reflects changes in both of these features. Thus the value index is not truly a price index. A price index should reflect changes in price alone, while a quantity index should reflect changes in quantities.

4.4.2 Laspeyres and Paasche indices

Two ways of calculating a price index will be described. In order to get an index which reflects changes in prices only, the quantities have to remain constant over the two time periods. This is achieved by using either the base year quantities to give the Laspeyres price index, or the current year quantities to give the Paasche price index.

The Laspeyres price index is calculated as

$$\text{Laspeyres price index} = \frac{\sum P_{ci} Q_{bi}}{\sum P_{bi} Q_{bi}} \times 100.$$

The denominator represents the value of the basket of goods in the base year, while the numerator is a fictitious value representing the value of the basket of goods in the current year assuming that the quantities have not changed. The index for our basket of UK goods has a value of 122.9:

Laspeyres	1995			1998			
	Quantity	Price	Value base	Quantity	Price	Value base	
UK							
Beer	10	1.5	15	15	1.8	18.00	
Cheese	1	3.76	3.76	0.8	4.4	4.40	
Bread	3	0.8	2.4	4	1.2	3.60	
			21.16			26.00	122.87

Shaded figures not used in the calculation.

The Paasche price index is based upon a real numerator representing the value of the basket of goods in the current year, but this time the denominator gives the fictitious value of the basket of goods in the base year assuming that the current year quantities are valid in the base year. In symbols,

$$\text{Paasche price index} = \frac{\sum P_{ci} Q_{ci}}{\sum P_{bi} Q_{ci}} \times 100.$$

For our UK basket, the value of the Paasche price index is 123.0 which in this case, but not all cases, is similar to the Laspeyres price index.

Paasche	1995			1998			
	Quantity	Price	Value current	Quantity	Price	Value current	
UK							
Beer	10	1.5	22.5	15	1.8	27.00	
Cheese	1	3.76	3.008	0.8	4.4	3.52	
Bread	3	0.8	3.2	4	1.2	4.80	
			28.708			35.32	123.03

Shaded figures not used in the calculation.

The calculation of the Laspeyres and Paasche price indices for the Italian data is shown in Table 4.6. The Laspeyres price index is 111.5 and the Paasche index is 111.4. Comparing the UK and Italian examples, it can be seen that according to the fictitious data here there was an 11.5% increase in the Laspeyres price index for Italy and a 22.9% increase for the UK for the same basket of goods. This implies a greater increase in prices in the UK than in Italy.

Table 4.6 Laspeyres and Paasche price indices for Italy

	1995			1998			
	Quantity	Price		Quantity	Price		
Laspeyres							
Beer	3	1 300	3 900	5	1 500	4 500	
Cheese	2	17 000	34 000	2	18 500	37 000	
Bread	5	3 000	15 000	4	3 500	17 500	
			52 900			59 000	111.53
Paasche							
Beer	3	1 300	6 500	5	1 500	7 500	
Cheese	2	17 000	34 000	2	18 500	37 000	
Bread	5	3 000	12 000	4	3 500	14 000	
			52 500			58 500	111.43

Shaded figures not used in the calculations.

To summarize, the Laspeyres price index is a comparison of values in two years assuming that the quantities in each year are the same as those in the base year. The Paasche price index is a comparison of values in two years assuming that the quantities in each year are the same as in the current year. As these two indices have constant quantities for each of the commodities, they are interpreted as price indices.

4.4.3 Quantity and value indices and Fisher's ideal index

If prices change then this can bring about a change in the quantities purchased. If prices increase dramatically then a reduction in the quantities purchased is to be expected. If prices fall then an increase in the quantities purchased may well follow. It is possible to calculate Laspeyres and Paasche quantity indices to reflect the changes in quantities assuming the prices remain the same. The Laspeyres quantity index calculates the change in the quantities purchased assuming that the prices remain fixed at base year prices, while the Paasche quantity index is based on the assumption that the prices are fixed at current year prices:

$$\text{Lapeyres quantity index} = \frac{\sum P_{bi}Q_{ci}}{\sum P_{bi}Q_{bi}} \times 100,$$

$$\text{Paasche quantity index} = \frac{\sum P_{ci}Q_{ci}}{\sum P_{ci}Q_{bi}} \times 100.$$

By comparing the equations for the value, price and quantity indices it can be seen that they are related to each other in that

Value index = Laspeyres price index × Paasche quantity index,

= Paasche price index   × Laspeyres quantity index.

In a sense the two price indices represent the two extremes. As the Laspeyres price index is based upon the base year quantities this index will tend to overestimate the effect of a price increase. With a price increase you would expect the quantity to decrease, but the Laspeyres price index keeps the quantity unchanged from the base year. This will tend to lead to an overestimation of the effect of the price increase as the quantity used will be larger than it should be. On the other hand, the Paasche price index will tend to underestimate the effect of a price increase for exactly the same reasons, as the quantity on which the calculation is based is the current year quantity which is likely to be the smaller than the base year quantity.

This is illustrated with a small example based upon the UK data in Table 4.5. Keep everything the same apart for the price and quantity of beer in the basket in the current year of 1998. The figures are manipulated to maintain a constant value of £15.00, which is the same as the 1995 value. As the price of beer in 1998 is steadily increased, with a corresponding reduction in the quantity sold, you can see that the Laspeyres and Paasche price indices start to diverge, with the Laspeyres index giving the higher values.

Price	0.5	0.75	1	1.25	1.5	1.7	2	2.5	3	4
Quantity	30	20	15	12	10	8.8	7.5	6	5	3.75
Value Index	110.2	110.2	110.2	110.2	110.2	110.2	110.2	110.2	110.2	110.2
Laspeyres price index	61.4	73.3	85.1	96.9	108.7	118.1	132.3	156.0	179.6	226.8
Paasche price index	45.5	64.4	81.2	96.3	110.0	120.0	133.6	153.3	170.1	197.1
Fishers ideal index	52.9	68.7	83.1	96.6	109.3	119.0	132.9	154.6	174.8	211.4

The status quo for the beer prices and quantities is at a price of £1.50 and a quantity of 10 bottles. In this case the Laspeyres price index is 108.7 and the Paasche price index 110.0; the differences are due to the changes in prices and quantities of the other two commodities. As the price of beer is steadily increased and the quantity decreased correspondingly to give a constant value, at £15.00, the Laspeyres price index increases at a faster rate than the Paasche price index. If the price is decreased and the quantity consumed increases then both indices fall, as there has been an overall price reduction, but the Paasche price index falls at a faster rate than the Laspeyres. This illustrates that the Laspeyres price index tends to overestimate the effect of a price increase and the Paasche price index to underestimate its effect. However, the price and quantity changes have to be quite dramatic for there to be an extremely large difference.

These two indices can be thought of as being at two extremes of under- and overestimating the effects of price changes. The differing effects of price changes on the quantities consumed have led to the use of a composite measure, Fisher's ideal index given by

$$\text{Fisher's ideal index} = \sqrt{\text{Laspeyres price index}} \times \sqrt{\text{Paasche price index},}$$

which will give a value between the two extremes. This index is illustrated in the last row of the table.

4.4.4 Relationship of price indices to weighted averages

The indices discussed in this section are all examples of weighted indices The price indices are calculated from a change in average price where the average price is weighted by the quantities. From Section 4.2, a weighted average is

$$\bar{x}_{\text{weight}} = \frac{\sum w_i x_i}{\sum w_i}.$$

Within the Laspeyres and Paasche price indices, the weighted 'average' prices are as follows:

	Laspeyres	Paasche
Weighted 'average' price base year	$\dfrac{\sum P_{bi} Q_{bi}}{\sum Q_{bi}}$	$\dfrac{\sum P_{bi} Q_{ci}}{\sum Q_{ci}}$
Weighted 'average' price current year	$\dfrac{\sum P_{ci} Q_{bi}}{\sum Q_{bi}}$	$\dfrac{\sum P_{ci} Q_{ci}}{\sum Q_{ci}}$

The indices are just the ratio of these two weighted 'average' prices multiplied by 100 as the denominator terms, $\sum Q_{bi}$ for the Laspeyres index and $\sum Q_{ci}$ for the Paasche index, cancel out. The word 'average' is used in inverted commas here because the quantity does not really have any physical meaning. The important point is that the quantities are weights. However, it is not necessary to use the exact quantities, and weights which reflect the relative importance of the items in the basket of goods may be used instead. Indeed, when indices are calculated in practice the weights used seldom correspond to exact quantities but are used to reflect the importance of the different commodities. This is discussed further in Section 4.5 in relation to the retail price index.

4.4.5 Comparison of Paasche and Laspeyres indices

There are advantages and disadvantages to the two basic price indices. The Laspeyres price index is based upon the baseline quantities, and once these have been obtained only the current year's prices are required for the calculation of the index. For the Paasche price index both the quantities and prices for the current year are required. This implies that the Laspeyres price index is slightly easier to calculate and can be obtained in situations where the current year's quantities are unknown.

As the Paasche price index uses current quantities, it can be argued that this index is more relevant and topical. This is certainly the case when the index is calculated over a long period of time where the quantities may have changed dramatically.

When using the Paasche price index over a long series of data, the whole index series has to be recalculated each period new data comes along as the current period will have changed. If there are substantial quantity changes then the previously calculated Paasche price indices may take substantially different values from the newly calculated ones based upon the new current year quantities. The Laspeyres price index does not need to be recalculated with the addition of extra periods as the base period remains the same.

The Paasche price index is slightly more relevant, with a better interpretation in terms of current quantities, but is slightly more inconvenient to calculate. If both quantities and prices are readily available then it is a simple matter to calculate both and then to work out Fisher's ideal index. If there are substantial differences between the Laspeyres and Paasche price indices, these differences convey information about the changes in both the prices and the quantities.

There are many modifications on these basic price indices which attempt to overcome some of the deficiencies in these indices. Some of these modifications include updating the base year repeatedly during the series. This is known as the chained base method. With this method, for the Laspeyres type of index, the year 2 index is calculated using year 1 as the base, the year 3 index is calculated using year 2 as the base, etc. There is a similar updating method for the Paasche index, where the current year 2 is the current year for the year 1 to year 2 index, year 3 is the current year for the year 2 to year 3 index, etc. This has an advantage for the Paasche index in that historical series no longer need to be recalculated each time data for a new period become available.

4.5 Retail price index

4.5.1 Calculation of the retail price index and related indices

The retail price index (RPI) is arguably the most important price index in the United Kingdom. It is calculated and published each month, and changes in the rates are regularly reported in the television news and newspapers. It is used as a measure of inflation in the economy, with the rate of change in the index from one year to the next regularly quoted as the rate of inflation. Thus the index is important in terms of economic policy. The changes in the index over time are also used in wage negotiations, personal savings schemes, and increases in state benefits such as pensions.

The rates for October 1997 until March 1998 are presented in Table 4.7, together with the percentage change in the rates over a 12-month period. The index is currently calculated relative to January 1987. In March 1998 the RPI was 160.8, which represents a 60.8% increase in prices since January 1987 but a 3.5% increase over the previous year.

A form of the retail price index has been calculated in the UK since 1914. The retail price index is calculated using data from two separate sources. The prices information comes from a survey of consumer prices at various locations around the UK carried out every month. The quantities information comes from the Family Expenditure Survey, a continuous government survey with a sample size of about 10 000 households each year.

The index is a chained Laspeyres price index covering 14 major categories of expenditure (see Table 4.8). Within each of these categories there are various subgroups – for example, alcoholic drink includes subgroups of beer, wines and spirits. The weights (quantities) are

Table 4.7 Retail prices index

Date	All items (RPI)		All items excluding mortgage interest payments (RPI-X)	
	Index (Jan. 1987 = 100)	Percentage change over 12 months	Index (Jan. 1987 = 100)	Percentage change over 12 months
October 1997	159.5	3.7	157.9	2.8
November 1997	159.6	3.7	158.0	2.8
December 1997	160.0	3.6	158.3	2.7
January 1998	159.5	3.3	157.7	2.5
February 1998	160.3	3.4	158.5	2.6
March 1998	160.8	3.5	158.9	2.6

All items (RPI) percentage change over 12 months
1995	3.5
1996	2.4
1997	3.1

Data published 22 April 1998.
© Crown Copyright.

revised each year in January. Thus the index for 1998 is based upon weights which are revised in January 1998. These are likely to be only minor modifications of the weights used in January 1996, which are illustrated in Table 4.8. For all of the indices published in 1998 the same weights are used, until January 1999 when new revised weights are calculated.

Table 4.8 Retail price index categories, weights and category indices, September 1998

Category	Weight (1996)	Value
Food	143	144.1
Catering	48	191.1
Alcoholic Drink	78	181.2
Tobacco	35	224.2
Housing	190	199.9
Lighting and Fuel	43	124.3
Household Goods	72	141.3
Household Services	48	179.8
Clothing and Footwear	54	142.5
Transport and Vehicles (Motoring)	124	174.5
Personal Goods and Services	38	171.5
Fares and Travel Costs	17	179.8
Leisure Goods	45	199.9
Leisure Services	65	192.5

Overall Index Value 164.4, base = 100, January 1987.

The prices of about 600 items, including services, are collected by obtaining price quotations monthly in about 180 locations throughout the UK. The items investigated are representative of all the items in the major categories. For example, in the fruits subsection of the food category the prices of apples, oranges and bananas will always be included, but not necessarily the less common seasonal fruits such as blackcurrants and redcurrants or the more recent imported fruits such as kumquats and passion fruit. The price quotes are obtained from different retail outlets, ranging from large supermarkets to chain stores and independent retailers. From this information the average price is calculated separately for 12 regions of the UK and for three separate retail outlets – independent stores, companies with multiple stores and large-scale co-operative stores.

The weights column in Table 4.8 represents the relative importance of the categories in 1996. These weights add up to 1000. From year to year these weights change in accordance with the relative importance of the categories. The RPI for September 1998 of 164.4 is calculated using 1998 weights. You can see that housing is the most important category, as it has the largest weight, followed by food and then motoring. The individual category with the largest increase in price since January 1987 is tobacco with an index of 224, while lighting and fuel have only increased by 24.3%. Both of these categories are relatively small and do not have as much influence on the overall index as the categories with the bigger weights.

The composition of the items which go into the RPI has been debated and a number of other indices are calculated. One is illustrated in Table 4.7. It is the retail price index with mortgage interest payments excluded. This index was derived because the RPI can increase or decrease as a result of changes to the mortgage interest rate. These changes are included in the housing category, which has the biggest weight. In the UK a large percentage of families own their own home and consequently changes to the interest rate have a large effect on the RPI.

The national price indices are used to make comparisons over countries, and one of the basic statistical principles of the comparisons is that, as far as possible, like should be compared with like. In other European countries home ownership is not so common and rents are controlled over a period of time. For example, in Italy rents for private apartments are controlled for a four-year period and increase at a rate of three-quarters of the rate of inflation each year, though there can be large changes at the end of this period. Mortgages are relatively rare, tend to be of shorter duration than in the UK and may be at fixed interest rates throughout the whole period. Consequently, mortgage rate changes in Italy may have a minimal effect, while in the UK they will have a larger effect. Thus the retail price index with mortgage interest payments excluded may be a better index to compare with the IPC (indice dei prezzi al consumo) in Italy.

The calculation of the RPI, like the calculation of many national economic indicators, has a political dimension as well as an economic and statistical one. The effect of mortgage rate changes on the RPI is unlikely to worry politicians in the UK if interest rates are going down, but does raise concern among them when interest rates are increasing. In the 1980s there was considerable pressure on the government statistical service as a result of essentially political changes to the components of some indicators. It is important to realize that the indices are not stable entities. They change in response to changes in society, and this has to be noted in the comparison of the indices over time.

A recent development is the introduction of the harmonized index of consumer prices (HICP), designed for use in the comparison of prices across European Union member states. This became necessary in the context of ensuring that member states met one of the convergence criteria for monetary union as required by the Maastricht Treaty. International comparisons became available after 8 May 1998 with the publication by Eurostat of figures for all member states. In the UK, the HICP does not replace the RPI, which remains the best indicator of UK consumer price inflation.

4.5.2 Interpretation and comparison of indices

A number of economic indicators of the standard of living for 1971–96 are presented in Table 4.9. Most, but not all, have been standardized to a base of January 1990. The average earnings figure is not an index number, as such, but has been standardized according to the value of the pound in 1990.

Table 4.9 Standard of living

	1971	1981	1992	1993	1994	1995	1996
Average earnings (GB only) (£ per week) (April) (Jan 1990=100)	28.7	124.9	304.6	316.9	325.7	336.3	351.5
Earnings index (annual average)	11.3	48.5	114.6	118.5	123.2	127.4	132.3
Retail prices (Jan 1987=100)	20.3	74.8	138.5	140.7	144.1	149.1	152.7
Real personal disposable income per head	59.8	69.7	101.2	101.1	101.8	104.3	106.9
Volume of consumer spending per head (1990=100)	60.3	72.5	97.0	99.0	101.2	101.4	106.1

Note: Some of the later figures are provisional. Most of the figures are rounded, leading sometimes to inconsistent totals.

The numerical comparison of values across these series is not really meaningful. For example, comparing the value of 11.3 for the earnings index in 1971 with the value of £28.7 is not valid as one is an index number (albeit with unknown base year) and the other is in pounds. Comparisons of the percentage changes within a series are valid. Between 1971 and 1996 average earnings rose by a factor of a 351.5/28.7 = 12.2, while retail prices rose by a factor of only 152.7/20.3 = 7.5. Real disposable income only increased 1.8 times over the same period.

In this example, the use of the appropriate summary quantities and the drawing of the appropriate comparisons are the domain of statistics. The interpretation and implications of the differences in the percentage changes in the indices are matter of economics.

Economic indicators are rough summaries measuring a wide variety of disparate information. They cannot be expected to be very specific. The retail price index measures the cost of a basket of goods designed to cover the broad spectrum of spending in the UK. There are a large number of disparate groups, and the RPI is not specific to the needs of all of then. Examples would include pensioners and the unemployed, who predominantly live on state benefits, and people living in isolated rural areas.

5

Large surveys and market research surveys

- Knowing and appreciating the importance of the various stages in the design of a survey
- Knowing the important points about the four main random sampling methods – simple random sampling, stratified random sampling, cluster random sampling and systematic random sampling
- Appreciating the differences between random and non-random samples
- Knowing how a quota sample is selected and the advantages and disadvantages of such a non-random sample
- Appreciating the importance of the large government-sponsored surveys in the United Kingdom and their impact on decisions made in society
- Realizing that these surveys have components in them which are common to all surveys
- Appreciating the differences between the main companies carrying out opinion polls and the ways in which they try to combat non-response
- Understanding the importance of good questionnaire design, particularly for surveys using self-completed questionnaires
- Realizing that non-response and bias are enormous problems in many surveys and understanding the ways in which these problems can be tackled during the design and analysis of the survey.

5.1 Planning a survey

In business and the social sciences data are generally collected through surveys of individuals in a predefined population. Much of our daily life is influenced by the information derived from such surveys. The General Household Survey and Family Expenditure Survey are used to construct the retail price index. This then forms the basis for wage claim discussions. The Labour Force Survey is used to provide an estimate of the number of people out of work and seeking employment. The British Crime Survey provides information on the level and type of crime in different regions of England and Wales.

If a survey is carried out well using a representative sample of the population, then the results can be very informative. If the study is not carried out to the highest standard, then the results can be misleading. Sampling incompletely from the population will result in a bias being introduced. Using an inappropriate questionnaire may lead to a large number of individuals refusing to respond to the whole survey, or parts of the questionnaire, and hence also to bias.

There are two important statistical properties of a sample survey which we will cover during this chapter. There are the concepts of **bias** and **precision**. Surveys are carried out to estimate some quantity in a population, for example average earnings. If the sample is representative of the population, then the estimated quantity, sample average earnings, will be close to the true value, population average earnings. If the sample contains too many wealthy people relative to the population, then the average earnings in the sample will be higher than in the population. The sample is biased as a result of the sample not being representative of the population. Precision is related to measures of variability such as the standard deviation through the standard error of the quantity being estimated in the survey. The standard error is a measure of the sampling variability in the estimate and, for example, the standard error of the sample mean is equal to the standard deviation divided by the square root of the sample size. Precision and standard error are concerned with how close the sample estimate is to the population value. Generally you can have a more precise survey by increasing the sample size or reducing the variability. Both of these lead to a reduction in the standard error. These concepts will be discussed in more detail in Chapter 10.

There are a number of different strands to carrying out surveys, and some of these will be discussed in this chapter.

5.1.1 Target and sampled population

Each study should have a **statement of objectives**. This needs to be concise and clear. The objectives should be simple enough to be understood by those working in the field and by those taking part in the study. If the study objectives are not clear at the outset then the wrong type of survey may be instigated, the wrong questions or types of questions asked, or the wrong population sampled.

The **target population** must be carefully defined so that a sample can be selected from it. The target population is the group of individuals about whom you wish to collect information. Different surveys have different target populations. The General Household Survey has all households in the United Kingdom as its target population, whereas a survey of crimes will have as the target population all crimes.

In planning a survey information is to be collected from the target population; however, the data collected may be representative of the **sampled population** which may not coincide with the target population. There is a distinction between the two. The sampled population represents the population about which information is available. This is the population to which the survey results will refer.

In a crime survey you may wish to survey all crimes, but if you sample from police records then information will only be available from crimes reported to the police. For serious crimes it is not likely that there is a big difference between the two as most serious crimes will be reported to the police, though there are cases of assault which are not reported. It is possible that there are many thefts and burglaries involving small amounts which are not reported to the police. There can be a large difference between the target population and the sampled population. Any survey results will pertain to the sampled population, and only to the target population if the two coincide.

The **sampling frame** should match the target population. Any major omissions could lead to bias in the results. The sampling frame is effectively a list of all individuals or items in the sampled population. For the General Household Survey and other government surveys the sampling frame is the list of all residential addresses in the United Kingdom. For a telephone survey the frame will be the list of telephone numbers. If the target population were all private individuals then the list would have to exclude business phone numbers. This could lead to a problem with individuals who have their business registered at home if they only have one telephone line for private and business use. Such people will be excluded from the target population. A second problem may arise with individuals who have more that one telephone line into the house. This is quite rare but is becoming more common. Such individuals will have a slightly greater chance of being selected into the survey.

Many surveys are carried out using a conveniently available sampling frame which may not match the target population exactly but is sufficiently close that the effort involved in getting a sampling frame to match the target population may not be worthwhile. In the UK, one way of sampling individuals in an area, such as an electoral ward, is to use the electoral register as the sample frame. This excludes people who have not registered to vote, and also people under 18. Since it is compiled once a year, it may also include people who have moved away from the area and exclude people who have recently moved into the area. Thus there may be a discrepancy between the sampling frame and the target population.

The **sample design** refers to the type of sample chosen and the required sample size. There are many different types of sample that can be selected. The main distinction is between random and non-random samples. Generally random sampling methods are to be preferred from a statistical point of view. They tend to be much more expensive than non-random samples, which are used extensively in market research. The common sampling schemes will be discussed in Section 5.2.

The sample size should be calculated to ensure that the main quantity of interest is estimated with sufficiently high precision. This generally requires expertise from statisticians and a brief discussion of the methodology and main points is given in Chapter 11. Many small-scale surveys provide little or no useful information because of a lack of a good design.

5.1.2 Methods of data collection

There are four main methods of **data collection**. For the large-scale government surveys in the United Kingdom personal interviews are used. Many market research surveys also use this method. Self-administered questionnaires are considered to be the easiest method of collecting information, and telephone interviews are becoming much more common. Direct observation is not often used, though it can be a successful method of collecting information about the functioning of a process such as interactions between individuals.

Personal interviews tend to have a high response rate of around 60–75% or higher. There is expected to be a high level of accuracy in the responses as the interviewer can probe the subject and make notes to supplement the answers. This is particularly useful if the subject requires clarification or if the questions require the subject to recall incidents in the past where much probing may be required to elicit the information. Direct personal interviews are costly and time-consuming. For the General Household Survey and Family Expenditure Survey the initial interview may last 2–3 hours. With the travel involved it may only be possible to interview two households per day.

It is very important to train the interviewers very well. In some smaller studies it is common for the same person to interview all the subjects in the sample. Interviewers who are not well trained may deviate from the survey protocol (i.e. the sampling plan and the interview procedure) and introduce an unmeasured bias. There is a small possibility of recording errors if the interviewer does not accurately write down the subject's response. There is also the possibility that the presence of the interviewer could lead the subject to change his or her response to something which is more acceptable to the interviewer. The interviewer may also ask leading questions which will induce the subject to respond in a certain way. Many surveys now combine a personal interview with a computer-assisted interview in which the individual being surveyed responds to the questions by keying a response directly into the survey database, with the interviewer only used for clarification. In this way the interviewer need not know the responses of the individual.

Telephone interviews are very common in the United States, but were not used so widely in the UK until recently They are much less expensive than personal direct interviews as there is no travel involved. Generally the interview will be much shorter than in the case of direct interviews or self-administered questionnaires. Tape-recording of the interview can permit monitoring to check the accuracy of the interview records. The sampling frame may not resemble the target population – for example you may get small businesses when looking for houses, and people with unlisted phone numbers will not be in the sample.

All questionnaires require careful design and all of the questions must be free of ambiguities. This is especially important for **self-administered questionnaires**. Generally these are posted out to the selected individuals and returned to the survey organization with no telephone or personal contact. The UK's 2001 Census is an example of a self-completed questionnaire, albeit with personal contact in the event of assistance being required and in the case of non-response. A well-designed covering letter should be sent with the questionnaire. This letter should be informative and at the same time written in such a style as to encourage the selected individuals to respond. Generally, there are no interview costs so it is a much cheaper method of collecting data than direct personal interviews.

There is a much lower response rate (30% on average, but this depends on the context and location of the survey) than personal interviews, and this can be a source of bias. Those who respond may not be a completely random selection of all those sampled, and if the reason for responding is related to the aims of the study a bias can result. If the study is looking at attitudes among individuals in a community to the provision of nursery facilities for single parents and you find that the proportion of single parents among the respondents is greater than the proportion in the sampling frame then the survey results are likely to be biased towards the opinion of single parents in the community rather than all individuals in the community. In order to combat non-response you need to use follow-up letters, telephone calls and interviews to try to encourage all individuals sampled to respond. A relatively large portion of the survey budget should be devoted to minimizing non-response.

Direct observation is common in traffic studies, supermarket check-out surveys and some psychological studies of interactions among individuals in groups. The use of hospital records, university records, school records and business information systems, where data are extracted from existing sources, can also be thought of as a method of direct observation. The process of extracting the data can be very time-consuming and you can often find that records are incomplete. This means that there may be a large number of missing values in the database. This is a problem of partial response and is a similar problem to that of non-response.

5.1.3 Questionnaire design and data management

Good **questionnaire design** is a crucial aspect of obtaining data. This is especially true for self-administered questionnaires but is also relevant to direct interviews, telephone interviews and direct observation. A well-designed questionnaire which looks good and is easy to use and understand, will help to minimize non-response, which in turn will help to reduce bias. This is discussed further in Section 5.5.

The successful completion of a study requires competent staff and the co-operation of everyone involved. This can be achieved by careful planning of the fieldwork and ensuring that all interviewers and data collectors are trained to the appropriate level. One of the aims of planning must be to ensure that all interviews are completed at much the same time. This can be very important in political opinion polls during elections, when opinions can change rapidly.

In some types of sampling design the interviewer is responsible for the selection of the sample. If this is the case then a sampling plan should be specified. This will give details of the procedure for selection. The sampling frame may be the household but the target population may be the individuals living in the house, only one of whom should be chosen. This can be done at random or according to a prespecified schedule. In either case the sampling plan needs to be specified and the interviewer trained to use it correctly. Interviewers have to be trained in the correct method of asking the questions, what they are permitted to say to prompt a response from a subject and how much information they are allowed to give the subject about the meaning of a particular question.

Many people think that statistics is just concerned with **data management.** They could not be more wrong. Statisticians have a great deal to offer at all stages of a survey, but the area where statistical analysis is most obvious is in the management of the data and in the preparation of the final report. Data management refers of the coding of the responses and entering the data into a computer database, dealing with non-response, data security and data confidentiality issues, data checking and verification, and quality control. This is a vital part of any survey. If there are gross errors in data entry then the results are unreliable.

Good questionnaire design will help to minimize any coding errors, which can be reduced by making sure that as many questions as possible have a fixed number of responses to be chosen from a list. There are many computer database packages, such as Paradox, dBase, Access and Oracle which can be used for data entry. A key principle is to make the data entry as easy and as error-free as possible. One way to achieve this is by devising special data entry forms for each survey to ensure that the data entry screen looks exactly like the questionnaire completed by the subject. In some instances it is possible to use scanners to read in the data. With direct personal interviews it is possible to use a portable computer in the subject's home to eliminate one layer of data transcription and also to speed up data entry.

Many checks can be built into the database to ensure that only valid data are entered and that the data are internally consistent. The former is achieved by specifying that only a fixed number of values are permitted in response to a particular question, If the subject's sex is to be recorded, then only 'male' or 'female' is permitted, possibly with a code for a missing value in case any subjects do not respond to the question. Internal consistency can be verified by making sure that the date of birth, date of interview and age tie up (age should be the difference between the two dates). Dates of employment and unemployment should also be internally consistent in that an individual should not be unemployed in the middle of a period of employment.

5.1.4 Pilot studies and final report

The **pilot study** is a vital part of any survey. This is used to check and modify the survey procedures. The questions are tested and any difficulties with the responses are noted. Any problems with the sampling plan should come to light here. Information may also be available on non-response. The main reason for this part of the study is to check that the survey procedures are all working well so that the main survey can proceed smoothly.

The survey will end with the data analysis, data summary tables and graphs, and the final report. Many of the statistical techniques described in this book will be of use here. A key principle of the report is to make the final conclusions of the survey stand out by using correct and striking graphical displays. The report should also contain details of the survey methodology, together with details of the statistical analysis.

5.2 Sampling schemes

The main random sampling schemes are: simple; stratified; cluster; and systematic. In each case, the individuals sampled are selected at random in some way from the sampling frame. The simple random sample is the simplest method, but is seldom used in practise.

Non-random samples, where the individuals sampled are not selected according to a random mechanism, are used though they do not have good statistical properties being of unknown reliability. The most important is quota sampling, and its use is common. Non-random samples are often used when a large sample is required but the study budget does not support the costs of a random sample.

5.2.1 Simple random sampling

A sample in which every individual in the population has the same chance of being included is a **random sample**. It may not, however, be a simple random sample, since the latter has a slightly more restrictive definition. A **simple random sample** is one in which every possible *sample* from the population has exactly the same chance of being selected.

This, in turn, implies that each individual in the population has exactly the same chance of being selected into the sample, but the reverse implication is not necessarily true.

The selection of a simple random sample begins with a sampling frame which matches the target population exactly. The sample is obtained by using a random selection procedure, for example involving a computer program or a table of random numbers, or drawing numbers out of a hat. Each unit in the sampling frame is assigned a unique identifier, usually a number, and these are selected at random.

Example 5.1: Selecting a simple random sample

Suppose a population consists of just 10 items and we are to select a simple random sample of size 2. The ten items and their values are:

Item	A	B	C	D	E	F	G	H	I	J
Value	18	17	19	18	17	18	20	19	19	25

By writing down all the possible samples, we can establish that there are 45 possible simple random samples of size 2 from 10. We assign identifier numbers to each item in the population, say 0 for A, 1 for B, . . . , 9 for J, and use a list of random numbers to give us a sample **without replacement**. The sample has to be selected without replacement because no item can appear more than once.

If the random numbers are 1 0, then this corresponds to sample B, A, with values 17 and 18 and a mean of 17.5. A second batch of random numbers might be 2 8, corresponding to C, I, with values 19 and 19 and a mean of 19.

If you selected every fifth letter from a random start then the only possible samples would be AF, BG, CH, DI and EJ. You would have a random sample, actually a systematic random sample but not a simple random sample. With a simple random sample there are 45 possible samples, representing all possible samples of size 2 from 10. With a systematic random sample there are only 5 possible samples of size 2 from 10, and, for example, the sample AB is not one of them. While there are many types of random sample, the simple random sample is the only one in which every possible sample has exactly the same chance of being selected.

The main benefits of simple random sampling are that it is a simple procedure for yielding a random sample. It is the optimal random sampling scheme for selecting a sample from a 'uniform' population – a population with no subgroups. The main limitations arise if the population is not 'uniform'. If the sample size is small then the procedure may result in an unrepresentative sample, by chance. Also, if there is a structure to the population which is related to the subject under study but has not been taken account of in the design, a bias can result from over- or under-representation. If the population does contain subgroups (e.g. males and females, geographic regions) then the precision of the estimators may be small with a simple random sample compared to a random sample which takes into account the different subgroups (a stratified random sample).

The sample selection is also time-consuming compared to a more systematic approach or a non-random method. When sampling individuals or households, the sample units may be widely dispersed and so lead to a costly survey. Furthermore, it is not particularly useful if the population is clustered (e.g. pupils within schools) as units within a cluster may be more similar to each other than units in different clusters. If the population has a structure, this should be reflected in the sampling procedure. The structure of the population dictates the type of sampling procedure, and stratified sampling and cluster sampling are commonly used to take into account different population structures.

5.2.2 Estimation in a simple random sample

The total number of individuals in the population is denoted N, and the number in the sample is n. The ratio of these two numbers is the sampling fraction,

$$f = \frac{n}{N}.$$

The mean for the population is estimated by the sample mean, $\bar{x}$, and the population proportion by the sample proportion, which is usually denoted $\hat{p}$. In fact, you can think of the sample proportion as a mean of binary values. If we are measuring the presence of an attribute in a population then the response of each individual in the sample can be denoted $x_i = 1$ if individual i has the attribute and $x_i = 0$ if the individual does not have the attribute. This type of variable is known as a binary variable as it takes only two values, 0 or 1. The total number of people in the sample with the attribute is given by $\sum x_i$ and the proportion by

$$\hat{p} = \frac{\sum x_i}{n},$$

which is the sample mean of a binary variable.

Samples are often used to estimate totals, rather than means or proportions. If you were carrying out a survey for a bank then you might wish to estimate the total amount in deposit accounts. If you take a sample of all deposit accounts then you have the sample mean balance, $\bar{x}$, and the estimated total over all N accounts is

$$x_T = N\bar{x}.$$

The proportion of accounts which have not been accessed in 6 months may be represented by $\hat{p}$, and so the estimated total number of accounts which have not been accessed in 6 months is $N\hat{p}$. Most of this is obvious, but the important point is that in order to estimate a total for the population you need to know the number of individuals in the population. When carrying out a random sample this information needs to be available, however, it is not necessary for a non-random sample.

5.2.3 Stratified random sampling

In many situations the population of interest consists of one or more sub-populations or **strata**. For example, students are: male or female; live at home or away from home; are mature students or have come to university straight from school. Another example might be the different types of customer who have an account with a wholesale firm: single individuals; small businesses; and large businesses. In such cases it is desirable to make sure that each stratum is represented in the sample. This is done to ensure that there is sufficient information on each subgroup in the sample so that estimates can be provided for the subgroups. With a simple random sample it is possible to find that a sample does not have sufficient observations in some of these subgroups.

Generally the units within each of the strata, will be more similar to each other than the units in different strata. For example, an account for a small business is likely to be more similar to the accounts for other small businesses than to that for a large business. This means that part of the variability in the population of accounts is going to be associated with the systematic differences among the accounts as a result of the different types of customer. Stratified random sampling is one way of taking into account this structure in the population. This is accomplished by taking simple random samples from each of the strata and then combining the estimates from each stratum into a single overall estimate.

Stratified random sampling, when appropriate, has the following major advantages over simple random sampling. The cost of collecting the data is often reduced by stratifying the data into homogeneous subgroups which are in the same geographic area. Regional location is often used as a stratifying factor as estimates are often required for different regions. Urban and rural locations constitute another important stratifying factor. The strata would then be in different locations from each other. The precision of the estimator is also increased by stratifying since the variation within the subgroups is usually smaller than the overall population variation.

A stratified random sample is obtained by dividing a population into non-overlapping strata and then selecting a simple random sample from each stratum. Stratified sampling provides estimators for parameters in each stratum. In general, a larger sample is taken in each stratum if (a) the stratum is large; (b) the variance within the stratum is large; or (c) the sampling costs are lower in the stratum.

Suppose that the population has L strata. Each unit is in one and only one stratum. Stratum i has N_i units and the total population has $N = N_1 + N_2 + \ldots + N_L$ units. A simple random sample of size n_i is selected from the N_i units in stratum i and the total sample size is $n = n_1 + n_2 + \ldots + n_L$ units.

There are many ways of allocating the numbers of items to be selected from each stratum; some are much more complicated than others. The simplest way, which is commonly used, is proportional allocation, where the sampling fraction in each stratum is the same:

$$\frac{n_1}{N_1} = \frac{n_2}{N_2} = \ldots = \frac{n_L}{N_L}.$$

This is achieved by setting the sample size in stratum i equal to

$$n_i = n \frac{N_i}{N}.$$

Example 5.2: Proportional allocation in a stratified random sample

In this example it is assumed that there are three strata ($L = 3$) with the following sizes: $N_1 = 100$, $N_2 = 200$, $N_3 = 700$. A sample of size $n = 50$ is to be selected, and this means that the following samples are selected from each of the strata using proportional allocation.

$$N = N_1 + N_2 + N_3 = 1000,$$

$$n_1 = 50 \frac{100}{1000} = 5$$

$$n_2 = 50 \frac{200}{1000} = 10,$$

$$n_3 = 50 \frac{700}{1000} = 35.$$

If a simple random sample were selected, then there is about a 10% chance that there would be 2 or fewer items from stratum 1 in the sample of size 50. Thus there is a possibility that a simple random sample would not be very representative.

As there is a simple random sample for each stratum, estimation of stratum means, totals and proportions proceeds in the same way as discussed in connection with simple random samples. The stratum estimates have to be combined to form an overall estimate for the population using weighted averages.

In stratum i a sample of size n_i is selected from the N_i units in the strata. The sample mean in this stratum is denoted $\bar{x}_i$ and the sample standard deviation is s_i. The estimate of the population mean based on a stratified sample is denoted $\bar{x}_{st}$ and is calculated by a weighted average of the individual stratum means where the weights are proportional to the stratum sizes N_i:

$$\bar{x}_{st} = \frac{1}{N} \sum_{i=1}^{L} N_i \bar{x}_i.$$

Example 5.3: Estimation in a stratified random sample

A stratified random sample of 40 households is selected from a population of 310 households in 3 strata (village A, village B, and a rural area) with a view to estimating the average numbers of hours per week that households in the population watch television. The data are as follows:

i		n_i	$\bar{x}_i$	s_i	N_i
Stratum 1	Village A	20	3.9	5.95	155
Stratum 2	Village B	8	25.125	5.25	62
Stratum 3	Rural	12	19.0	9.36	93

Observe that proportional allocation has been used:

$$\frac{n_1}{N_1} = \frac{n_2}{N_2} = \frac{n_3}{N_3}.$$

The estimated average time spent watching TV is:

$$\bar{x}_{st} = \frac{1}{310} (155 \times 3.9 + 62 \times 25.125 + 93 \times 19.0) = 12.675 \text{ hours per week.}$$

There are differences in the standard deviations in the three strata, s_i, of Example 5.3. This is information that stratified samples give but not simple random samples. In a simple random sample the investigators would be completely unaware of the differences between the three strata.

As the sample means are different in the three strata the precision from a simple random sample from this population would be much smaller than the precision from the stratified sample. The technical, algebraic, details will not be dealt with here but the reason hinges on the fact that some of the variability among the 40 observations in the sample is associated with systematic differences among the three strata.

There are substantial differences in means across the three strata. This implies that stratification has been useful because there will be an increase in precision. Even if there were not large differences in means across the strata, stratification can still be useful as it provides information on the individual strata. As a simple random sample is selected in each stratum we can provide estimates for the stratum means. This is information which could not be obtained by selecting a simple random sample from the whole population of 310 households.

Exactly the same procedure is used to calculate the estimate of a population proportion based on a stratified random sample.

Example 5.4: Estimating a proportion in s stratified random sample
The numbers of households watching programme X in each of the strata are noted. It is required to estimate the proportion watching X in the whole population.

	N_{ii}	n_i	r_i	$\hat{p}_i$
Stratum 1	155	20	16	0.80
Stratum 2	62	8	2	0.25
Stratum 3	93	12	6	0.50

From the data given,

$$\hat{p}_{st} = \frac{1}{N}\sum_{i=1}^{L} N_i \hat{p}_i = \frac{1}{310}(155 \times 0.80 + 62 \times 0.25 + 93 \times 0.50) = 0.60.$$

5.2.4 Cluster sampling

Cluster sampling is used when the units in the population are a collection or cluster of elements. A random sample of clusters is selected. This type of design is very common as all that is needed is a sampling frame of clusters, together with a list of units within the clusters selected. It is the type of design used when you want information about individuals but only have addresses or home telephone numbers as a sampling frame. The household is the cluster, and this can have 1, 2, 3, . . . individuals in it.

Suppose a survey of teenage children under 18 is to be carried out. No sampling frame exists. However, a list of household addresses is available, and each household forms a cluster with 0, 1, 2, 3, 4, . . . teenage children. One appropriate way to select a random sample is to select households at random and interview all the teenage children in the selected households, if any. An extension of this is to select just one of the teenage children at random from the sampled households. The former method is known as **single-stage cluster sampling** and the latter as **two-stage cluster sampling**.

Suppose a survey of school teachers' opinions is to be carried out. A complete list of teachers is difficult to get, but a list of schools is readily available. The school is the cluster and a random sample of schools could be selected and all teachers in a school interviewed. Alternatively, a random sample of teachers in a school could be selected to give a two-stage cluster sample.

The essential difference between stratified and cluster sampling is that in a stratified sample some elements are sampled from each stratum, whereas in single-stage cluster sampling a sample of clusters is selected and all units in these clusters are sampled. In all cluster samples there is some sample selection at the first stage of clusters. Generally speaking, there are a few strata but a large number of clusters. Usually cluster sampling is less costly than stratified sampling, which in turn is less costly than a simple random sample.

The difficulty with cluster sampling is that the units within a cluster are not likely to be independent of each other. One of the most common cluster samples is to select households and then interview the individuals within this household. Quite clearly you would expect there to be an association between the responses of the different household members. They are likely to eat the same types of food, spend money on the same types of activities, go on holiday to the same places. This means that there is likely to be some relationship between the responses of one unit in the cluster and another unit in the same cluster. This is known as the **intra-class correlation** or **within-cluster correlation**. This correlation has to be taken into account when estimating the population quantities. The methods developed for simple random samples are not appropriate for cluster samples as they assume that the units in the sample act independently of each other.

As a result of the intra-class correlation (which is usually positive) the estimates from a cluster sample are usually less precise (have a larger standard error) than a simple random sample of the same size. Contrast this with a stratified random sample which will generally be more precise (have a smaller standard error) than a simple random sample of the same size.

In single-stage cluster sampling all units in the cluster are sampled. A random sample of units within a cluster units can be selected, and this is a two-stage cluster sample. This can be extended to three- and four-stage cluster sampling. Estimation is complicated and requires care. In a sense, the ease of selecting the sample is paid for by difficulties in analysis.

Two-stage cluster sampling gives a random sample of units in the population without the need to construct a sampling frame from all the units in the population. All that is needed is a frame for the primary sampling units, the clusters; once a random sample of these is selected, the frames of the secondary sampling units for the clusters are needed. It is not necessary to enumerate the whole population, and it is possible to obtain a cluster sample without knowing the total size of the population.

5.2.5 Systematic random sample

This is a very quick method of selecting a random sample. It is used when the N units in the population are, in effect, arranged in a long list. A random unit is selected from the first k elements and every kth element is selected thereafter. The sample size is $n = N/k$.

By way of illustration, assume that there is a population of 9 units and that every third item is to be sampled. Thus $N = 9$, and $k = 3$, leading to a sample size of $n = 3$. With a systematic random sample, there are only three possible samples, as illustrated below: {A, D, G}, {B, E, H} and {C, F, I}.

A	B	C	D	E	F	G	H	I
x			x			x		
	x			x			x	
		x			x			x

Systematic sampling is easy and quick. If the population is listed in a random order then systematic random sampling is equivalent to simple random sampling, and the same procedures for estimation can be used. Systematic sampling is not identical to simple random sampling as not all samples are equally likely – in the above illustration {A, E, I} and {A, B, C} among others, are impossible. With the 9 units in the illustration above, there are 84 possible simple random samples and only 3 systematic random samples, so it is clear that systematic sampling is much more restrictive than simple random sampling.

Systematic sampling is, in fact, a special case of cluster sampling. The population in the illustration above is arranged in 3 clusters, and one of these is selected at random as the sample and all the items in the cluster are sampled. This is a single-stage cluster sample with only one cluster selected.

If the population is ordered in such a way that a periodic pattern is present in the data, then systematic sampling can lead to serious bias. Suppose that we are interested in estimating the daily revenue in a shop, which is given for the six shopping days in a week. The sampling frame is thus a list:

$$M, T, W, Th, F, S, M, T, W, Th, F, S, M, T, W, \ldots.$$

If $k = 6$, then the sample will be all Mondays, all Tuesdays, etc. Thus the sample contains only data from one day of the week and is clearly biased. If $k = 2$ or 3 then the possible samples will still only contain data from certain days of the week and not others, and the same serious bias will be present as daily revenue is expected to follow a periodic pattern. In fact it is very difficult to select an unbiased systematic sample for a sampling frame which is ordered over time.

Often the patterns present in the population list are not known before the systematic sample is selected. Thus, while systematic sampling is attractive it is potentially dangerous unless you are sure that the sampling frame is in a completely random order

5.2.6 Non-probability samples

In the discussion of survey sampling methods attention has been focused solely on probability or random samples. These are the only sampling schemes which permit valid statistical inferences about the population under study. Non-probability samples are common, particularly in market research, and the advantages and disadvantages will now be discussed. There are three main types: convenience samples; judgement samples, and quota samples. Of these quota sampling is the most highly regarded and is the technique of choice for most market research surveys.

A **convenience sample** consists of a group of units which are readily available. Examples include patients in a hospital ward, members of a class, people in a street, volunteers for a study. There is no theoretical justification for such a sample. It is extremely likely to be biased, but the magnitude of this bias is unknown. As there is no random selection of units from the population no standard errors can be calculated, and consequently the investigator can have no idea of the precision of the estimate. Often with a convenience sample there is no clear idea of the population under study. Convenience samples are most useful in pilot studies to check the sampling procedure and questionnaire design.

In experiments **judgement samples**, where the opinion and experience of an expert is used to select a sample, have been shown to be seriously biased despite being supposedly representative. If you try to estimate the mean weight of a group of individuals with a judgement sample then bias can result by always selecting a large person and a corresponding small person to balance out the large one. In this case while the average might be unbiased the variation will tend to be overestimated as there will tend to be extreme individuals in the sample. The converse can occur by selecting only individuals who are near the average weight into the sample, giving a sample with a reduced variation.

5.2.7 Quota sampling

This is a non-random version of stratified random sampling. It was devised in an attempt to ensure that simple random samples yielded representative samples at all times by making sure that individuals for the subgroups of the population are always sampled.

Suppose we have a population of 10 000 people: 5500 men and 4500 women. A simple random sample of 100 individuals could quite easily yield 45 men and 55 women. Such a sample might be considered unrepresentative as 55% of the population are men but only 45% of the sample. Stratifying would overcome this problem, and the stratified random sample would be designed to select $n_m = 55$ men and $n_w = 45$ women, using proportional allocation.

Quota samples also ensure that $n_m = 55$ men and $n_w = 45$ women are selected, but accomplish this by a non-random selection process in that no sample frame is necessary and the interviewer decides which people to include in the survey and which to exclude. In order to collect the data in a quota sample the interviewer simply goes out into the population, either into a high street, a housing estate, or by telephone, and picks people until the quotas of men and women are both filled up.

Quota samples are cheap and quick. They will always result in a sample which is representative of the population on the quota characteristics. As the sample selection is non-random there is the likelihood of unmeasured bias, and also no standard errors can theoretically be calculated. It is common practice to use the same process as for a stratified random sample, but this is not correct.

One of the main sources of bias in any survey arises from non-response. In a quota sample there may not be a record of the number, or type, of people who refused to participate in the survey. If someone refuses to answer the questions the interviewer simply goes and asks someone else. This may lead to a selection bias.

Another source of bias is best seen from examples. If an interviewer is required to fill a quota of 20 women then she may go into the street one morning. She is then likely to meet mainly women who do not have a full-time job, and this will lead to a biased sample. Alternatively, suppose an interviewer has to fill a quota of 20 men, with five in each age range 18–30, 31–45, 46–64 and 65+. It has been shown that typically people in the middle of the age ranges (22–25, 37–40, 50–54, 70–74) are selected, and people at the extremes of the age ranges tend not to be represented.

Quota sampling can be quite complicated as gender, age and social class are commonly used as controlling variables. This means, for example, that the interviewer can be out looking for a man aged 30–34 in social class AB, or a woman aged 65+ in social class C, etc. Examples of quota samples are discussed in Section 11.5.2 and the file web-Analysis.doc on the book's website.

Virtually all opinion polls are quota samples though this changed in the mid-1990s in the United Kingdom with the advent of telephone interviews. Collection of the data for a quota sample is a relatively quick procedure: opinion polls are generally taken over a period of 2–3 days. A random sampling procedure could not be gathered together so quickly. There are about 1500 voters in an opinion poll, and a stratified random sample of 1500 people can take about 2–3 weeks to gather, using a similar number of interviewers and taking into account call-backs and travel.

5.2.8 Summary

Non-probability samples are likely to be subject to bias which cannot be assessed. Random samples may also be biased due to non-response. However, the non-responders are known and further study can be carried out to see if the non-response is random or if it leads to bias. This means that there is the possibility of quantifying the bias and correcting the sample estimates to take this into account. Such a procedure is known as 'weighting the sample'. Also, as there is no random selection of units from the population in non-probability samples there is no mechanism for the calculation of the standard errors of the estimates. As a consequence the precision is unknown so the investigator has no idea of how good the estimate is.

Non-probability samples have their uses (pilot studies) and are quick and cheap. They are no substitute for random samples, and well-run random samples, with adequate checks on non-response and bias, are to be preferred in virtually all cases. The conclusions based on random samples can be defended on statistical grounds, but any results from non-random samples can always be attacked on the grounds of subjective collection of data and unmeasured response bias.

5.3 United Kingdom large government-sponsored surveys

There are a large number of government-sponsored surveys in the United Kingdom. These have an impact on many walks of life. The major features of some of the more important surveys will be highlighted in this section. The Office for National Statistics website has a great deal of information about these surveys (http://www.statistics.gov.uk/ census2001; http://www.statistics.gov.uk/themes). One important feature of these surveys is their continuity over a long period of time, which gives a way to monitor the changes in society. Many of these changes are documented in an annual publication, *Social Trends*, which gives a very good description of the changing face of society and attitudes in Britain. In this section we will review some of the more important surveys, but we begin with what is arguably the most important survey, though it is not in fact a sample – the Census.

5.3.1 Population Census

The Population Census of the United Kingdom has been held every 10 years since 1801, except for 1941 when it was missed because of World War II. The variables which are recorded include a head count, age, employment status, sex, marital status, ethnic origin, housing, and ownership of domestic goods. It is difficult and expensive to conduct an accurate census of a large population. A decision has to be taken on how and when to collect the data. In the United Kingdom the collection is done on a house-to-house basis by paid enumerators. The form requires the head of the household to enter details of all those who are actually present in the house on Census night – a *de facto* basis. Census night is a Sunday night in April chosen to avoid Easter Day. In the 2001 Census it was 29 April.

The United States Census requires reporting of all those who are normally resident in the house – a *de jure* basis. Experience has shown that information collected on a *de facto* basis yields numbers which are about 0.2% above the normal population because of foreign visitors.

On previous occasions in the UK an enumerator would assist the head of each household in filling out the census form and then collect it. In the 2001 Census, while the enumerator called on each household, the form could be returned by post and so the Census has become a bit like a postal survey – except that there is a legal obligation on the head of each household to complete the form, with a penalty of a large fine in the event of refusal to comply. In the Spanish Census of 2001 it was possible for some households to complete their form online through an internet site.

The 1991 Census in the United Kingdom had a problem with undercount. A substantial number of people were not included in the population count. Many of these were people who lived in non-standard housing, and enumerators are now required to include tents and caravans as well as fixed dwellings. Even this may not reach the homeless or others of no fixed abode. There is also always a perennial problem of people who have two 'usual' addresses, such as students who have a college address and a parental address.

5.3.2 Family Expenditure Survey

The UK Family Expenditure Survey (FES) is carried out by the Office for National Statistics and the Northern Ireland Statistics and Research Agency. It has been conducted annually since 1957, and about 7000 private households are surveyed by personal, computer-assisted interview. The data are collected throughout the year and throughout the United Kingdom to allow for seasonal and regional variation in income and spending patterns.

The primary use of the FES is to provide information about spending patterns for the retail price index, although over recent years the uses of the FES have been extended. The survey concentrates on household personal income and expenditure on goods and services. Information is also collected on the socio-economic characteristics of households – composition, size, social class, occupation, and age of the head of household.

The FES consists of a comprehensive household questionnaire which asks about regular household bills and expenditure on major but infrequent purchases. Each adult (aged 16 or over) in the household is also asked detailed questions about their income and is asked to keep a two-week diary of all personal expenditure; and children aged 7–15 are also invited to keep a diary of their own expenditure. The target sample size for the FES is 11 400 addresses per year; these are selected from the Postcode Address File, which is a comprehensive list of all postal delivery addresses in the UK.

In 1999–2000 full response was achieved from 63% of households that were eligible for the survey. This is quite low for an interviewer survey. However, the FES operates vary strict response rules – households count as responding only if the household expenditure questionnaire is complete, and all adults complete an income questionnaire (without refusing any item of information) and keep a two-week diary of all their expenditure. It only needs one adult to refuse to contribute one source of income or expenditure for the household to be deemed a non-responding household. This is a strict criterion as the more common procedure would be to only exclude households where all members refuse all information.

5.3.3 General Household Survey

Each year, approximately 10 000 households, and about 20 000 adults aged 16 or over, are surveyed in the General Household Survey (GHS). This is a multi-purpose survey which collects information on a range of topics from people living in private households in Great Britain. Data are collected for all of Great Britain and some regional information is available. Approximately 13 000 addresses are selected each year from the Postcode Address File. All adults aged 16 and over are interviewed in each responding household. The response rate is 72%.

This survey began in 1971 and focuses on five core areas: family composition, housing, employment, health and education. It has been used to fill in the gaps between censuses. Although a census is carried out every 10 years the numbers of people living are required for every year and the data from the GHS are used to adjust the census figures taking into account the numbers of births and deaths as well. In 1997, the Office for National Statistics carried out a review of the GHS, which concluded that the survey should be relaunched with a different design. The new survey will consist of a continuous section, focusing on the major aims of the survey above, which will remain unchanged for five years. In addition, additional special surveys will be included on an *ad hoc* basis.

5.3.4 Labour Force Survey

The UK Labour Force Survey (LFS) is carried out for the Office of National Statistics Socio-economic Division. This survey is a continuous household survey, which provides a wide range of data on labour market statistics and related topics such as training, qualifications, income and disability. The UK is required by European Union regulations to carry out a labour force survey annually. Results from the spring quarter of the LFS are supplied to Eurostat each year to meet this requirement.

The LFS focuses on the economic activity of those aged 16 and over, those in employment, occupation, industry worked in, job-related training, hours of work, those unemployed (under the International Labour Organization definition) and the length of time they had been without a job, redundancies, those economically inactive and the reasons for this. Information is also collected on trade union membership, disabilities, sickness absence, journey to work time and distance, and income. Some demographic characteristics are also collected.

In previous years this was a controversial survey as it yielded a much higher estimate of unemployment than the official government figures, which were based upon the number of claimants of unemployment benefit. However, since April 1998 the LFS figures have provided headline UK unemployment and employment figures each month for the preceding quarter.

The survey sample of addresses is taken from the Postcode Address File. In addition, a small sample of addresses of NHS and health trust accommodation is included in the survey and anyone aged 16 or over and at boarding school or living in a hall of residence is included in their parental household. The survey has a stratified random sample, and within any continuous 13-week period at least one household from every postcode sector is sampled. This feature allows representative results to be produced for any 13-week period. The survey has a panel design where each sampled address is interviewed on five occasions. As the interviews take place at three-month intervals the fifth interview at an address takes place one year after the first. Data from around 59 000 households, including about 138 000 individuals, are collected, and a 79% response rate is achieved.

All first interviews (with the exception of a very small sample located north of the Caledonian Canal) are carried out by a team of face-to-face interviewers who work exclusively on the LFS. Subsequent interviewers are carried out, where the informant is willing, by telephone. Over 60% of all LFS interviews are now conducted by telephone.

Data are produced separately for England, Scotland, Wales and Northern Ireland, as well as for regions and counties. This is one of the reasons for the extremely large sample size. Initially, the LFS was conducted every two years from 1973. From 1984 it was conducted annually. Then, in 1992 it became a quarterly survey, except in Northern Ireland, where it remained an annual spring survey until winter 1994–95, when it too became quarterly. The main results for the LFS for Great Britain are available six weeks after the end of the reference period, with full UK results published three months after the reference period. For further details, see http://www.statistics.gov.uk/themes/labour_market/surveys/labour_force_text.asp.

5.3.5 British Crime Survey

The British Crime Survey (BCS) measures the amount of crime in England and Wales (Scotland and Northern Ireland have their own surveys). This includes crimes which may not have been reported to or recorded by the police. It thus provides an important alternative to police-recorded crime statistics. Without the BCS the government would have no information on unreported crimes. The BCS also helps identify those most at risk of different types of crime. This is used in designing and informing crime prevention programmes. It is also used to assess people's attitudes to crime and towards the criminal justice system. It is one of the major sources of information about levels of and public attitudes to crime.

The BCS started in 1982. The 2000 BCS, the eighth survey, was carried out by Social Survey Division of the ONS and the National Centre for Social Research. It covered a sample of 20000 individuals, with a boost of 4000 individuals from ethnic minorities. The achieved response rate was over 80%. One person (aged 16 or over) is selected, at random, for interview at each selected address. This is a cluster sample with the household forming the cluster and one individual selected from each cluster. The interview is administered using a computer-assisted personal interview. There are two self-completion sections at the end of the questionnaire, and the respondent is encouraged to do these on the laptop themselves. This is thought to be a more impersonal way of obtaining some information about crime which may in some cases be quite sensitive.

Topic areas include: experiences of crime, both property and personal; attitudes to the criminal justice system, including the police and the courts; worries and fears about crime; violence at work; perceptions of equality and prejudice; illegal drug use; and sexual victimization.

5.4 Differences between major election opinion poll groups

The 1992 general election in the United Kingdom yielded a Conservative victory with a lead of 7.6% over Labour. The five major opinion poll companies – Gallup, Harris, International Communications and Marketing (ICM), Market and Opinion Research International (MORI), and National Opinion Polls (NOP) – all predicted a Labour lead of 1.3% averaged over the different polls. The margin of error in the opinion polls was wider than can be accounted for by sampling error, and the polls were criticized for their inability to perform up to expectations.

Before 1992, there were 13 general elections in the United Kingdom with 'eve of poll' forecasts, and the record of the opinion polls was generally quite good. In these elections there were 51 election forecasts based on opinion polls and in about a quarter of these the share of the vote for the two main parties, Labour and Conservative, was estimated to within 1%, and in over half of the polls the share of the vote was estimated correctly to within 2%. Thus the forecasts had been quite accurate.

Election opinion polls in the United Kingdom were generally carried out using a quota sample with personal interviewers in a number of locations throughout the country. In the run-up to the 1992 general election the sample sizes of the final eve of poll predictions ranged from 2478 for a Gallup poll to 1731 for a MORI one, though there had earlier been an ICM poll which sampled 10 460 individuals in 330 sampling points and took nearly a week to complete. Different companies use different locations and different numbers of locations, ranging from 52 for ICM and 54 for NOP and MORI to 110 for Harris and Gallup, though there are deviations from these numbers for specific surveys.

The Market Research Society produced a report on the opinion polls leading up to the 1992 general election. The first conclusion was that the media and the public tend to expect a greater degree of accuracy from polls than can be delivered. A poll of 1000 individuals has a precision of about ±3% in the percentage support for the parties, see Chapter 10 (sampling distributions, proportions). Many in the media and the general public do not understand the implications of sampling variability. A precision of 3% implies that the difference between 47% and 44% could simply be due to sampling variability. Essentially the percentages are indistinguishable within the limits of sampling variation because they are based upon samples, and different samples will give slightly different results. The accuracy of the polls is related, among other things, to the sample size. In the absence of bias, increasing the sample size increases the accuracy.

The failure of the 1992 polls was attributed to: (1) a late swing to the Conservatives in the last 24 hours before the election, which cannot be covered by an opinion poll; (2) differential response to the interviews by party of preference, such that individuals who were intending to vote Conservative were more likely to give the wrong response, for example by replying 'don't know', than those not intending to vote Conservative, (3) differential non-response, such that individuals who were intending to vote Conservative were more likely to refuse to participate in the interview than voters from the other parties; and (4) an effect due to quota sampling.

Point (1), an extremely late swing in voting intentions, is very difficult to deal with by any opinion poll used as a forecast. Opinion polls are best viewed as cross-sectional snapshots summarizing opinion at specified time points. The other three points are all statistical in nature and all contributed to a bias in the polls in favour of Labour. Points (2) and (3) come under the general heading of response and non-response bias, which will be discussed in Section 5.6. Response bias brought about by the failure of a subject to answer a question correctly can be addressed by good questionnaire design, though this may not totally remove the problem.

As to point (4), a quota sample is supposed to result in a representative sample from the population under study because individuals are sampled according to predefined quotas derived from variables such that the distribution of these variables in the sample matches the distribution in the population. A bias will be introduced if these quotas do not reflect sufficiently accurately the social profile of the population. One variable used is the proportion of council tenants. In the 1991 Census, which was not available at the time of the 1992 general election, 19.5% of individuals were council house tenants. As the Census results were unavailable the opinion poll companies used the results of the National Readership Survey of 1991 as the basis for forming the quotas, and this gave an estimate of 21.5% of individuals in council housing. Information on the percentage of council tenants was available for 22 of the 1992 election polls carried out in the month before the election, and this ranged from 17% to 25% with a median of 24%. Thus the quota samples tended to produce a bias towards council tenants, who tend to be more likely to vote Labour. This in itself is not a failure of the method of quota sampling so much as a failure of the implementation of the method. The target distribution of the quota factor (household tenancy) did not match up to the population.

As a result of reviewing the opinion polls, the five main companies modified the way in which they were to conduct their polls in the future. The main points are listed in Table 5.1. In the 1992 election there were no substantial differences among the five companies as regards the sampling and estimation methodology, whereas the responses to the deficiencies by the different companies have produced methodological differences.

Table 5.1 Characteristics of opinion polls

Characteristic	MORI	HARRIS	NOP	ICM	Gallup
Sampling	Quota, in home	Quota, in home	Quota, in home or on street	List based in home	Random, in home
Mode	Face-to-face	Face-to-face	Face-to-face	Phone	Phone
Don't knows	No imputation	Past vote	Party identification, the economy	Past vote	Past vote, best prime minister, the economy
Weight by past vote	No	No	Yes	Yes	No

Source: O'Muircheartaigh and Lynn (1997, Table 1).

The most important difference is in the sampling method, with two companies adopting telephone interviews. Gallup used random digit dialling and an objective method of selecting individuals within a household (nearest birthday) to give a probability sampling method. ICM used a quota sample based upon the telephone directories. The other three companies remained with the face-to-face interviews, but with improvements to the method of selecting quota controls.

The other major difference among the companies was in the treatment of the 'don't knows'. MORI made no adjustment, while Harris and ICM made an adjustment on the basis of the past vote. One way this would work is as follows. An individual is asked which party he or she will vote for. If the response is 'don't know' then a subsequent question is asked about what party the subject voted for at the last election. When it comes to estimating the percentage of individuals intending to vote Conservative this is based upon the numbers responding Conservative to the first question and those 'don't knows' who voted Conservative last time. This technique is known as imputation. Clearly this procedure will have no effect for people who did not vote at the last election perhaps because they were too young, but NOP and Gallup base their imputation on other variables such as the participant's opinion on the economy.

In their review of the performance of the polls in the 1997 election, O'Muircheartaigh and Lynn (1997) concluded that the ICM poll appeared to have performed best, though Gallup, because it used a random sampling method, had a better coverage of the whole electorate. Both polls using telephone dialling had information about non-response bias on an area basis. For the other three companies using traditional quota sampling methods, any unit non-response bias (where a person refuses to answer any questions) is inseparable from coverage and sampling bias.

5.5 Questionnaire designs

The questionnaire is the means by which information is obtained in a survey, consequently it is one of the most important aspects of survey design. A poor questionnaire design may make it difficult for individuals to complete the questionnaire and may lead to an increase in refusals to return it, if it is a self-administered questionnaire. If a direct interview is being carried out, poor questionnaire design may lead to an increase in item non-response for questions asked during the latter part of the interview. A well-designed questionnaire will have an attractive layout, with questions which are unambiguous and easy to understand, thus leading to an increase in the response rates.

Much questionnaire design is concerned with worrying about getting the correct information from what may seem to be relatively uncommon subgroups of society. With a direct interview in person or over the telephone clarification can always be given by the interviewer, but with self-administered questionnaires it is vital that all questions are very clear and unambiguous.

There are four points to consider in the design of a questionnaire. These apply to self-completed questionnaires as well as ones used by interviewers. The most important is the information being sought, why is it needed and how the results will be analysed. The purpose of the survey dictates the questions which will be asked. The response scale for the questions has an influence on the type of analysis which can be carried out on the data collected. There are four types of variables, measured on nominal, ordinal, interval and ratio scales. Most social and business surveys deal with the first two. We begin this section with an example of good questionnaire design and testing, and close it with examples of different types of questions.

5.5.1 Case study: the ethnic group question from the 1991 Census

Designing good questions and testing them can be a very time-consuming process. Sillitoe and White (1992) provide details of the search for a question about ethnic group for use in the UK Census. A question was introduced for the first time in the 1991 Census, and the first field trials began in 1975. Between 1975 and 1979 there were three separate field trials of four question designs and the question used in the main census trial is shown in Figure 5.1. The results of the trials and the main census test in April 1979 were poor, with a response rate of 54% compared to a response rate of over 70% normally achieved in similar trials. Furthermore, there was a low percentage of correct answers, with only 14% of West Indian and 34% of Asian households answering the question correctly. As many as 32% of the Asian and West Indian households objected in principle to the inclusion of a question on ethnicity in the Census, and even more objected to the question about birthplace. As a result of these trials, no ethnic question was included in the 1981 Census.

The absence of a question about ethnic origin made the results of the 1981 Census much less useful as there was no national information about ethnic groups. With a view

V(A) Parent's country of birth	
Write the country of birth of	
a the person's father	a Father born in (country)
b the person's mother	b Mother born in (country)
This question should be answered even if the person's father or mother is no longer alive. (If country not known write 'NOT KNOWN'). Give the name by which the country is known today	

V(B) Racial or Ethnic Group	
Please tick the appropriate box to show the racial group to which the person belongs. If the person was born in the United Kingdom of West Indian, African, Asian, Arab, Chinese, or 'Other European' descent, please tick one of the boxes numbered 2 to 10 to show the group from which the person is descended.	1 ☐ English, Welsh, Scottish or Irish 2 ☐ Other European 3 ☐ West Indian or Guyanese 4 ☐ African 5 ☐ Indian 6 ☐ Pakistani 7 ☐ Bangladeshi 8 ☐ Arab 9 ☐ Chinese 10 ☐ Any other racial or ethnic group, or if mixed racial or ethnic descent (please describe below)

Source: Sillitoe and White (1992, Figure 2)

Figure 5.1 Ethnicity question trialled in 1979.

to including an ethnicity question in the 1991 Census, a second series of field trials were carried out in 1985–86, with a major test of the whole census in April 1989. The ethnicity question tested in 1989 is given in Figure 5.2, and a number of differences from the 1979 version can be seen.

Approximately 90 000 households took part in the 1989 Census test in six locations in England and Scotland. The response rate for the test ranged from 42% in Birmingham to 94% in Berwickshire, with an overall response rate of 60%. There was a post-enumeration survey of 2322 households, of whom 81% took part in the survey. Less than 0.5% of those who did not return the census test gave the ethnicity question as the reason for not doing so. The main reasons people gave for not taking part in the census test were that they forgot about it, were too busy, or could not be bothered, accounting for over 50% of the refusals. The accuracy of the question was tested by comparing the results of the post-

11 Ethnic Group		
Please tick the appropriate box		
	White	1
	Black Caribbean	2
	Black African	3
	Black Other (*please describe*)	4
If the person is descended from more than one ethnic or racial group, please tick the group to which the person considers he/she belongs, or tick box 9 and describe the person's ancestry in the space provided.	Indian	5
	Pakistani	6
	Bangladeshi	7
	Chinese	8
	Any other ethnic group (*please describe*)	9

Source: Sillitoe and White (1992), Figure 9

Figure 5.2 Ethnicity question trialled in 1989.

enumeration survey with the census test, and this question was found to be 90% accurate for whites, 86% accurate for blacks and 89% accurate for Asians.

Sillitoe and White (1992) round off their discussion of the testing and development of the ethnicity question by stating that 'the question which was included in the 1991 Census is a compromise between obtaining the type and detail of information that users require and devising a question which members of the public understand and will answer'. Although it is likely to attract criticism from individuals at either extreme, it was based upon empirical evidence and with the co-operation of the groups most closely involved with the issue such as the Commission for Racial Equality.

The illustration of the process of developing and field-testing the ethnicity question for the 1991 Census shows great care and detail. Admittedly the Census is the most important survey, and one would expect a greater attention to detail there than in any other survey. The two main points to note are the large number of revisions of the questions tested and the length of time before the Census the field-testing took place. In most sample surveys there are not so many revisions, nor does the pilot study take place so long before the main survey. The key aspect of the illustration is that the aim was to develop a question which would obtain accurate information and would be understood and answered by the public. This question was further revised for the 2001 Census.

5.5.2 Quantitative, interval or ratio scale variables

For questions which have a response on an interval or ratio scale it is usual to ask for a value such as:

What is your age? _____ ☐☐ (years)

What is your height?   ☐☐☐   *or*   ☐   ☐☐
                                             Centimetres   Feet   Inches

These are relatively straightforward, though you need to consider the units of recording carefully. Most people in mainland Europe and young people in the United Kingdom will know their height in metres or centimetres, whereas older people in the United Kingdom and people in the United States and Canada will not know their height in centimetres but will only know it in feet and inches. Similarly for weight.

Even asking a woman a simple question such as

How many children do you have?   ☐☐

can lead to a number of difficulties depending on the purpose for which you wish to use the information. Imagine a woman who has a rather unfortunate history of births, with two sons aged 12 (twins), a daughter aged 10 and an adopted daughter aged 4. This woman also had a miscarriage before the birth of her twin sons, a child who died at 3 months and has a 14-year-old niece living with her. She will answer 3 or 4 or 5 to the question, depending on how she interprets the meaning of the question in relation to the baby who died, her adopted daughter and the niece currently living with her. She has three natural children still alive, she has given birth to four children, she currently has four children including one adopted, and she is currently looking after five children.

If the main purpose of asking the question is to obtain information about family size, then you need to ask

How many children do you currently have living with you?   ☐☐

You will also have to specify details about the ages of the children. In some instances you may also want to know if they are financially dependent upon the woman. If you wish to know about family composition with regard to adoption and fostering, then you need to ask details about the individual components. For example:

How many children have you given birth to?   ☐☐

How many of your natural children live with you?   ☐☐

How many legally adopted children live with you?   ☐☐

How many other children live with you?   ☐☐

You may find that you need to have a more detailed set of questions, but for the woman in the example these will suffice. She will answer 4 to the first, 3 to the second, 1 to the third and 1 to the fourth. We will still not extract any information about the miscarriage or the twins.

5.5.3 Qualitative, ordinal and nominal scale variables

Most social surveys collect nominal and ordinal data, and there are a number of type of questions for extracting the relevant responses. Ordinal data are usually ranked on a scale of 4–7 points, though other scales may be used. Such response variables are normally gathered from closed questions where the respondent has to choose from a limited number of alternative responses. Information on nominal variables can be collected using closed questions or open questions where any response is permitted.

The example below is of a question which results in an ordinal response. It is a closed question as there are only four possible response categories.

How would you describe your level of physical activity during leisure time? (*Please tick one box.*)

Mostly sedentary, such as reading and watching TV ☐

Involved in some walking or light exercise ☐

Involved in regular walking or moderate exercise ☐

Engaged in regular, vigorous training or sport ☐

This question is trying to obtain information about the level of physical activity during leisure time, and it does so in the context of examples of the type of levels of activity that presumably are important in the study. An alternative way of recording the amount of physical activity would be on a ranked scale.

How would you describe your level of physical activity during leisure time? (*Please tick one box.*)

No physical activity ☐

Mild activity ☐

Moderate activity ☐

A great deal of physical activity ☐

This is probably not as good question as the first one as the amount of physical activity in each of the responses is not very well defined. Thus someone who takes a walk every day may consider themselves to have moderate or mild activity depending on how they viewed the scale.

Another version of the same type of question is:

How often do you engage in regular, vigorous training or sport? (*Please tick one box.*)

Never ☐

About once a month ☐

About once a week ☐

About two or three times a week ☐

Daily ☐

Here there is a clear and unambiguous ranking to the responses. There is not likely to be much confusion about the meaning of the grades, though to be absolutely clear the grades corresponding to 'about once a month' and 'about once a week' may have to be tightened up to 'Not more than once a month' and 'Between twice a month and once a week' so that there is no potential confusion for a person who takes a five-mile run every two weeks.

While the vigorous physical activity question has a better set of response grades and is less ambiguous, there is still some doubt as to the exact meaning of vigorous physical activity, and it may be necessary to include some examples. Also in terms of the original question about level of activity during leisure time only one aspect is covered, and to capture fully someone's level of activity during leisure time it will probably be necessary to ask similar questions about watching television and walking, though with different response levels.

For variables which are recorded on nominal scales where there is no clear order, there is a choice between an open-ended question and a closed question. Generally closed questions have many advantages if there are only a limited number of responses because the coding of the responses and the analysis is easier. Questions about occupation are generally open-ended questions.

> What is your occupation or main activity? If you are not currently employed please write down your last occupation.

This question will have to be coded at the data entry stage using, for example, the National Statistics Socio-economic Classification (NS-SEC). The NS-SEC (see http://www.statistics.gov.uk/nsbase/methods_quality/ns_sec/default.asp) is an occupationally based classification but has rules to provide coverage of the whole adult population. The information required to create the NS-SEC is occupation coded to the unit groups of the Standard Occupational Classification 2000 (SOC2000) and details of employment status (whether an employer, self-employed or employee; whether a supervisor; number of employees at the workplace). This is a considerable amount of work, but is a more efficient way of collecting the information than a closed question which may run into several pages with the different types of occupations. The Labour Force Survey and the General Household Survey, which both deal with occupation, have considerably more detailed questions about occupation.

When there is a choice it is better to use a closed question, as this controls the possible responses. Also you know the responses you will get and this then allows you to focus on the methods of analysis at the design stage.

> Please indicate your current employment status. (*Please select only one category.*)

Work more than 20 hours per week	☐
Work less than 20 hours per week	☐
Retired	☐
Retired, but work part-time	☐
Looking for work	☐
Unemployed, not looking for work	☐
Unable to work for health reasons (disability)	☐
Other	☐

This is a closed question with various categories which are supposed to be mutually exclusive, though in practice may not be. Someone could be working and at the same time looking for work, so the ambiguity could be cleared up by replacing the 'looking for work' response with 'Unemployed and looking for work'. If someone responds by ticking the 'Other' box then there is a loss of information as there is no scope to write down the reason for this response. The question could be improved by replacing the last option with:

Other, *please specify*: _____ ☐

If an open-ended question were used,

Please write down your current employment status

_____

then there would no control over the responses. Individuals might reply 'working', 'at home', 'not working', 'retired', 'part-time', 'don't know', 'plumber', 'nurse', etc., and there would be much confusion and many coding problems. Also no information would be obtained on the number of hours worked per week, and a separate question would be required for that.

5.5.4 Question clarity and impartiality

A general rule is to set questions which are straightforward and clear, with easily understood response options. If there are filter questions then clear instructions have to be given as to the appropriate steps to be taken on answering it. Usually filter questions have a yes/no answer where if you answer 'yes' you answer one set of questions and if you answer 'no' you go on to a different set. For example:

13. Have you left school?	Yes ☐	
		To to Question 15, page 2
	No ☐	
Answer Question 14		

Questions should be framed in as neutral, or impartial, a way as possible. Loaded words or phrases should be avoided, such as:

Many people do not agree with the introduction of job creation schemes for young school leavers. Do you agree with the following statement?
    The introduction of job creation schemes for young people is a waste of public money.
(*Please select only one category.*)

Strongly agree ☐

Agree ☐

Neither agree nor disagree ☐

Disagree ☐

Strongly disagree ☐

In this example the use of the phrase 'Many people do not agree with' carries a loading with it such that the respondent may feel obliged to agree with the statement. The phrase is unnecessary and a neutral question would be much better.

With questions which are perceived to be embarrassing there tends to be a great deal of item non-response, and there have been many studies of ways to improve response and response accuracy for such questions. Questions dealing with sexuality are notoriously difficult to get responses to, accurate or otherwise. In a study I analysed recently, more than 30% of men aged 60 or over did not answer questions concerned with sexuality even although they answered questions concerned with other aspects of their quality of life.

In the National Survey of Sexual Attitudes and Lifestyles, item non-response was low for the face-to-face interviews: less than 2% for the question about age at first heterosexual intercourse. The self-completion questionnaire from this survey on sexual partners and practices had a refusal rate of 4%, but once an individual had agreed to complete the booklet item non-response was low at less than 5%. The item non-response rates quoted for the National Survey of Sexual Attitudes and Lifestyles refer to those who actually participated in the survey. Overall there was a 63.3% response rate. This is a similar response rate to other surveys of sexual behaviour. For more details, see http://www.natcen.ac.uk/research/surveys/index.htm.

Questions should normally not be phrased using emotive language which might provoke extreme responses. In English words like 'racist', 'fascist' and 'aristocrat' tend to provoke a hostile reception in certain quarters, and their use may influence the responses. That said, there may be situations where the use of some emotive words is unavoidable: if the study is about attitudes to the public schools, then there may be no option but to use terms such as 'aristocrat'.

Many of the considerations about questionnaire design are just a matter of common sense. Questions should not have double negatives in them. 'Do you not agree that many people are not helpful? is a poorer question than 'Do you agree that many people are helpful?' Even then the question is still rather vague as there is no context. There should only be one point to each question.

Are you in favour of a parliament in Scotland or

should Scotland remain in the United Kingdom? Yes ☐   No ☐

The above is a bad question because it is really two questions which are partly contradictory. In this instance it would be much better to set two separate questions.

Questions with extremely long lists in them tend not work too well as response tails off. So a question such as

Which towns in Britain have you visited during the last year

(*Please tick the appropriate boxes from the following list of 100 towns.*)

will be a waste of space and is not likely to lead to any more accurate response than just asking people to write down a list of towns that they have visited during the last year which can then be coded later. One reason for giving a list is to prompt the subject's memory. You may forget that you spent a day in Bangor in the rain but if you see the name on a list then you may recognize it and recall that you had been there. However, a list of 100 is not likely to be read thoroughly.

Any questions which deal with recall entail a loss of accuracy the longer a subject is asked to recall from the past. One of the intriguing difficulties with the National Survey of Sexual Attitudes and Lifestyles is the discrepancy between the average number of heterosexual partners reported by men and women. Men in the survey reported an average number of 2.6 partners in the last 5 years, whereas women reported an average of 1.5. This type of discrepancy is common in all surveys of sexual lifestyle and one reason for it is over-reporting by men and under-reporting by women. This is related to the length of time the subject is asked to recall in the past. Over the last two years men report an average of 1.6 partners and women and average of 1.1; over the last year the corresponding averages are 1.2 and 1.0, while over the last four weeks they are 0.75 and 0.74. Generally you would expect these two averages to be equal, and they are over a short recall period but not over a much longer period.

Questionnaires which deal specifically with events in the past – such as periods of employment and unemployment, periods of cohabitation and marriages in the British Household Panel Study, and past diet, as in studies of the effect of diet and alcohol consumption on the risk of disease – all use special techniques such as lifetime diaries and specially trained interviewers to help prompt recall and increase accuracy.

5.5.5 Question and response order

Question order and response order can influence the response. Even a simple thing such as changing the order of the response options in a question can have some influence. You may find that you get different responses to the following, even though the question is essentially the same.

In general, would you say that you have confidence in the ability of the Prime Minister to govern the country?

☐	☐	☐	☐	☐
Strongly Agree	Agree	Neither Agree nor Disagree	Disagree	Strongly Disagree

In general, would you say that you **do not** have confidence in the ability of the Prime Minister to govern the country?

☐	☐	☐	☐	☐
Strongly Agree	Agree	Neither Agree nor Disagree	Disagree	Strongly Disagree

In general, would you say that you have confidence in the ability of the Prime Minister to govern the country?

☐	☐	☐	☐	☐
Strongly Disagree	Disagree	Neither Agree nor Disagree	Agree	Strongly Agree

Furthermore, asking the same question immediately after a sequence of questions about the monarchy could yield different responses compared to the same question after a sequence of questions about Britain's international affairs.

Question order and response option order can influence the level of item non-response and the accuracy of response. There is no single correct order for listing questions in a questionnaire. In general, it is a good idea to have factual questions first and attitudinal questions later, with questions which are likely to influence non-response last, i.e. embarrassing questions. Another overriding principle is that the questions which are crucial to the study are asked near the beginning, particularly if there is a direct or telephone interview. In this way you can often get responses to the key issues before the subject decides that he or she has had enough and refuses to answer any more questions.

5.6 Bias and representative studies

5.6.1 Examples of bias

There are a number of biases which need to be considered in any survey. The key aim in a sample survey is to achieve a representative sample of the population which will give you accurate information about the population. A sample is representative of the population if the characteristics of the sample match the known characteristics of the population.

The National Survey of Sexual Attitudes and Lifestyles was a mixture of a cluster and stratified sample. The primary sampling units in the first stage of the cluster sample were the electoral wards, and each electoral ward was located within strata based upon region, population density and housing tenure profile. The electoral wards were selected with probability proportional to size, and 1000 were selected at random. From each ward 50 addresses were selected. Each address was a cluster of individuals of varying numbers. The target population was all individuals in Great Britain aged 16–59.

The overall response rate was 63.3%. It is thus necessary to investigate non-response and the possible effects that it may have. One question to address is whether the sample is representative. This is answered by comparing the sample distribution with the population distribution on certain key variables such as age, sex and housing tenure which are available in the survey as well as in other large surveys and the Census. The figures from the preliminary analysis of the 1991 Census show that 49.7% of the sample are expected to be men. In fact only 44.4% were men, so men are under-represented in the sample. Furthermore, response was also related to age. Among men aged 50–59 and women aged

55–59 there was a deficit of 2 percentage points from the census figures which was balanced out by a slight over-representation of younger people. Thus we can see that there is evidence that this sample is not completely representative of the population.

If the sample is not representative then this may be associated with a bias in the estimates. Generally a bias means that the estimated quantity from the sample is too large or too small compared to the true value in the population. So as a result of refusal to participate by men, the National Survey of Sexual Attitudes and Lifestyles gives a biased estimate of the number of men in Britain aged 16–59. This is not a problem as the survey is not designed to estimate this quantity – the Census is supposed to provide accurate information on this. Where it does lead to a problem is in the effect it may have on the main aims of the study, namely to estimate sexual behaviour.

Bias may arise from reporting bias or item non-response bias. Recall bias is one aspect of reporting bias. The tendency to over- or under-estimate the occurrence of an event is also an example of reporting bias. Surveys of alcohol intake are plagued with under-reporting. Everyone does it to some extent. People who maybe have a glass of wine every couple of weeks call themselves non-drinkers, people who have a pint of beer every night tend to forget about the occasional glass of wine at lunchtime, and people who drink very heavily may under-report by as much as 50%. Much of the time it is not deliberate under-reporting, just a result of forgetting. People are also likely to over-report the amount they donate to charities. If an act is perceived to be good there is a tendency to over-report, and if it is bad to under-report. An exception may be income, which tends to be under-reported.

Item non-response arises when a subject answers most questions but refuses to answer some questions. Occasionally this is just an oversight, particularly if the question is at the end of a page. Most often it is a deliberate decision to refuse. In this case the subject usually has a reason for not answering the question and this can also result in a bias.

5.6.2 Using sample weights to compensate for bias

The best way to try to overcome bias is to have a well-designed questionnaire with appropriate techniques for collecting the data, by direct interview, telephone interview or self-completed questionnaire as appropriate, and for there to be some mechanism to assess the bias. In many surveys the results will be weighted to try to correct for bias induced by a refusal to participate in the study. This will be illustrated with an example.

Recall that the sample mean is

$$\bar{x} = \frac{1}{n}\sum x_i.$$

A convenient method of estimating the proportion of a population with some attribute is to consider each individual taking a binary measurement $x_i = 1$ if the subject has the attribute and $x_i = 0$ if the subject does not. In this way the sample mean of the binary variable will give the proportion of individuals with the attribute in the sample. A weighted mean is

$$\bar{x}_w = \frac{1}{n_w}\sum w_i x_i,$$

where $n_w = \sum w_i$. Generally it is convenient to ensure that $n = n_w$.

In the National Survey of Sexual Attitudes and Lifestyles the sample size was 18 876, of whom 8378 (44.4%) were men. In the population there are 49.7% men, and we can use the ratio of 49.7% to 44.4% to obtain weights which will 'correct' the sample estimates for the under-representation of men. The weight for men is going to be proportional to 49.7/44.4 = 1.1194 and the weight for women proportional to 51.3/55.6 = 0.9227. As there are 8378 men and 10 498 women the sum of the weights is

$$n_w = \sum w_i = 8378 \times 1.1194 + 10\,498 \times 0.9227 = 19\,064.8378.$$

This should be equal to $n = 18\,876$, and so this means that the weights should be chosen to be equal to

$$1.1194 \times \frac{18\,876}{19\,064.8378} = 1.1085$$

for men and

$$0.9227 \times \frac{18\,876}{19\,064.8378} = 0.9136$$

for women. Using these weights we can check that

$$n_w = \sum w_i = 8378 \times 1.1085 + 10\,498 \times 0.9136 = 18\,875.8.$$

With these weights the percentage of men in the sample is estimated as

$$\frac{8378 \times 1.1085}{18\,876} = 49.2\%$$

The difference of 0.5% from the population figure is due to rounding errors.

Roughly speaking, the weights imply that every man who replies to the survey counts as 1.1 men, while every woman who responds counts as 0.9 women.

The calculation of the sample weights has been illustrated only for a simple case of gender. Often there are many other variables which go into the calculation of the sample weights such as age, housing tenure, area and cluster size. This weighting is a device to correct for over- or under-representation in the sample. Such weights are only applicable to the estimation of the quantities on which they were based – here, the percentage of men and women – but they are often used as general weights over the whole range of variables recorded under the assumption that, having taken into account the information to calculate the weights, individuals fail to respond at random so we can extrapolate from those who do respond.

5.7 Data presentation, tables and bar charts

When presenting the results of a survey it is vital to include details of the sampling scheme and the way in which the information was obtained. For random samples the number of units, individuals or households contacted, together with the number responding, will also be required. A detailed analysis of the characteristics of the non-responders is needed. This gives the reader an idea of the reliability of the study and the likely representativeness of the sample. If weighting has been performed then it is necessary to provide details of how the weights were derived.

Item non-response is also an issue when it comes to reporting the results for specific questions or groups of questions. The correct approach is to report the numbers of individuals not responding to these specific questions to give the reader a fair idea of the results. Also, a comparison of the percentage of missing information across categories – for example, men and women – is often appropriate.

Generally with social surveys the results are expressed as percentages of individuals with a certain characteristic. This is often stratified by smaller subgroups of the population. A good general rule is never to quote a percentage without also giving the sample size on which it is based. Certainly all tables of percentages should include the sample sizes on which these percentages are based.

The tables can be cumbersome to use and graphs such as bar charts and pie charts can be used to convey the information pictorially. This can help immensely in getting the main point across.

The results of surveys should be presented clearly. It is a good investment to pay considerable attention to the formation of the appropriate tables and the construction of appropriate graphs. It is better to use two tables or two graphs to show two points rather than to construct a single table or graph with the two separate bits of information on them. Clarity is the key to a good presentation of results.

The web page for the book has examples of a survey report and presentation of diagrams and tables.

6

Investigating relationships

- Realizing that the association between two quantitative variables can take many forms but that a straight-line association is the easiest one to describe
- Understanding that by transforming the variable some curved associations can be made into straight-line ones
- Appreciating that there are some complex associations which cannot easily be described
- Realizing that there is no causal interpretation to any observed statistical association
- Appreciating that a statistical model is a concise description of a general pattern in the data
- Knowing how to calculate the correlation coefficient and when its use is appropriate
- Knowing how to calculate the slope and intercept of a straight-line relationship and how to interpret the parameters of this model
- Knowing how to display inequality through the Lorenz plot and measure it through the Gini coefficient.

In the graphical displays of Chapter 3 we were primarily concerned with investigating the distribution of a variable, and the plots focused on that aspect of the data. The important features were the location and shape of the distribution as well as, ranges, peaks and outliers. No one plot covered all aspects equally well, but taken together the plots covered all the important aspects of understanding and presenting distributions. We also looked at plots for presenting relationships between two variables, the scatter plot and the time series plot.

In this chapter we will look in more detail at investigating and describing relationships between two quantitative variables. The most important plots, the scatter plot and the scatter plot matrix, have already been described. In Chapter 3 we also saw that boxplots were very useful displays for investigating distributions in a number of different subgroups (see Figure 3.12). This type of plot is investigating the relationship between one quantitative variable and one qualitative variable. Relationships between two qualitative variables are often presented in tables, and bar charts can then be used to display the relationships. Examples are given in Chapter 3 and the Sites Web-Surveys.doc and Web-Analysis.doc on the book's website.

In Section 6.1 we will see that there are many different types of relationships between two variables, each with their own shape. The most important association is one in which there is a linear relationship between the two variables. Curved relationships are also common and in Section 6.2 we consider common transformations which can sometimes be used to make a curved association into a linear one, as linear associations are easier to describe than curved ones. We then go on to see how we can measure the strength of the linear association and how to estimate the linear relationship. We look at how to describe and summarize curved relationships on the book's website. Finally, we look at a plot of the association between two sets of cumulative proportions which is used to describe inequality.

6.1 Housing data – different types of associations

In this section we will look at the scatter plot matrix of some data published by the Office for National Statistics in the United Kingdom. These data were collected every month over a 10-year period from January 1985 until December 1994 and refer to the building industry. The variables and the labels used in the plot are shown in Table 6.1. These variables all measure some aspects of the state of the house building industry in Great Britain. Some refer to the value of orders and commitments to mortgages, others to the number of houses started and completed in different sectors, and one to the average house price. We would expect to see some associations among these variables, and the purpose of showing this example is to see the shapes that these associations can take. The matrix of scatter plots is shown in Figure 6.1.

Table 6.1 Variables in the housing data

Lebel	Variable
FCAS	Orders received by contractors for new houses (GB) £m (1990 prices), from May 1992 only
FCAT	Private sector: housing starts (GB) – thousands
CTOQ	Housing starts: housing associations (GB) – thousands
CTOU	Housing starts: LA's new towns & govt. depts (GB) – thousands
FCAV	Private sector: housing completions (GB) – thousands
CTOS	Housing completions: housing associations (GB) – thousands
CTOW	Housing completions: LA's new towns & govt. depts: (GB) – thousands
AHLO	OFI: building societies: commitments on new dwellings, £m
AHLS	OFI: building societies: advances on new dwellings, £m
BAZH	Average price new dwellings: building society mortgages approved, exc. Abbey National (£ thousands), from July 1989 only

Source: http://ons.gov.uk October 1998.

Some of the associations among the housing variables take the form of straight-line associations. The best example is that between AHLS and AHLO. Both of these refer to moneys committed or advanced by building societies in the same month, and it is not surprising that they are associated. FCAV and FCAT also have a linear association, and as they are starts and completions of houses in the private sector we can again see that this association is not unexpected. In months where a large number of houses are completed a larger number are also started. CTOS and CTOQ have a weaker linear association which appears to be slightly more variable at higher values than at lower values as there is a greater spread of CTOS at higher CTOQ.

The association between the local authority housing completions and starts (CTOU and CTOW) is not linear but curved. There is a positive association in that high starts and

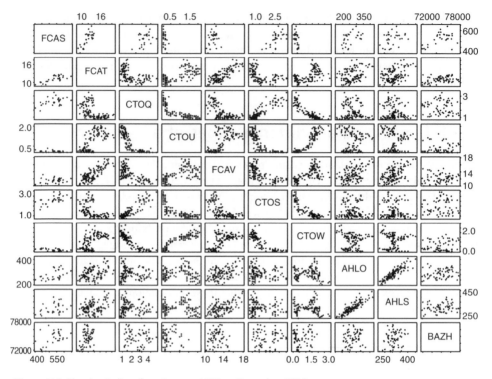

Figure 6.1 Housing indices from January 1985 to December 1994.

high completions go together as do low starts and low completions, but a straight line does not describe it. There are other curved relationships, notably between both CTOQ and CTOS, on the one hand, and both CTOU and CTOW, on the other. These four associations are all negative in that high values of CTOQ and CTOS tend to be associated with low values of CTOU and CTOW.

AHLS and AHLO both have fairly weak positive associations, which might be described as linear, with FCAV, and this again has an interpretation as money lent by building societies will be used to buy the newly completed houses. There are strong patterns in the plots of AHLS and AHLO with CTOW but these could not be described as linear or curved, and indeed almost defy description. There is a similar shape in the plots involving AHLS and AHLO, on the one hand, and CTOU, CTOQ and CTOS, on the other, but the pattern is not quite as obvious. Most of the plots involving the average house price (BAZH) do not exhibit any strong associations. FCAS and FCAT may have a linear association, but it is hard to see clearly as FCAS is only observed for the last two and a half years of the series.

This example has shown straight-line and curved relationships both of which could be described mathematically. This type of mathematical model forms the basis of an important area of the application of statistics in business and commerce called regression analysis which will be discussed in Chapter 15. In some of the plots there was what appeared to be a random scatter of points, implying that there is no association between the variables. In other plots there was evidence of a relationship because the points were not in a complete random scatter, but the relationship could not easily be described by a mathematical equation. Although it is not present in Figure 6.1, one type of relationship which is not well described by a mathematical equation occurs when the points fall into distinct clusters or subgroups.

In the discussion of mathematical models you may look at Figure 6.1 and think that there is no exact straight line, no exact curve, with all the points lying on it. If there were an exact mathematical relationship then all the points would lie exactly on the line or curve. Statistical modelling acknowledges the presence of variability, and while a mathematical curve or line underlies the statistical model we are mainly interested in the general shape or trend. So when we say that a relationship could be described by a curve or a straight line, this refers to the general shape in the scatter plot, allowing for variability.

A final point to make about Figure 6.1 is that the data were collected every month over a 10-year spell. None of the plots take this information into account. In fact some of the peculiar patterns in Figure 6.1 may arise because of the separate association between two variables with time. Methods of presenting and modelling data which arise as a sequence of observations in time, a **time series**, are presented in Section 3.8 and Chapter 7.

6.2 Common transformations to linearity

The simplest relationship between two variables is a straight line, which is represented by the equation $y = a + bx$. In Figure 6.1 a number of the relationships between the variables had an approximate linear relationship, but the vast majority did not. In this section we will look at ways of making curved relationships appear linear by making some changes to the variables. This is known as **transforming** the variable and is accomplished by using a mathematical function. Arguably the most common transformation is the logarithmic transformation which is written $\log_{10}(x)$ or $\log_e(x)$, depending on whether you take logarithms to base 10 or to base e – it doesn't really matter which one you use; $\log_e(x)$ is sometimes written $\ln(x)$. The square root transformation, $\sqrt{x}$, and the inverse transformation, $1/x$, are also common transformations to use.

The graphs in Figure 6.2 illustrate how you can straighten out a relationship by taking a transformation of either x or y. Each row of plots corresponds to the same set of data, and the first column gives the original plot without any transformation. Among the plots in the first row you can see that the relationship between y and x is a concave curve. Taking the log of y gives a plot which is a good straight-line relationship. The square root plot is straighter than the original but not as good as the log plot, and the inverse plot does not get rid of the curvature.

In the second row the original relationship is also curved though not as much as in the first row. This time the square root transformation works to give a straight line, while the log transformation results in a convex curve. The third row of plots has a concave relationship between x and y, but y decreases as x increases. Neither the log nor the square root completely get rid of the curvature, but the inverse transformation of y works in this instance.

In the fourth row the relationship between y and x is very difficult to see as most of the points are crowded down in the bottom left-hand corner, with a few outliers. You may also notice that there is a tendency for the points to spread out in a fan shape from bottom left

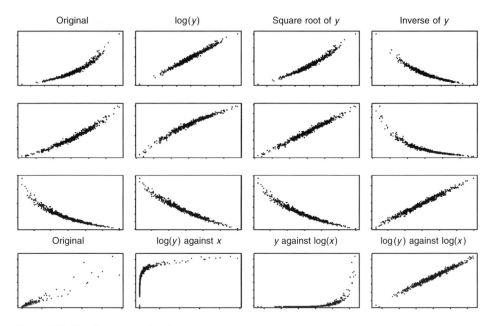

Figure 6.2 Transformations of variables.

to top right. Plots of log y against x and log x against y both give very curved relationships, but a plot of log y against log x gives a good straight line relationship.

6.3 What is a relationship?

In this section we will look at what a relationship is in statistical terms. If you think about a relationship in human terms, perhaps with your parents, brothers and sisters, boyfriends or girlfriends, then you may have the idea of two people doing something in harmony some of the time and doing other things in conflict at other times. You may also think of decisions taken by one person influencing actions of a second. When we look at the statistical aspects of a relationship, both of these ideas have their place. We look to see if two variables move in harmony, or if they move in opposite directions. This can be accomplished by examining the patterns in graph.

The simplest type of relationship is a straight line, and this can either have positive slope, indicating that the two variables tend to move in harmony, or a negative slope, where one variable tends to increase while the other decreases. Other type of relationships involve curves, and we have seen above that if there is just one bend to the curve then it is often possible to make the relationship into a straight line by taking a transformation of one or other of the variables. The reason why transformations are attractive is that they turn a more complex relationship into a familiar simple one.

The main message of this section is that statistical relationships need not have any causal link to the variables. This is where the analogy with family relationships breaks down, as there is much emphasis on cause and effect there – you do or say something as a direct result of something someone else has said or done. Statistical relationships can still be useful even in the absence of cause and effect. In this sense a relationship is often just described as an **association**.

The idea of cause and effect will figure prominently in the rest of this chapter. In Section 6.4 a measure of the strength of a linear association, indicating the degree to which two variables are related, will be introduced – the correlation coefficient. If two variables are related in such a way that the relationship can be described by a straight line, then you can use this straight line to summarize the relationship. Calculating the equation of the line is discussed in Section 6.5. Statistical work cannot give you any information about cause and effect. A statistical analysis of relationships is only based upon associations, and not upon a cause and effect model.

The *Glasgow Herald* of 26 September 1991 featured an article about the high correlation between school clothing grants and exam results. This relationship may be of use statistically, but there cannot be a direct cause and effect relationship between the percentage of pupils receiving a clothing grant and the percentage of pupils passing exams. If there is a causal relationship then it has to be indirect.

It may be possible to make a case for the influence of a third variable on both of these which is leading to the correlation. One candidate might be poverty, which you might expect to have a direct relationship with school clothing grants; another might be social class. However, it is difficult to make a strong case that either of these two variables is directly related to exam performance in a cause and effect fashion. The lack of poverty does not cause good exam results, neither does being in an advantaged social class.

The message of this example is that statistical models involving variables which are associated with each other are virtually all based upon variables which do not have a direct causal relationship to each other. In this way statistical models are vastly different from physical models, where there is a great dependency on causality. This does not invalidate the statistical model, but it does mean that you should have a healthy critical attitude towards them. You should not attach more significance to them than they deserve.

A key feature of a causal model is that if you change one of the variables then you anticipate a corresponding change in the value of the other variable as a direct result. With a statistical model based upon associations or relationships, changing one variable may be associated with changes in the other variable but the direction of the causality may be unknown – and, indeed, direct causality may not exist. The existence of an equation describing the relationship does not imply a causal relationship. Statistical models are only descriptive of an association, and are not proof of the existence of causality.

6.4 Correlation

6.4.1 Calculation

The numerical measurement of the linear association between two quantitative variables x and y is known as the **correlation** between them. If we denote the sets of observations in the sample by (x_i, y_i), $i = 1, \ldots, n$, then the sample correlation coefficient is defined to be

$$r_{xy} = \frac{\text{covariance between } x \text{ and } y}{(\text{standard deviation of } x)(\text{standard deviation of } y)} = \frac{s_{xy}}{s_x s_y},$$

where the **covariance** between x and y is calculated as

$$s_{xy} = \frac{1}{n-1} \sum_{i=1}^{n} (x_i - \bar{x})(y_i - \bar{y}) = \frac{1}{n-1} \left(\sum_{i=1}^{n} x_i y_i - \frac{\sum x_i \sum y_i}{n} \right).$$

The standard deviation of a variable has already been discussed in Chapter 2.

The calculation of the correlation coefficient is illustrated using the data in Figure 3.25 which shows the relationship between energy consumption and growth. The data extracted from the graph are shown in Table 6.2, together with the three columns of squares and products needed in the calculation of the standard deviation and the covariance.

Table 6.2 Energy and growth data from the *Guardian*, 25 April 1997

Country	Energy intensity use, index	Growth, per annum (%)			
	x	y	x^2	xy	y^2
Japan	0.28	3.7	0.0784	1.036	13.69
Italy	0.33	2.9	0.1089	0.957	8.41
UK	0.41	2.2	0.1681	0.902	4.84
W. Germany	0.41	2.4	0.1681	0.984	5.76
USA	0.43	2.1	0.1849	0.903	4.41
Spain	0.44	1.6	0.1936	0.704	2.56
Australia	0.49	1.8	0.2401	0.882	3.24
New Zealand	0.71	0.3	0.5041	0.213	0.09
Totals	3.50	17.0	1.6462	6.581	43.00

The variance of the energy index is

$$s_x^2 = \frac{1}{7} \left(1.6462 - \frac{3.5^2}{8} \right) = 0.016\,421,$$

giving a standard deviation of $s_x = \sqrt{0.016\,421} = 0.128\,144$. For growth the variance is $0.982\,143$, so that $s_y = 0.991\,031$. The covariance between energy and growth is calculated as

$$s_{xy} = \frac{1}{7} \left(6.581 - \frac{3.5 \times 17.0}{8} \right) = \frac{-0.8565}{7} = -0.122\,357,$$

and so the correlation is given by

$$r_{xy} = \frac{-0122\,357}{0.128\,144 \times 0.991\,031} = -0.9635.$$

For these data we would report that the growth per annum and energy intensity use index are negatively associated with a correlation coefficient of –0.96.

6.4.2 Interpretation of the correlation coefficient

As with every other statistical summary, the correlation coefficient is useless if we cannot interpret it. The sample correlation coefficient is also known as the Pearson correlation coefficient, after the statistician E.S. Pearson, who first proposed its use in the early 1900s. It is a measure of the strength of the **linear association** between two variables, and the emphasis is clearly on the word **linear**. The mathematical properties of the sample correlation coefficient show that $-1 \le r_{xy} \le 1$, and we use these limits to put an interpretation on the value of the correlation:

- $r_{xy} = 0$ means that there is no linear association between x and y. This does not mean that there is absolutely no association between the two variables, only that there is no linear relationship. It is possible, but quite rare, for there to be a non-linear association which has a correlation of zero.
- $r_{xy} = 1$ means that all the points lie on a straight line with a positive slope. This is the strongest form of direct positive linear association between two variables.
- $r_{xy} = -1$ means that all the points lie on a straight line with a negative slope. This is the strongest form of direct negative linear association between two variables.
- $r_{xy} > 0$ is interpreted by saying that there is a positive linear association between x and y. This means that when the value of one variable is large then the value of the other is also large; when one is small then the other is also small.
- $r_{xy} < 0$ is interpreted by saying that there is a negative linear association between x and y. This means that when the value of one variable is large then the value of the other is small, and vice versa.

The interpretation of the Pearson correlation coefficient is always in terms of a linear association. In order to assess if the relationship between two variables is linear you really need to look at the scatter plot. There is not much point in just calculating the correlation coefficient as you can have a non-linear relationship, as with some of the curves in Figure 6.1. In that graph the correlation between CTOQ and CTOU is -0.83, yet the relationship is strongly curved and so is non-linear, while the correlation between AHLS and FCAV is smaller at 0.54 for a relationship which is approximately linear.

The scatter plots in Figure 6.3 are presented so that you can see the value of the correlation coefficient and the pattern in the plot for some linear associations. These are all based upon samples with 100 observations in them. In the plots which have a correlation with an absolute value bigger than 0.7 you can easily see a linear association in all the graphs. The larger the correlation, in absolute terms, the easier it is to see the linear association. When the correlation is 0.5 you can still see the linear trend. However, when the correlation is between -0.3 and 0.3 it is very difficult to see any linear association and most of the scatter plots in the middle row just look like random scatters of points.

The Pearson correlation coefficient should not be used if there is a clear non-linear association between the variables, with outliers, curves or evidence of subgroups of observations (see Figure 6.4). The correlation coefficient should be used in conjunction with a scatter plot as an aid to the interpretation of the plot and an assessment of the strength of the association.

Returning to the data in Table 6.2, we saw that the correlation coefficient has a value

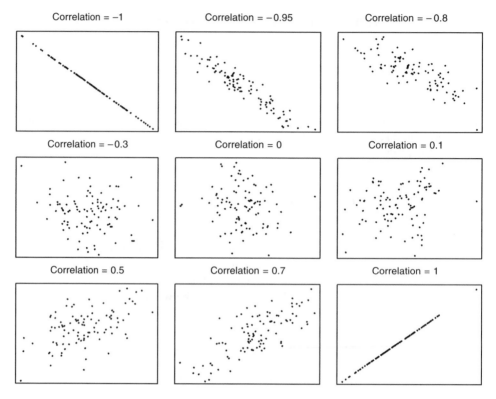

Figure 6.3 Scatter plots and correlation coefficients.

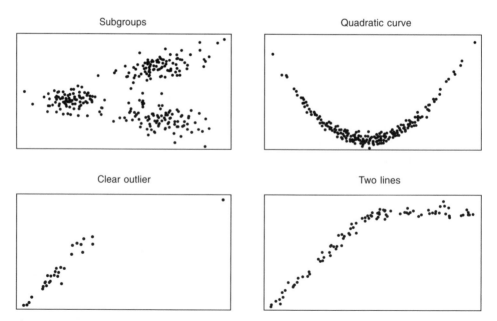

Figure 6.4 Scatter plots where a correlation coefficient should not be calculated.

of −0.96. This, in conjunction with the scatter plot, indicates a very strong negative linear association between growth and energy. More statistical problems arise when you come to think about the implications of this result. Firstly, you need to think about the data – where they came from and whether they can be checked. In fact, this information is available on the graph in Figure 3.25. Secondly, you need to worry about the countries which are included. They are obviously not a random selection of countries and the guidelines given above for the interpretation of the correlation coefficient are really based upon the assumption that the data are a random sample from a population. There is certainly a population of countries, but the eight countries in the data set are not a random sample. They are all economically well-developed countries but they are probably not even a random sample of all economically well-developed countries, nor are they all economically developed countries, nor are they all countries in the G8 group – France is missing and New Zealand included.

6.5 Regression lines

6.5.1 Parameters

In the previous section we saw how to measure the degree of linear association between two variables in a scatter diagram using the correlation coefficient. In this section we look at how to summarize the relationship in terms of a straight line. The equation of a straight line can be written as

$$y = a + bx,$$

where a is the **intercept** and represents the value of y when $x = 0$. The **slope** or **gradient** of the relationship is denoted by b, and this measures the rate of change in y for a unit change in x. These two parameters, the intercept and the gradient, are all that you need to summarize a straight-line relationship.

One of the main uses of this summary is in the prediction of the value of y given a value of x. This is illustrated in Figure 6.5 using the data from the schools in Surrey, first discussed in Chapter 1. The average A-level points score of a school is plotted against the percentage of pupils in the school who obtained 5 or more GCSE passes at grades A*–C and the straight-line summarizing the observed relationship is drawn. Each school is identified by a number. The two arrows illustrate the prediction of the average A-level points score for schools in which 60% and 85% of the pupils obtained 5 or more GCSE passes. The predicted values are 17 points and 21 points, respectively. These are derived from the equation of the straight line, which in this case is

$$\text{A-level points} = 7.735 + 0.154 \times \text{GCSE percentage},$$

by substituting for the GCSE percentage.

In this example the slope of 0.15 can be interpreted as the predicted increase in the average A-level points score for the school for every 1% increase in the percentage of pupils obtaining five or more GCSE passes. So if there is an increase in this percentage of 10% then the average A-level points score in the school is predicted to increase by 1.5 points.

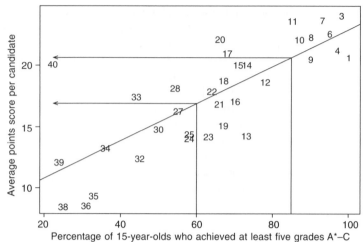

The numbers in the plot identify the different schools, see Table 1.2

Figure 6.5 A-level points scores and GCSE passes, Surrey.

The intercept of 7.74 corresponds to a predicted A-level points score for a school where no pupils passed five or more GCSE examinations. This is not of much interest in this example, as you can see from the graph that there are no schools near this point. In all the schools at least 20% of the pupils passed at least 5 GCSE, and 0% is far outside the range of the observed percentages.

It is quite common that the intercept does not have a very useful interpretation and for this reason many people prefer to use a model where the x-values are replaced by their deviations from the mean of the x-values. This gives

$$y = a + b(x - \bar{x})$$

as the model. In this case the interpretation of the slope b is exactly the same but the interpretation of the intercept term is now the predicted value of y when x is at its mean value. We will see below that in this case the estimated value of a is $\bar{y}$, the mean of the y-values.

6.5.2 Estimation of the parameters

The estimation of the slope and intercept involves a considerable amount of calculation if you have to do it by hand. The formulae are quite similar to the ones for the correlation coefficient and use the same sums of squares and products:

$$b = \frac{S_{xy}}{S_x}, \quad a = \bar{y} - b\bar{x}.$$

If the mean of the x-values is zero then it is easy to see that the intercept a is equal to $\bar{y}$. We illustrate the calculation of these two terms using the small set of data in Table 6.2.

From Section 6.4 we can see that as $s_{xy} = -0.122\,357$ and $s_x = -0.128\,144$ the slope is going to be

$$b = \frac{s_{xy}}{s_x^2} = \frac{-0.122\,357}{0.016\,421} = -7.4513.$$

We also have $\sum x = 3.5$, $\bar{x} = 0.4375$, and $\sum y = 17.0$, $\bar{y} = 2.125$, which gives the intercept as

$$a = \bar{y} - b\bar{x} = 2.125 - (-7.4513) \times 0.4375 = 5.3849.$$

Thus the equation describing the relationship between energy consumption and growth based upon the data in Table 6.2 is

$$\text{Growth} = 5.38 - 7.45 \ \text{Energy}.$$

The line and data are shown in Figure 6.6.

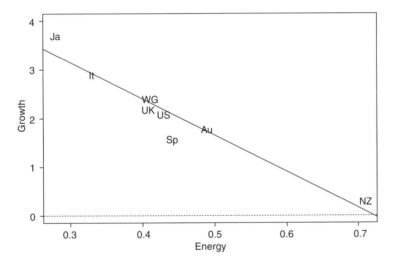

Figure 6.6 Relationship between energy and growth for selected countries.

All the points lie close to the line, as is to be expected as the correlation is very close to −1. However, you need to be very careful about the interpretation of the coefficients and the utility of the straight-line model. There is a negative slope, which means that growth decreases as energy consumption increases. Ultimately this means that there will come at time when the model predicts a negative growth. This may not be sensible, and you should be very concerned about making predictions from straight-line models, or indeed any statistical models, outside the range of the observed data. The intercept of 5.38 corresponds to a predicted growth when the energy intensity index is at 0. This is really meaningless as a country's energy index is never likely to be zero.

6.5.3 Interpretation of the model

These linear models, describing approximate linear relationships, are the most important type of statistical model. Despite the drawbacks mentioned above, they are extremely useful and give quantitative predictions to assist business decisions. Obviously, the closer the points lie to the line the smaller will be the error in the predictions. Also, you should not be tempted to take seriously any predictions made outside the range of data observed – for example, in Figure 6.6 predictions for energy levels less than 0.3 and greater than 0.7.

The deviations of the observations from the fitted straight line are known as the **residuals**. In Figure 6.6 Spain has a negative residual as the predicted value from the straight line is above the observed value. Japan has a positive residual.

In Figure 6.5 we described the relationship between a school's average A-level points score and the percentage of pupils with five or more GCSE passes. In terms of business decisions, the headmaster or board of governors might be tempted to see how the pupils in one school perform relative to other schools.

Over all the schools in Surrey the average A-level points score is 16.9. The comparison of one school with the mean leads to labelling of schools as 'above' or 'below' average. So, in simple terms, you might label the A-level performance in the school with a points score in excess of 16.9 as 'above' average with all the rest 'below' average. Essentially this is what is done in the publication of the school league tables. So, for example, schools 1, 4, 9, 40 would all be labelled above average, while 34, 35, 36, 38, 39 are all below average.

From the graph you can see that there is a strong relationship between GCSE and A-level scores so that it is necessary to modify what is meant by a good A-level performance in the school. The correlation coefficient is high at 0.79, so the percentage of pupils obtaining five or more GCSE passes can be used to predict the A-level points score. Thus all the schools which lie on the fitted line in Figure 6.5, for example schools 27 and 34, obtain the A-level points scores predicted by their GCSE score. Schools which are above this line, and have positive residuals, do better than predicted and schools which are below it, and have negative residuals, do worse than predicted.

This classification of schools, adjusting for GCSE performance, gives different results than for unadjusted comparison. For example, schools 11, 20 and 40 are all better than predicted while 1, 4, 9, 13, 19, 23, 35, 36, 38 all have lower A-level scores than predicted.

With this investigation we have the idea that the deviation of a school from its predicted value is going to give you some idea about its performance. Exactly the same type of analysis can be made about supermarkets or other retail outlets. A statistical model like that in Figure 6.5 may be developed to predict the profits or turnover using the floor space as the predictor variable to see if the store is performing up to expectations. Among investment managers with a number of different portfolios the rate of return may be predicted from the size of the portfolio with a view to seeing which managers have a higher rate of return than expected.

The idea of using statistical models to assess performance is now quite widespread in business and commerce. These models are used to help make decisions. If the models are not valid then the decisions may be compromised. All statistical models need to be investigated and tested before being used, and some methods for doing so will be investigated in Chapter 15, where we will also look in detail at some of the statistical properties of the straight-line model.

It is vitally important to look at these properties as the models can be used for decision-making. If you are going to label a school as being below average, or are going to close down a retail outlet on the basis of a poor turnover, then you need to be very sure that the model is correct. In both instances this will occur because the school or retail outlet is far below the predicted value. This requires a judgement about how far below the prediction you need to go before there is cause for concern. In Figure 6.5, is the distance of school 1 below the line a cause for concern? How about schools 13 and 32? In order to answer these questions we need to take into account the variability of the points about the line and also variability in the estimation of the line itself. There are sound statistical techniques for dealing with these points, and we will return to them in Chapter 15.

6.6 Lorenz plot

6.6.1 Income inequality

In 1971 the 7:84 Theatre Company was founded by a playwright, John McGrath. One of the first plays they performed was *The Cheviot, the Stag and the Black, Black Oil*, which charted the social history of Scotland from the Highland clearances and the introduction of sheep, through the use of the land as hunting grounds for absentee landlords, to the use of oil profits by multinational companies. The numbers in the company's name come from an often quoted statistic that 7% of the population of Britain own 84% of the wealth. Similar statistics could be quoted about most countries. In some the inequality will be more extreme and in others it will be less extreme. A Lorenz plot can be used to display and assess the degree of inequality.

The data in Table 6.3 are concerned with the distribution of pre- and post-tax income among taxpayers in the United Kingdom in 1995/96. There is also information on the distribution of the numbers of individuals who have some tax liability. In the first row of this table we have information on all individuals who have an income of between £3525 and £3999 before tax. There are 547 000 of them, out of a total 25 820 000 taxpayers. The total income of these people before tax is £2070 million and after tax it is £2040 million. As you would expect, there is very little tax collected from the group of people, £30 million in all.

The next three columns give the cumulative distributions for the preceding three variables. Thus there are 1 129 000 individuals with an income before tax between £3525 and £4499. Their total income before tax is £4530 million and after tax £4430 million. The final three columns give the cumulative relative frequencies or the cumulative proportions, which could also have been expressed as percentages. Thus we see that 4.4% of taxpayers have an income between £3525 and £4499 before tax but that this represents only 1.1% of the total pre-tax income and 1.3% of the total post-tax income. These figures give you some idea of the inequality in the income distribution, and we will use all of these data to illustrate the Lorenz plot and how it can be interpreted.

Table 6.3 Distribution of pre- and post-tax income in the United Kingdom among taxpayers, 1995/96

	Number (1000s)	Total income before tax (£m)	Total income after tax (£m)	Cumulative distribution functions			Cumulative relative frequencies		
	A	B	C	A	B	C	A	B	C
3 525	547	2 070	2 040	547	2 070	2 040	0.021	0.005	0.006
4 000	582	2 460	2 390	1 129	4 530	4 430	0.044	0.011	0.013
4 500	714	3 410	3 290	1 843	7 940	7 720	0.071	0.019	0.023
5 000	885	4 640	4 440	2 728	12 580	12 160	0.106	0.031	0.036
5 500	861	4 950	4 670	3 589	17 530	16 830	0.139	0.043	0.050
6 000	1 610	10 400	9 680	5 199	27 930	26 510	0.201	0.069	0.079
7 000	1 730	12 900	11 900	6 929	40 830	38 410	0.268	0.100	0.115
8 000	3 040	27 300	24 300	9 969	68 130	62 710	0.386	0.167	0.188
10 000	2 720	29 800	25 900	12 689	97 930	88 610	0.491	0.240	0.266
12 000	3 470	46 500	39 500	16 159	144 430	128 110	0.626	0.354	0.384
15 000	3 930	67 800	56 200	20 089	212 230	184 310	0.778	0.521	0.553
20 000	3 790	90 700	73 700	23 879	302 930	258 010	0.925	0.743	0.774
30 000	1 360	50 300	38 600	25 239	353 230	296 610	0.977	0.867	0.889
50 000	455	29 700	21 000	25 694	382 930	317 610	0.995	0.939	0.952
100 000	126	24 700	15 950	25 820	407 630	333 560	1.000	1.000	1.000
	25 820	407 630	333 560						

6.6.2 Visualising inequality

A Lorenz plot is like a scatter plot in that it is used to investigate the relationship between two variables. This time the variables are two sets of cumulative relative frequencies. In the example we are looking at the relationship between the cumulative distribution of the number of taxpayers and the cumulative distribution of the total value of the income of these taxpayers both before tax and after tax. This plot is shown in Figure 6.7, where the cumulative relative frequencies of the populations are plotted on the x axis and the cumulative relative frequencies of the income distributions on the y axis.

From the graph you can read off that 50% of the taxpaying population have 25% of the pre-tax income and 27% of the post-tax income. This is obtained by going vertically from 0.5 on the x axis until you reach the two curves and reading off the corresponding values on the y axis. Reading from the y axis to the x axis, we can see that 87% of the population have 70% of the total post-tax income and 89% have 70% of the total pre-tax income. Alternatively, the richest 11% have 30% of the pre-tax income and the richest 13% have 30% of the post-tax income.

If there were no inequality between these two distributions then we would find that, for example, if 30% of the taxpaying population had an income below a certain level then 30% of the income would also be among those individuals. The percentages would match at every possible level and the points would lie on a line connecting 0, 0 to 1, 1 in the diagram. This line is drawn in the plot and represents the line of complete equality.

In Figure 6.7 the solid line corresponds to the relationship between the cumulative relative frequency for the population of taxpayers and the cumulative relative frequency of the pre-tax income, while the dotted line gives the relationship for the post-tax income. As the dotted line is uniformly closer to the diagonal line, which represents equality, there is less inequality in the post-tax income than in the pre-tax income. As one of the goals of a taxation system is to redistribute income, the reduction in inequality is anticipated.

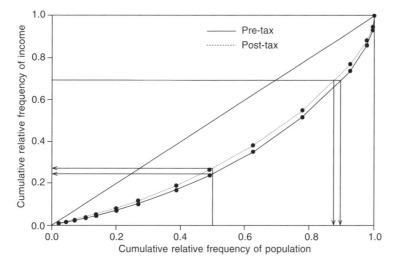

Figure 6.7 Lorenz plot of income in the United Kingdom.

6.6.3 Measuring inequality – Gini coefficient

If we were to calculated the area between the curves and the diagonal line in Figure 6.7 then we would have a measure of inequality. This measure is known as the Gini coefficient. If there are k classes then the coefficient is given by

$$\sum_{i=1}^{k}(x_{i-1}y_i - x_iy_{i-1}),$$

where the x_i and the y_i are the two sets of cumulative relative frequencies with $x_0 = y_0 = 0$ and $x_1 = y_1 = 1$. This coefficient will lie between 0 and 1, with larger values indicating greater inequality. The formula for this coefficient can be derived by working out the area under the curve in Figure 6.7 as a series of rectangles and right-angled triangles and subtracting this from the area of the triangle under the straight line. This will give a value ranging between 0 and 0.5, as the area of the square in the Lorenz plot is 1, which is then multiplied by 2 to give the Gini coefficient above. It is quite possible to draw a Lorenz curve as a convex curve, above the diagonal line, rather than the concave curve in Figure 6.7. If you do so the Gini coefficient will turn out to be negative so you need to use the absolute value.

The calculations for the Gini coefficients for the pre- and post-tax incomes are set out in Table 6.4. We see that the coefficient for the pre-tax income is 0.38 and for the post-tax income 0.33. As the post-tax coefficient is smaller than the pre-tax coefficient there is less inequality in the distribution of post-tax income than of pre-tax income, but there is still considerable residual inequality. For most countries the Gini coefficient for income inequalities lies between 0.2 and 0.4.

Table 6.4 Calculation of Gini coefficients

Group	Population	Pre-tax	Post-tax	Gini pre	Gini post
	x_i	y_i	y_i	$(x_{i-1}y_i - x_iy_{i-1})$	$(x_{i-1}y_i - x_iy_{i-1})$
0	0	0	0		
1	0.021	0.005	0.006	0	0
2	0.044	0.011	0.013	0.000011	0.000009
3	0.071	0.019	0.023	0.000055	0.000089
4	0.106	0.031	0.036	0.000187	0.000118
5	0.139	0.043	0.050	0.000249	0.000296
6	0.201	0.069	0.079	0.000948	0.000931
7	0.268	0.100	0.115	0.001608	0.001943
8	0.386	0.167	0.188	0.006156	0.005994
9	0.491	0.240	0.266	0.010643	0.010368
10	0.626	0.354	0.384	0.023574	0.022028
11	0.778	0.521	0.553	0.050734	0.047426
12	0.925	0.743	0.774	0.096129	0.090647
13	0.977	0.867	0.889	0.076064	0.066127
14	0.995	0.939	0.952	0.054738	0.045549
15	1.000	1.000	1.000	0.056	0.043
Sum				0.377096	0.334525

6.7 Summary

In this section we have looked at relationships between two quantitative variables and how this relationship might be displayed and summarized. Much statistical work is concerned with the investigation and use of relationships in order to make predictions. A fundamental aspect of the use of a statistical model to make predictions is looking at plots of the data. One of the crucial aspects of most simple predictions is the existence of a straight-line relationship. You can use the plots to check up on the validity of this in any particular case.

We have also looked at the use of the correlation coefficient to summarize the strength of a linear association. This is a useful summary measure but you should be wary of using and interpreting it without having seen the plot. A high coefficient does not necessarily imply a linear relationship. Also there could be a strong relationship which is not well described by a straight line and which has a low correlation.

The straight-line relationship is the most important one in elementary statistics. It is a relatively simple tool and is used to provide an approximate summary of a relationship which has a considerable amount of variation in it. In none of the scatter plots shown in this chapter was there a strong relationship without some variability. One of the main aims of statistics is to summarize the important information in the presence of variation, and finding a straight-line relationship among a cloud of points is one good example of this principle.

Many things have not been covered in this chapter, and many more details of fitting and interpreting straight lines will be covered in detail later. You can see from any of the plots and fitted lines in this chapter that there is normally a great deal of variation about the line. We need to know how to take this variation into account when making predictions from the models. In order to understand how to do this it is necessary to study probability and probability models so that you can appreciate the reliable and unreliable aspects of the predictions.

Probably the most important warning to give about the straight-line models is that there is no evidence from the statistical analysis that there is any causal relationship in the model. Obviously the model will have more significance and you will have more faith in it if you can establish a causal link. There is no direct causal link in any of the graphs shown in this chapter, although for most of them you can make a case for there being a link. Earnings per share should be affected by profits; GCSE exams precede A levels and require the same type of skills, so should be predictive. Statistical associations alone do not imply causal relationships.

7

Relationships with time

- Knowing the various components of a time series model
- Knowing how to estimate the trend in a time series using a moving average
- Knowing how to estimate the trend in a time series using exponential smoothing
- Knowing how to estimate the seasonal components of a time series and hence provide a seasonally adjusted time series
- Knowing how to use exponential smoothing to forecast future values of the time series
- Appreciating the difficulties in forecasting future values of a time series.

7.1 Introduction

Almost any business statistic which is collected regularly may be of interest as a **time series**: the retail price index, the FTSE-100, the Royal Bank oil index, a company's annual profit figures, weekly sales volume, share price and salary costs are all examples of statistics which are routinely examined for their behaviour with relation to time. A time series is any sequence of numbers where the position in the sequence specifies a point in time.

There are two main goals in working with time series: we may want to describe or summarize the key features of the data, or we may want to predict what happens after the end of the available observation time (this is called forecasting in the statistical literature); we will see in this chapter that some of the descriptive techniques are also useful for making forecasts. These goals are very similar to the goals of describing the relationship between two variables as a straight line which was covered in Chapter 6. There are a number of special features about time series which mean that different techniques are required. However, the basic features of statistical modelling – summarizing and describing a relationship, and predicting future values – are all present.

The essential idea of both time series description (developed in Section 7.5) and forecasting (Section 7.6) is that we model the data by ascribing features to one of five sources:

1. **Trend.** This is the long-term behaviour of the variable of interest. Here 'long-term' means a period of time covering at least the length of the time series. Saying that there is an upward trend in some or all of the time series *does* not mean that the values observed in that section always increase with time: it just means that there is, taking the whole section into account, a tendency to increase; both upward and downward

variation will occur. The major stock market indices such as the FTSE are good examples. Despite recessions, increasing globalization, mergers and demergers, these indices have shown a consistent pattern of increase over the decades. We look at some ways to estimate trend in Sections 7.3 and 7.4.

2. **Seasonality.** This is fluctuation in the data which occurs in a regular and repetitive pattern. In many business statistics there is an annual 'cycle' of variation, where certain times of the year typically have low values, and other months typically have high values. Electricity and sunscreen sales are two good examples where seasonal variability is well known (of course, the highs and lows in these variables occur at different times of the year!). Section 7.5 covers some methods of extracting seasonality from a time series.

3. **Cyclical patterns.** These are similar to seasonal effects in that they occur in a more or less repetitive pattern over the period of the time series. Seasonal effects are considered to be associated with observable temporal patterns such as days of the weeks, months of the year and quarters of the year, while cyclical effects are often associated with cycles of longer duration, such as a two-year business cycle. If the time series is collected on a daily basis then a systematic pattern associated with the days of the week would be considered as a seasonal effect, while a repetitive pattern every 10 days would be described as a cyclical rather than seasonal effect.

4. **External factors.** There will always be factors affecting a given variable which are external to the context in which it is viewed. The global economy in general has an impact on even local, small-scale businesses; the consequences of extensive and aggressive currency targeting, which occurs from time to time, can be profound in the country whose currency suffers; changes in the law can affect some businesses to the extent of destroying them (or creating new opportunities). Such factors can produce changes in the behaviour of a given business time series, for example by introducing large changes over a short period of time. Although we may attribute some features in a time series to such external factors, there is little we can do about modelling them. In this chapter when we encounter features that look to have such external origins, we will note them, but take analysis no further than this.

5. **Randomness.** Even if we could take complete account of trends, seasonality, cycles and external factors, there would still be variation left in the time series which was not completely predictable. The combination of all the small decisions, transactions and environmental effects going on all the time, all about us, means that detailed prediction of the overall result is effectively random. In modelling a time series, we try to take account of the trend and any seasonality, and treat the remainder as random 'noise', predictable only as a random variable (see Chapter 8).

7.2 Simple time series models

Now that we have considered the uses of a time series, its various components, and how best to display it, we shall turn to the statistical description of such a series. It seems natural in light of the enumeration of time series components of this chapter to phrase our model in the same terms.

Let us denote the series of observed values as X_t, where $t = 1, 2, 3, \ldots, N$. As mentioned in the previous section, this arbitrary use of time t presents no difficulties, as the difference in time between t and $t + 1$ (say) is exactly that between two successive observations, measured in the appropriate units. For many time series in business these are usually days, weeks, months or quarters. We write our model for the data as

$$X_t = T_t + S_t + C_t + E_t + R_t,$$

where T represents the contribution from the trend, S the contribution from seasonal effects, C any cyclical effect, E the effect of external factors, and R the random effects. It is worth noting that in many economic time series the above description best applies to the logarithm of the variable of interest; in that case we are saying that our model is

$$X_t = T_t \times S_t \times C_t \times E_t \times R_t,$$

so that

$$\log(X_t) = \log(T_t) + \log(S_t) + \log(C_t) + \log(E_t) + \log(R_t).$$

In a case where we believe that the structure is multiplicative rather than additive, we can just work with the logarithm of the variable, and otherwise proceed as in the rest of the chapter to examine trend, seasonality, etc.

In general, we try first to extract any trend, then take account of any seasonality we may suspect, then any cyclical patterns, then make what adjustment we can for any external factors which show up in the data, and hope that this will leave us with an essentially random set of small effects which we call the **residuals**. In other words, when we have extracted and explained all that we can, we should be left with a variable which exhibits random behaviour:

$$R_t = X_t - T_t - S_t - C_t - E_t.$$

The point of this is first that we have constructed a model of the average behaviour of the variable of interest which is useful in itself for understanding the observed patterns in the time series, and second that many of the methods of estimation and inference under conditions of random variability found elsewhere in this book may be used to aid forecasting.

The importance of the residuals in time series analysis, or indeed any statistical analysis, cannot be over-stressed. It is only by looking at the residual variation after trying to extract some other contributory component that we can see whether further analysis is necessary, or if we may stop at that point. The methods used to test residuals are many, varied, and often statistically complicated. In this book we shall discuss only graphical methods. The use of residuals is seen as a method of checking the validity of the model that we have adopted. If the residuals exhibit any systematic behaviour then the model is deemed to be inadequate. If the residuals exhibit random behaviour then we would conclude that the model is valid.

The first plot is simply a histogram or boxplot of the residuals, looking for any tendency to deviate from a moderately symmetrical form. Too much deviation may mean that a systematic component of the model has not been taken into account – for example an external factor, or just outliers. The second kind of plot is simply that of the residual values as a time series in their own right. Here we are looking for a series which shows no particular pattern: specifically, there should be no trend or seasonality or cyclical pattern, or we have not extracted these from the original time series!

In the next sections we shall look at two methods for extracting the trend from a time series. The use of residual analysis will be illustrated throughout. As the methods used to estimate cyclical effects are similar to the methods used for seasonal effects, and as seasonal effects are much more common and important than cyclical effects, we will not discuss cyclical effects any more. Also, we will not dwell further on external factors, other than to note possible effects in the context of some case studies.

7.3 Moving averages

Apart from correcting for seasonality (the subject of Section 7.5), we may wish to highlight any underlying trend in the data that random effects may be masking. One way to do this is to use a moving average to smooth the time series.

The idea behind the moving average approach is that the trend may be visible over groups of data points (eight or ten successive ones, say) while the random effects should be contributing equally to any point. By taking the average of the eight or ten points we should get a better estimate of what is happening in that chunk of time, because we have averaged out the effects of the random components. By performing this calculation for all sets of eight or ten points successively through the time series, we might hope to extract an idea of how the average changes with time, having smoothed out the random effects.

7.3.1 Moving averages with an odd number of points

The easiest form of moving average uses an odd number of points, centred on each data point. For example, suppose we want to use a five-point moving average; then the average is calculated according to

$$\bar{x}_t = \frac{1}{5}(x_{t-2} + x_{t-1} + x_t + x_{t+1} + x_{t+2}).$$

The general equation for an m-point moving average, when m is an odd number, is

$$\bar{x}_t = \frac{1}{m} \sum_{k=t-(m-1)/2}^{t+(m-1)/2} x_k.$$

As soon as we attempt to implement such a moving average, however, we run into a problem – the first and last few points in the series do not have a sufficient number of neighbours to use the formula! In fact all we can reasonably do is to calculate the moving average only for the range of time where sufficient points are available for the appropriate average. In other words if we calculate a five-point moving average, we can only get estimates of the average trend for points ranging from the third in the series, to the third from the end of the series. This is a limitation of the moving average method.

7.3.2 Moving averages with an even number of points

As we shall see in Section 7.5, there are occasions when we need an even number of points in the average. There is no problem in calculating a moving average with, say, 4 points, but we are left with values that do not correspond to any time point where there is an observation, as the 'centre' of the average lies midway between two time points. To get over this nuisance, we generally use the average of pairs of adjacent moving averages, and use this as the estimate of trend.

We do this in two stages. First if m is the order of moving average we actually want, we obtain one 'raw' m-point moving average centred halfway between t and $t-1$ ($\bar{x}_t^{\text{raw}}$), and then average successive pairs of these to get a trend estimate centred on t. For $m = 4$ these would be

$$\bar{x}_t^{\text{raw}} = \frac{1}{4}(x_{t-2} + x_{t-1} + x_t + x_{t+1}),$$

with the general equation being

$$\bar{x}_t^{\text{raw}} = \frac{1}{m} \sum_{k=t-m/2}^{t+m/2-1} x_k.$$

Then we take the average of two successive estimates, to get a trend estimate centred halfway between them, that is, at time point t:

$$\bar{x}_t = \frac{1}{2}(\bar{x}_t^{\text{raw}} + \bar{x}_{t+1}^{\text{raw}}).$$

7.3.3 FTSE-100 index to illustrate moving averages

We will use a short data set to illustrate the method. The following table gives the closing values of the FTSE-100 index for 20 trading days of the month of April 1996:

1–10	3718.4	3728.5	3725.1	3755.6	3755.3	3767.4	3744.2	3766.8	3790.5	3825.3
11–20	3805.6	3820.7	3857.1	3852.7	3833.0	3817.6	3819.3	3832.8	3809.2	3817.2

If we plot these data as a time series (Figure 7.1), we see that there is a general trend towards increasing values day by day until the 13th working day followed by a decrease. The amount of increase is not constant.

Now we will calculate a five-point moving average to try and extract the trend from the rather fluctuating data. We must start with day 3, and use the points ranging from 1 to 5; so our estimate will be

$$\bar{x}_3 = \frac{1}{5}(3718.4 + 3728.5 + 3725.1 + 3755.6 + 3755.3) = 3736.58.$$

Then we calculate the estimate at the next day, day 4, using the points from day 2 to day 6. This gives us

$$\bar{x}_4 = \frac{1}{5}(3728.5 + 3725.1 + 3755.6 + 3755.3 + 3767.4) = 3746.38.$$

We continue in like manner for the other moving average estimates, stopping at day 18, the last for which five points are available. When we plot the five-point moving average trend along with the original data in Figure 7.1a we must be careful to remember that the trend estimates start from day 3, not day 1.

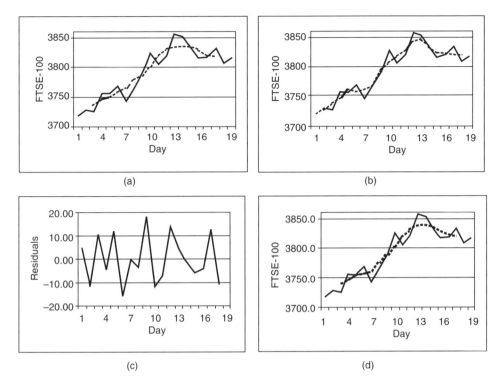

Figure 7.1 FTSE-100, April 1996: (a) original data (solid) and five-point moving average (dotted); (b) original data (solid) and three-point moving average (dotted); (c) residuals from three-point moving average; (d) original data (solid) and four-point moving average (dotted).

The trend line in Figure 7.1a is much smoother than the original series. But the loss of two points from each end of the series, the inevitable consequence of using the five-point moving average, means we are unable to assess the trend for a fraction of the time covered by the data. One way to reduce this difficulty is to use a smaller number of points in the moving average. The only possibility is 3, so we plot the three-point moving average in Figure 7.1b. We can see that the three-point average is only a little less smooth than the five-point one was, so the former is still a useful trend estimate. In addition, we now have estimates of trend for days 2 and 19, leaving us with just two days without trend estimates.

Our next step is always to look at the residuals, the differences between the actual and the trend values at each time point. This allows us to detect any remaining trend or non-randomness. In the table below we give the three-point moving average estimates, together with the residual values for the first six days:

Day	1	2	3	4	5	6
FTSE-100	3718.4	3728.5	3725.1	3755.6	3755.3	3767.4
Trend		3724.0	3736.4	3745.3	3759.4	3755.6
Residual		4.5	−11.3	10.3	−4.1	11.8

The residuals show no particular pattern in a time series plot in Figure 7.1c. They appear to be equally distributed on either side of zero. As a result we can be reasonably content with our moving average model.

We shall also illustrate the calculation of moving averages where the number of points is even using the same data. In this case we shall use the four-point average. The first of the 'raw' averages is

$$\bar{x}_3^{\text{raw}} = \frac{1}{4}(3718.4 + 3728.5 + 3725.1 + 3755.6) = 3731.9,$$

and this is centred at 2.5. The second one is

$$\bar{x}_4^{\text{raw}} = \frac{1}{4}(3728.5 + 3725.1 + 3755.6 + 3755.3) = 3741.1,$$

which is centred at 3.5. This means that the first centred average, centred at 3, is

$$\bar{x}_3 = \frac{1}{2}(3731.9 + 3741.1) = 3736.5.$$

The following table lists the complete calculations for the four-point average for the first 6 days:

Day	1	2	3	4	5	6	
FTSE-100	3718.4	3728.5	3725.1	3755.6	3755.3	3767.4	
Raw		3731.9	3741.1	3750.9	3755.6	3758.4	
Trend			3736.5	3746.0	3753.2	3757.0	

In the third row we have offset the raw estimates from the other rows to underline the fact that they are not centred at the equivalent time point, but halfway between it and the preceding one. Figure 7.1d shows the original data and the four-point moving average estimate of the trend.

In Section 7.5 we shall look at methods of removing seasonality from a time series. Moving averages have a role to play in this process.

7.3.4 Case study: Irish gross agricultural output

The time series of the gross agricultural output (GAO) for the Republic of Ireland in millions of dollars in shown in Figure 7.2. Each point represents one year, and the series runs from 1973 to 1992, giving 20 data points. This is a time series illustrating some cyclical behaviour.

Superimposed on a prevailing trend of increase in output there are several episodes of decline. These may be part of the larger economic pattern, but we will not attempt to describe this. We will use a moving average method to extract the underlying trend.

The results for the five-point moving average are shown in Figure 7.2. The overall trend seems to be well captured by the five-point model. However, if we look at the residual plot (Figure 7.3), we see that by smoothing out the three periods of decline, we are left with residuals which display a considerable degree of remaining pattern. The residual series has a clear cyclical behaviour, with peaks at points 6 (1978), 12 (1984),

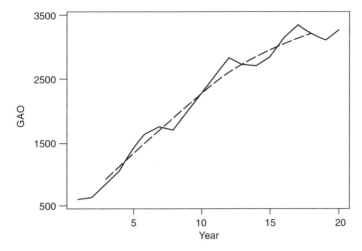

Figure 7.2 The gross agricultural output for the Republic of Ireland in millions of dollars. Each point represents one year, and the series runs from 1973 to 1992. GAO data (solid line) and the five-point moving average (dashed line).

and 17 (1989) and troughs at 4 (1976), 8 (1980) and 14 (1986). This corresponds to a cycle of 5–6 years.

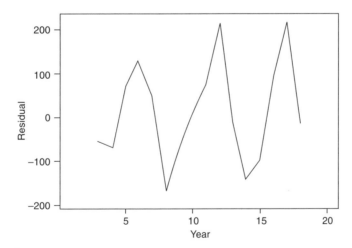

Figure 7.3 Residuals between the data and five-point moving average of the Irish GAO data.

Dealing effectively with this kind of problem involves methods for modelling the residuals themselves, repeating the procedure until we are left with a set of residuals which have the desired randomness properties. Such methods are outside of the scope of this book; but for the purpose of extracting the underlying trend, the moving average method has performed well. The methods used to estimate cyclical components have strong similarities to the methods for estimating seasonality in Section 7.5.

7.4 Exponential smoothing

Smoothing with a moving average is not by any means the only method available. A common alternative is the **exponential smoothing** technique. Here we create a new smoothed time series (y) from the original one (x) as follows. Let $y_1 = x_1$ so that the first point in each series is the same. Then for each successive time point, calculate the new series according to

$$y_t = \alpha x_t + (1 - \alpha)y_{t-1}.$$

Here α is a value between zero and one, known as the **smoothing parameter**. The first few values of the smoothed series are

$$y_1 = x_1,$$

$$y_2 = \alpha x_2 + (1 - \alpha)y_1 = \alpha x_2 + (1 - \alpha)x_1,$$

$$y_3 = \alpha x_3 + (1 - \alpha)y_2 = \alpha x_3 + (1 - \alpha)(\alpha x_2 + (1 - \alpha)x_1)$$

$$= \alpha x_3 + (1 - \alpha)\alpha x_2 + (1 - \alpha)^2 x_1,$$

$$y_4 = \alpha x_4 + (1 - \alpha)y_3 = \alpha x_4 + (1 - \alpha)(\alpha x_3 + (1 - \alpha)\alpha x_2 + (1 - \alpha)^2 x_1)$$

$$= \alpha x_4 + (1 - \alpha)\alpha x_3 + (1 - \alpha)^2 \alpha x_2 + (1 - \alpha)^3 x_1.$$

As $0 < \alpha < 1$ the powers of $(1 - \alpha)$ will gradually become smaller and smaller and the earlier values in the series will have less influence on the smoothed values later on in the series.

What this approach does is to include some information on all of the historical values before the current one, with older values having less weight. To see how this works, consider the calculation with $\alpha = 1$, giving

$$y_t = 1 \times x_t + (1 - 1)y_{t-1} = x_t.$$

The new series is identical to the original one, so there is no smoothing. Now consider what happens with $\alpha = 0$. In this case

$$y_t = 0 \times x_t + (1 - 0)y_{t-1} = y_{t-1},$$

and as $y_1 = x_1$, $y_t = x_1$ and each value is the same as the previous one, so that the new data series is a horizontal line. This is a very extreme form of smoothing!

When exponential smoothing is used the smoothed value at time t is a weighted average of the observed value at time t and the smoothed value at time $t - 1$. If $\alpha = 1$ all the weight is placed upon the observed value at time t and so there is no smoothing as there is no contribution from the previous values of the series. If $\alpha = 0$ all the weight is placed upon the smoothed value at time $t - 1$ and there is no contribution of the observed value at time t. Generally, values of α in the range 0.1 to 0.3 are used to place more emphasis on the past observations. Larger values of α are seldom used.

Clearly the trick here is to choose an appropriate value for α that smoothes enough to remove the unwanted fluctuations in the time series, but does not remove everything of interest. There is no simple way to go about choosing the value of α (although many statistical packages will offer a means for finding a value of α which satisfies certain conditions on the residuals), and some trial and error is often required.

We use the same data on the closing values of the FTSE-100 index for 20 trading days in April 1996 as in Section 7.3. The same information is presented as a time series plot in Figure 7.4 along with various exponentially smoothed series. In the original series we see a gradual increase in the index for the first part of the month to a peak on day 13 followed by a decrease towards the end. There are many fluctuations around this trend, and we will use exponential smoothing to try and extract a sense of the typical behaviour throughout the month.

The method will be illustrated for the first few data points for the particular case of α = 0.35. The first trend value is y_1, which is set to x_1 = 3718.4. Next we have

$$y_2 = \alpha x_2 + (1 - \alpha)y_1 = 0.35 \times 3728.5 + 0.65 \times 3718.4 = 3721.94$$

and

$$y_3 = \alpha x_3 + (1 - \alpha)y_2 = 0.35 \times 3725.1 + 0.65 \times 3721.94 = 3723.04.$$

The other values are calculated in the same way up to y_{20}. The effect of different choices of α on the exponential smoothing is illustrated in Figure 7.4.

As expected, the two time series become very similar as α approaches 1. We can also see that the effect of small α is to underestimate the time series; this arises because the two series start with the same value, but for small α there is little input from the original data, and the smoothed version relates primarily to itself. Either $\alpha = 0.25$ or $\alpha = 0.5$ would

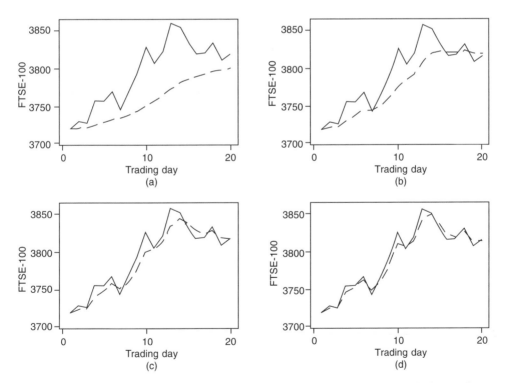

Figure 7.4 Exponential smoothing (dashed lines) of the FTSE data of Figure 7.1 (solid lines). The α-values used were (a) 0.1, (b) 0.25, (c) 0.5 and (d) 0.7.

probably be an adequate illustration of what is happening to the FTSE during this month; the former value is smoother but underestimates the data consistently in the middle of the month, while the latter value matches the data better as it is more variable. On balance, we should choose the $\alpha = 0.5$ version, because it is does not tend to underestimate the data.

This is rather a short time series, however. In the following case study we will look at a much longer series of FTSE closing values, and apply exponential smoothing.

7.4.1 Case study: FTSE-100 share index

The time series of the FTSE-100 index of leading shares from 3 January 1984 to 17 May 1996 in shown in Figure 7.5. The overall consistent increase is evident, as are the two stock market crashes. Superimposed on these there is a lot of random fluctuation which we wish to smooth out.

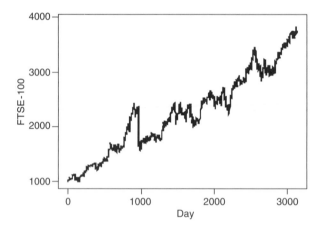

Figure 7.5 Daily closing price of the FTSE-100 index of leading shares from 3 January 1984 to 17 May 1996.

We will first try exponential smoothing with $\alpha = 0.1$. This is illustrated in Figure 7.6. A lot of the random fluctuations have been removed, leaving both trend and crashes visible. But there is still a lot of fluctuation in the time series, and we might perhaps try an even smaller value of α.

In Figure 7.7 we show what happens if we set $\alpha = 0.01$. This is much smoother, but the underestimation is starting to get serious, and the stock market crashes, particularly the second one, have become rather seriously smeared out. Which we choose really depends on what we want the smoothed estimate for: if all we care about is the overall growth of the FTSE value, then $\alpha = 0.01$ is probably the better; if we want more specific information then an intermediate value (say $\alpha = 0.05$) might be useful.

7.5 Correcting for seasonality

If there is seasonality in a time series, this manifests as a more or less cyclic fluctuation around the trend line. This can tend to hide the underlying behaviour, making comparison

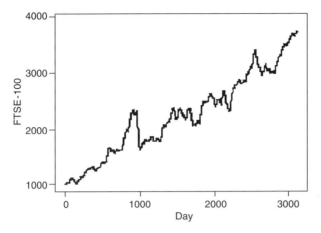

Figure 7.6 Exponential smoothing of the FTSE data with $\alpha = 0.1$.

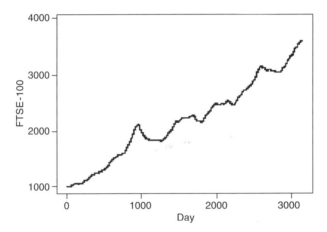

Figure 7.7 Exponential smoothing of the FTSE data with $\alpha = 0.01$.

with other similar series, or long-term assessment, difficult. For this reason we may often want to remove the effects of seasonality, a process called **seasonally adjusting**. The resulting time series is then called **seasonally adjusted**; for example, national unemployment figures are usually presented in seasonally adjusted form, to compensate for the large seasonal rise at the end of the school year and in winter.

On the other hand, there are many cases where the seasonal information is useful (planning retail sales, for example), so it is clear that we must consider both extraction of seasonal information and also properly incorporating it into a model of the time series data. We shall first show how to produce a trend line when there is seasonal variation, then illustrate how seasonal adjustment may be achieved, at the same time showing how seasonal information may be usefully presented.

The key to taking proper account of seasonality is to use a moving average with the same number of points as the length of a seasonal cycle. To see how this works, consider the following artificial time series where there is a perfect cycle of length 3 time units:

t	1	2	3	4	5	6	7	8	9	...
x_t	4	6	5	4	6	5	4	6	5	...

The three-point moving averages for this series, starting at time $t = 2$, are

$$\bar{x}_2 = \frac{1}{3}(4 + 5 + 6) = 5, \qquad \bar{x}_3 = \frac{1}{3}(6 + 5 + 4) = 5, \ldots,$$

so that $\bar{x}_t$ always equals 5, for all t. In other words, for a perfectly cyclic time series, which repeats itself exactly after K time units ($K = 3$ in the above example), a moving average of length exactly K will always involve all of the numbers in a complete cycle, so that the average trend in the time series will be a constant. Choosing a $(K + 1)$-point moving average will not have this useful effect.

Of course in a real time series there will not be such perfect cyclic repetition, and there will usually be some underlying trend, but the choice of a K-point moving average, where K is our best estimate (or guess!) of the period of the seasonality, allows us to remove a very large part of the seasonal contribution to the series. We will illustrate this with the spirits sales series first shown in Chapter 3 (Figure 3.22). In this case the unit of time is a month, and the period is annual, so that the appropriate moving average is a 12-point one. This is obviously an even-numbered average, so we need to use the two-stage averaging method of Section 7.3.2.

First we must calculate the raw averages $(\bar{x}_t^{\text{raw}})$, which for a 12-point average will be

$$\bar{x}_t^{\text{raw}} = \frac{1}{12}(x_{t-6} + x_{t-5} + x_{t-4} + x_{t-3} + x_{t-2} + x_{t-1} + x_t + x_{t+1} + x_{t+2} + x_{t+3} + x_{t+4} + x_{t+5}).$$

Clearly we can only calculate this average for t starting at $t = 7$, so that the first raw average is centred at $6\frac{1}{2}$, and the second is centred at $7\frac{1}{2}$; hence the average of the first two raw averages is centred at $t = 7$, and may be compared with the original data point for that time.

The original data (solid line) and the 12-point moving average (dashed line) are plotted in Figure 7.8. We can see that the large fluctuations which appear throughout the year

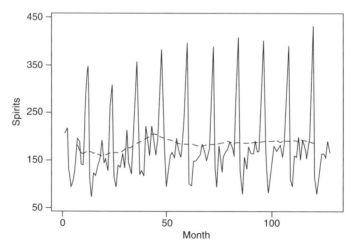

Figure 7.8 Original spirits data (solid line) and a 12-point moving average (dashed line).

have been removed, leaving us with a very smooth trend line. From around month 50 (early 1986) there is little change in the average.

The next stage in seasonal adjustment is to calculate the seasonal differences. This means that for each time period for which we have both a trend value and the original data value, we calculate

$$x_t - \bar{x}_t.$$

These deviations from the trend are plotted in Figure 7.9. Not surprisingly, this series bears a strong resemblance to the original data (the trend line is very flat).

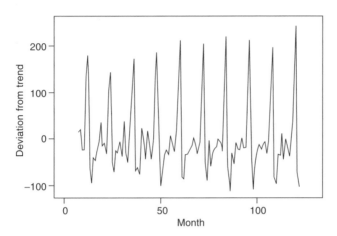

Figure 7.9 Deviations of monthly spirits sales from the moving average trend.

We now assume that the deviations we see are entirely due to a regular seasonal fluctuation, plus some random effects. Then we can calculate the average amount of deviation from the trend for each month of the year simply by, for each month, adding up all of the deviations from the trend across all years, and dividing by the number of years where data and trend for that month are both available. So for the spirits data we create a table of average deviation by month:

Month	J	F	M	A	M	J	J	A	S	O	N	D
Mean deviation	−57.95	−92.46	−40.51	−33.11	−13.56	−27.05	5.90	−8.61	−28.06	−10.90	111.21	198.84

So we can see how much deviation there is on average from the trend (and the direction of the deviation) for each month of the year. Now, if we were to add these 12 values up, we would find that their sum is 3.74; we would rather that their average be zero, so we impose a small adjustment by subtracting 3.74/12 from each value (this says that we spread our deviation of 3.74 over all 12 months equally). Then, to two decimal places (as in the original data), our seasonal deviations become

Month	J	F	M	A	M	J	J	A	S	O	N	D
Mean deviation	−58.26	−92.77	−40.82	−33.42	−13.87	−27.36	5.59	−8.92	−28.37	−11.21	110.90	198.53

This monthly mean deviation in sales from the overall trend line is graphed in Figure 7.10.

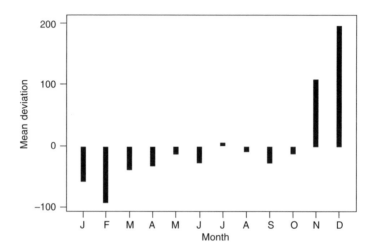

Figure 7.10 Adjusted seasonal (monthly) deviations from the 12-point moving average trend of spirits sales.

Although these deviations are interesting in their own right, we may also use them to complete the process of creating a seasonally adjusted time series. Where the monthly deviation is (say) W, we must remove the value W from the data value for the appropriate month in *every* year in the series. When we perform this calculation for the original spirits time series, we obtain Figure 7.11.

In Figure 7.11 the thick line is the trend as estimated above, while the thin jagged line is the seasonally adjusted result. Except near the start of the series (where of course the trend is not available anyway) these lines keep in fairly good agreement. There is quite a lot more variability in the seasonally adjusted line, which we can attribute to random differences.

We would conclude that we have a fairly good model of the data, particularly over the last 4–5 years, that there is little evidence of any change in the long-term trend, and that seasonal variation and random year-by-year variability account for most of the pattern we see.

7.6 Forecasting

7.6.1 Exponential smoothing forecasts

Having found various ways in which a time series model can be used to extract information on long-term trend and seasonal effects, it is natural to want to consider what the future

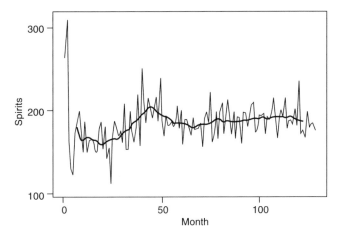

Figure 7.11 Seasonally adjusted spirits sales (thin line) and the moving average trend line (thick line) for the Republic of Ireland spirits sales data.

behaviour of these components of a time series might be. Calculating numerical values for future behaviour is known as **forecasting**.

Forecasting is an especially difficult area statistically, because by its nature it involves inferences for time periods for which no data are available. Indeed, in Chapter 6 it was stressed that predictions should only be made within the range of the observed data. This is impossible in time series analysis as the only predictions of any use are forecasts of future events. There are many, often highly sophisticated, statistical methods designed to help in the making of forecasts. In this book we will concentrate on a simple technique which should provide the reader with a tool with which forecasts may be made in some circumstances.

If there is no significant seasonal variation and no trend, then the exponential smoothing model may be used for forecasting. Recall that the smoothed estimate of each time series value is given by $y_t = \alpha x_t + (1 - \alpha)y_{t-1}$. In effect the y_t-values so produced are a kind of estimate of the x_t-values, taking into account the past values of the time series (via the α parameter). Given this, it seems quite reasonable to employ the same method to go about predicting values of x_t for time t beyond the observed series. If we write $\hat{x}_t$ as the predicted (forecast) value of x at time t, then we use

$$\hat{x}_{t+1} = y_t.$$

But we have an equation for y_t, and if we substitute this into the above equation, we get

$$\hat{x}_{t+1} = \alpha x_t + (1 - \alpha)y_{t-1}.$$

Rearranging this, we get the convenient equation for the forecast

$$\hat{x}_{t+1} = y_{t-1} + \alpha(x_t - y_{t-1}) = y_{t-1} + \alpha e_t$$

where $e_t = x_t - y_{t-1}$ is the prediction error at time t. We can use this to forecast the next value after the end a time series. Suppose $t = N$ is the last time for which we have data; then

$$\hat{x}_{N+1} = y_N + \alpha(x_N - y_N)$$

is our forecast. For $t = N + 1$, the only data we have is what we have just predicted, so that effectively $x_{N+1} = y_{N+1}$, which gives $\hat{x}_{N+2} = \hat{x}_{N+1}$. This holds true for $t = N + 2$, or $N + 3$, or indeed any t greater than N. In other words, our one-step-ahead prediction is not just a forecast of x at time $N + 1$, but at *all* future times, $N + 2, N + 3, \ldots$.

We illustrate the technique with some of the data on housing starts used in Chapter 6. The number of housing authority housing starts per month for three years from January 1985 until December 1987 are given in Table 7.1 and plotted in Figure 7.12. This is a series with no marked seasonal factors or trend, and so is suitable for forecasting with exponential smoothing.

Table 7.1 Housing authority housing starts, thousands

	J	F	M	A	M	J	J	A	S	O	N	D
1985	1.3	0.8	1.2	0.9	0.9	1.1	0.8	0.9	0.9	1.1	0.9	1.3
1986	1.0	1.0	1.1	1.6	1.1	1.0	1.2	1.0	0.9	1.0	1.0	1.1
1987	0.8	1.0	1.0	1.2	1.1	1.3	1.1	1.1	1.3	1.2	1.3	1.1

Source: http://www.ons.org.uk, October 1998.

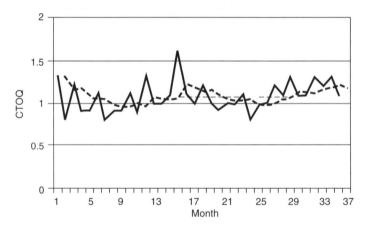

Figure 7.12 Housing authority housing starts (solid line) and exponential smoothing forecast using $\alpha = 0.3$ (dotted line).

The calculations of the forecasts from exponential smoothing are illustrated in Table 7.2. From there it is clear that the forecast for time $t + 1$, $\hat{x}_{t+1}$, is just the smoothed value at the previous time, y_t. This shows you that exponential smoothing is one-step-ahead forecasting. The errors in the prediction, $e_t = x_t - y_{t-1}$, are the differences between the observed value at time t, x_t, and the prediction based upon all values up until time $t - 1$, y_{t-1}. The percentage error measures the relative prediction error and is given by

$$\% \text{ Error} = \frac{x_t - \hat{x}_t}{x_t} \times 100.$$

Table 7.2 Forecasts from exponential smoothing, $\alpha = 0.3$, for 6 months

t	x_t	y_t	$\hat{x}_t$	e_t	% Error	e_t^2
Jan 1985	1.3	1.30				
Feb 1985	0.8	1.15	1.30	−0.50	−62.50	0.2500
Mar 1985	1.2	1.17	1.15	0.05	4.17	0.0025
Apr 1985	0.9	1.09	1.17	−0.27	−29.44	0.0702
May 1985	0.9	1.03	1.09	−0.19	−20.61	0.0344
Jun 1985	1.1	1.05	1.03	0.07	6.38	0.0049

The forecasted series is also shown in Figure 7.12, and you can see that the forecast for the second month is just the observed value at the first month. The forecast series starts at month 2 and ends at month 37, where the forecast is the smoothed value of the series at month 36. If you look carefully at the forecasts relative to the original series you will see that the forecasts lag one step behind the series. For example, the sharp drops in the series and months 7 and 20 are followed by drops in the forecasted series at months 8 and 21, and the sharp increases in the series at months 12 and 16 are followed by increases in the forecasts at months 13 and 17. The percentage errors are plotted in Figure 7.13, and these hover around 0, almost always in the range ±20%, from month 6 onwards.

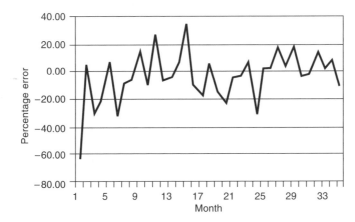

Figure 7.13 Percentage errors in the forecast.

7.6.2 Choosing α

The last column in Table 7.2 gives the squares of the prediction errors. The sum of squares of these errors, $\sum_{t=2}^{N} e_t^2$, is a measure of how far the one-step-ahead predictions are from the observed values in the series. Note that this sum begins at $t = 2$ as there is no prediction of the first point of the series. Obviously, if the prediction errors were all zero then the sum of squares of the errors would be zero and we would have perfect prediction. One way of choosing the value of α is to calculate the sum of squares of the errors for various values of α and choose the optimal value as the one where the sum of squares is minimal.

The sums of squares of the prediction errors are presented in Table 7.3 for various values of α, and you can see that the minimum value is going to occur around $\alpha = 0.3$. This was the value used in forecasting this series. Obviously we could try to take a finer grading of values around 0.3 such as 0.26, 0.27, . . . , 0.39, to get a more accurate value for α. This is seldom required in practice as there is usually very little difference in $\sum_{t=2}^{N} e_t^2$ over a wide range of values. This is the case here, where values of α in the range 0.25 to 0.4 have, more or less the same prediction error.

Table 7.3 Sum of squares of the prediction error and α

α	0.1	0.2	0.25	0.3	0.4	0.5
$\sum_{t=2}^{N} e_t^2$	1.44	1.29	1.26	1.24	1.25	1.30

7.6.3 Exponential smoothing forecasting and trend

To assess the effects of a trend on this forecasting technique, let us return to the small FTSE data set of Section 7.3.3. Using the technique in Section 7.6.2, we estimate that the optimal value of α is 0.95. This indicates that there is very little contribution of the past history of the series to the forecast and effectively we are just using the observation on day t as the forecast for the value of day $t + 1$.

In Figure 7.14 we plot both the original series and the forecast series. Note how the forecast 'lags' behind the real time series. The final forecast for day 21 is about 3816.9, which is only slightly below the series value on day 20 of 3817.2. This illustrates a problem with exponential smoothing technique of forecasting, namely that pronounced trends are not well dealt with. If you look back to Figure 7.4 you will see that with small values of α there is an enormous underestimation of the observed series, and this will lead to a systematic bias in the forecasts. When there is an increasing trend simple exponential

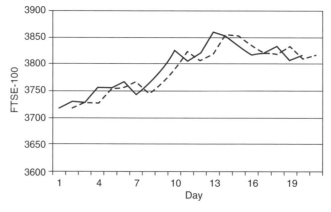

Figure 7.14 Original data (solid line) and forecast (dotted line) for the small FTSE data set.

smoothing, as discussed in this section, is not a good forecasting method as it will always tend to underforecast. If the trend is decreasing then there will be an overprediction.

There are other methods of forecasting, but most are beyond the scope of this textbook. It is certainly possible to use the linear model of Chapter 6 to fit a straight-line model of the form $x_t = a + bt$ between the time series x_t and time t. This is only valid if there is indeed a straight-line relationship, but as you will have noticed from the graphs this is very seldom the case. Consequently, this method of forecasting is not recommended. The moving average trend is not good for forecasting as you do not have moving average trends for the last few observations. Consequently, to make a forecast you need to guess values for the moving average trend, and this is also unsatisfactory.

7.7 Summary

In this chapter we have seen how to decompose a time series into three of its main components – the trend, the seasonal effects and the random components. The random components are used to assess whether the time series model is adequate. Essentially, if the residuals for a time series have no systematic structure then we conclude that we have extracted all the relevant systematic effects. Investigating the residuals to see if they exhibit any systematic structure is a common feature of statistical models and one that shall be pursued in other chapters.

Both moving averages and exponential smoothing are means of estimating the long-term trend in the series. Moving averages are a smoothing technique very closely related to the locally weighted least-squares procedure discussed on the book's website. Within time series analysis moving averages and exponential smoothing are preferred to a straight-line relationship for the estimation of the trend as it is a rare time series which has a linear trend over time.

Seasonal adjustment is an important technique in time series analysis, as many economic series are presented in an unadjusted and seasonally adjusted form. The methods used to seasonally adjust the time series of economic indices published by the Office for National Statistics are similar, in principle, to those described in this chapter. Seasonal adjustment is necessary to make sure that the figures are comparable, taking into account known seasonal effects. In this way you would hope to notice short-term trends in the series.

The methods discussed in this chapter have been a blend of graphical presentations and data summaries of the important features. The summaries are used to describe the main features of the series, and the residuals are used to check that all major systematic effects have been taken into account. The close association between graphical presentation and data summary is a feature of statistical analyses.

8

Probability as a model for random events

- Knowing about events and sample spaces and the three definitions of probability – equally likely outcomes, long-run relative frequency and subjective probability
- Knowing how to provide an interpretation of a probability and a conditional probability
- Understanding mutually exclusive events and independent events and the difference between them
- Knowing the laws of probability – addition law, complementary law and multiplication law
- Knowing how to manipulate probabilities using these laws
- Understanding the effects of mutually exclusive events on the addition law and independent events on the multiplication law
- Knowing how to use tree diagrams as an aid to working with the laws of probability
- Knowing about Bayes' theorem, its uses and its interpretation.

8.1 Introduction

The study of statistics is very closely linked with the study of probability. While it is possible to use descriptive statistical techniques, such as means and standard deviations, histograms and bar charts, without worrying too much about probability, taking the trouble to understand probability will bring with it a fuller understanding of statistical ideas.

Probability theory has a long history dating, in part, from the study of games of chance. The study of such games, such as tossing a coin, throwing a die, card games and, recently in the United Kingdom, the National Lottery, still provides much opportunity for the use of probability. For example, it is possible to evaluate the probability of an event occurring from knowledge of the chance mechanism of the game.

The study of probability is a means of studying how to model chance or random events. Statistics deals with the interpretation of the data arising from experiments which have some random component. Essentially, the use of a descriptive statistical technique to summarize some feature of data from an experiment is an attempt to understand the random process by which the data arose. Thus, when we calculate the mean and standard deviation of house prices for bungalows in a certain area we believe that the mean provides information about the average house price and the standard deviation provides information about the price variation, which comes about, in part, through some random mechanism.

One of the goals of a statistical analysis is to describe the process by which the data arose and then make predictions about future data which are likely to arise in similar situations. Suppose a camping store manager, studies the weekly demand for a product, say Vango Force 10 two-person tents, and finds that over a period of 15 weeks the sales are 1, 2, 2, 0, 3, 1, 2, 1, 4, 2, 0, 2, 1, 0, 2. As far as the manager is concerned the sales are a random event from week to week and are not predictable, though in the long term he may expect a seasonal trend with greater sales in spring and summer.

If the manager decides that she will only keep three tents in stock at the beginning of each week, then the past results suggest that only in one week out of 15 will she not have enough tents to satisfy demand. Suppose that she keeps four tents in stock; does this mean that she will always be able to satisfy demand? On the basis of the sample, yes; but in reality, clearly not. There is a chance, albeit a small one, that in one week she may sell five or more tents. Such an event was not observed in the 15 weeks she studied, but this does not mean that it can never occur.

It is possible to develop a probability model for the weekly sales of tents and use this model, together with the data collected, to estimate the probability of demand being five tents or more in a week. This is actually a better way of estimating the probability of demand being four tents or more in a week than using the observed number in the 15 weeks. If a different 15 weeks were chosen then it is likely that a different pattern of weekly sales would be observed and so the number of weeks with four or more sales would be different.

Probability and statistics are linked. Probability, by itself, can be an interesting but relatively abstract subject, where most of the easier applications are on relatively unimportant examples, such as games of chance. When linked with statistics, probability becomes an important and powerful tool for making inferences about future situations based on a careful statistical analysis of data. This is mainly through the use of probability models which allow you to describe how data were generated and, subsequently, to make predications about the likely events in the future.

The example of the tents discussed above is similar to many stock control problems. For this problem the weekly sales are not very high and a Poisson distribution is often a good probability model for this type of data (see book website). Before discussing specific probability models, we shall look at elementary probability and see how to manipulate and interpret probabilities.

8.2 Events and sample spaces

8.2.1 Sample space

A probability experiment consists of a number of components. Each experiment has a number of **elementary outcomes**, which are completely distinct such that no two can occur at the same time. The collection of all possible outcomes for an experiment is known as the **sample space**. A key feature of a probability experiment is that the outcome of the experiment is subject to chance. At any one time, you may know the possible outcomes, but which one will occur is the result of a random process.

The random nature of the outcomes distinguishes probability experiments from more traditional scientific experiments. If you consider an experiment to heat up water held in a container at room temperature, then application of sufficient heat to the water will eventually lead to the water boiling. The outcome of the experiment is water boiling and this outcome occurs every time the experiment is repeated under identical situations.

Even within well-controlled scientific experiments the outcome can still be the result of effectively random processes. If you consider measuring the length of time for 0.5 litres of water to boil using a 3-kilowatt water heater then you will not get exactly the same time each time the experiment is repeated under identical conditions. This occurs because there are certain features of the experiment which may not be completely in the control of the experimenter, such as the room temperature and small variations in the output of the water heater.

Experiments in business are not as easy to control as scientific experiments. In many of them you are looking at the behaviour of individuals, and this can be very unpredictable. Consequently, there is a great need for understanding probability experiments within a business setting. A marketing experiment is assessing the acceptability of the packaging of a particular product, say a brand of toothpaste. There are many different brands, and the company may change the packaging with a view to trying to increase the market share for theirs. A study may be carried out to monitor the acceptability of the new packaging on a number of features using the grades 1 to 5, with 1 indicating completely unacceptable and 5 very acceptable. This defines the sample space to be {1, 2, 3, 4, 5}, where the numbers themselves only denote a ranked position on the acceptability scale. A particular individual may give the background colour scheme a grade of 5, while a second individual may give it 4. Furthermore, it is entirely possible that the second individual may give the colour scheme a grade of 5 on second sight of the packaging the next day. Thus the outcome of the experiment is subject to a random process both between individuals and within an individual on separate occasions.

In simple probability experiments it is normally a simple matter to write down the sample space. In the toothpaste example above the sample space for package colour is {1, 2, 3, 4, 5}. If a second attribute is added such as text colour, then the joint sample space for the combination of acceptability of background colour and acceptability of text colour is:

1,1	1,2	1,3	1,4	1,5
2,1	2,2	2,3	2,4	2,5
3,1	3,2	3,3	3,4	3,5
4,1	4,2	4,3	4,4	4,5
5,1	5,2	5,3	5,4	5,5

There are 25 elementary outcomes in the sample space corresponding to every possible combination of acceptability of background colour and acceptability of text colour. Adding a third attribute such as acceptability of the size of the company logo would lead to a sample space of $5 \times 5 \times 5 = 125$ outcomes. In the tent example of Section 8.1, the sample space for the number of tents sold per week is {0, 1, 2, 3, . . .}. This has no theoretical upper bound, and while sales of 537 tents in a week are highly unlikely they are not, strictly speaking, totally impossible.

8.2.2 Events

In a probability experiment an **event** is defined to be one or more of the elementary outcomes. The event 'less than two tents' comprises the outcomes {0, 1} and the event 'more than four tents' consists of {5, 6, 7, ...}. An event is a number of outcomes, and events can be combined together. For example, we can make up a compound event 'four or fewer tents' and 'more than two tents' which contains two outcomes {3, 4}. A second example is 'less than 2' or '4 tents' which has {0, 1, 2, 4} as outcomes. Events are combined, using **or** and **and** in much the same way as numbers are combined using + and ×. In order to do this we need to study how to add and multiply probabilities, and this will be tackled in Sections 8.4 and 8.5.

In many respects the study of probability through games of chance paints an unrealistic picture of probability for the business world as the experiments are relatively contrived compared to experiments in marketing and economics. Consequently, we prefer to study probability and the laws of probability using examples taken from surveys, where the data can be tabulated. In this chapter we use data from a survey of young people in Scotland which was designed to gather information about young people as they moved from secondary school to college, university or employment.

Example 8.1: Scottish Young Persons Survey: events
The data in Table 8.1 were collected from a random sample of 1979 pupils who were in fifth year of secondary school (S5) in Scotland during the session 1986–87. At this time, the school pupils have their first opportunity to sit the Higher examination, which is the main determinant, in Scotland, for entry to higher education. Pupils in S5 can take Higher examinations in any number of subjects. Five subjects is the usual maximum in many schools, but some pupils can take more.

Table 8.1 Number of Highers passed, by gender, for a random sample of pupils in S5 in Scotland in session 1986–87

	0	1	2	3	4	5	6+	Total
Male	353	157	83	91	100	123	15	922
Female	376	164	140	132	121	113	11	1057
Total	729	321	223	223	221	236	26	1979

Source: Scottish Young Persons Survey

In Table 8.1 two variables are recorded: gender, where the sample space is {M, F}; and the number of Highers passed in S5, where the sample space is {0, 1, 2, 3, 4, 5, 6+}. We can define some events based upon the sample space for the number of Highers passed in S5 to illustrate the construction of events from elementary outcomes:

Event	Description	Outcomes
E_1	Highers in S5	{1, 2, 3, 4, 5, 6+}
E_2	No Highers in S5	{0}
E_3	3+ Highers in S5	{3, 4, 5, 6+}

Thus the event denoted E_3 corresponds to three or more Highers passed in S5, and the elementary outcomes {3, 4, 5, 6+} are all associated with this event. In Scotland, at that time, 3 Highers was generally regarded as the minimum number required for entry to higher education.

If we look at the whole of Table 8.1 rather than just the last row of it, then the elementary outcomes consist of the cross-classification of gender and the number of Highers passed in S5. The sample space is {M0, M1, M2, M3, M4, M5, M6+, F0, F1, F2, F3, F4, F5, F6+}, where M0 corresponds to a male pupil who achieved no Highers, and so on. There are 12 elementary outcomes all of which are distinct from each other, or **mutually exclusive**.

We can begin to define more complex events by combining the two variables. Also, the existing events can be redefined in terms of the two variables:

Event	Description	Outcomes
E_3	3+ Highers in S5	{M3, M4, M5, M6+, F3, F4, F5, F6+}
Male or E_3	Male or 3+ Highers in S5	{M0, M1, M2, M3, M4, M5, M6+, F3, F4, F5, F6+}
Female and E_2	Female and No Highers in S5	{F0}
Male and E_3	Male and 3+ Highers in S5	{M3, M4, M5, M6+}
E_2 and E_3	No Highers in S5 and 3+ Highers in S5	{ }

With these examples, it can be seen that the same event can be defined in terms of different outcomes in different sample spaces. Thus when we were only interested in the number of Highers passed in S5, E_3 was defined to have the outcomes {3, 4, 5, 6+}. When there are two variables involved in the sample space then E_3 is defined as having the outcomes {M3, M4, M5, M6+, F3, F4, F5, F6+}. It is easy to see in Table 8.1 that exactly the same number of pupils are involved in both definitions, 706. The sample space depends on the variables under study. Different events can have different representations, in terms of elementary outcomes, under different sample spaces. Whatever the representation, the probability of the event will be exactly the same.

All complex events are made up of a number of elementary outcomes. The event, E_2 and E_3, has no elementary outcomes associated with it. This occurs because the event is completely impossible. Other events, such as E_1 or E_2, will contain the whole sample space as the event is certain to occur.

The important lesson from this section is that all events are made up of a collection of outcomes of the experiment. Some events are impossible (no outcomes are associated with the event), others are certain to occur (all outcomes are associated with the event). Most lie somewhere in between (some outcomes are associated with the event and the rest are not).

8.3 Probabilities and their interpretation

The probability of an event is a measure of the likelihood that the event will occur. This ranges from impossible to certain, and the probability expresses this likelihood on a numerical scale. A probability of zero indicates that the event is impossible and will never occur. A probability of one indicates that the event is certain to occur.

Assigning a numerical value to the probability of an event is not a particularly easy task. There are three main ways of doing this. One occurs in special situations where the elementary outcomes of the probability experiment are all equally likely. This is known as the **equally likely outcomes** method. A second is the use of the **long-run relative frequency** of the event as a measure of probability. This is the most common method in practice. The third way is known as **subjective probability**, and is based on the idea that probability measures the likelihood of an event for a particular individual. In situations where sufficient experimental evidence can be collected the three definitions of probability will tend to coincide as the experimental evidence will influence the subjective measures of probability.

8.3.1 Equally likely outcomes

The equally likely outcomes definition of probability has its roots in games of chance, where there are a fixed, usually small, number of possible outcomes, all of which are equally likely. In such a case the probability of an event E is defined as

$$P(E) = \frac{\text{Number of outcomes favourable to event } E}{\text{Total number of outcomes in the sample space}}.$$

In order to work out a probability, all that is necessary is to count the number of possible outcomes for the experiment and then the number of outcomes in which event E occurs. We will look at some simple examples from games of chance to see how the calculation operates.

Example 8.2: Throwing a die

A die has six sides and so the sample space is $\{1, 2, 3, 4, 5, 6\}$. If the die is unbiased then each of these elementary outcomes is equally likely. The probability of the event E_1, 'a 2 lying face up', is simply 1/6 as there are 6 outcomes, all of which are equally likely, and one of these outcomes corresponds to the event E_1, namely $\{2\}$.

The probability of the event E_2, 'at least a 4' is 3/6 as E_2 comprises the 3 elementary outcomes $\{4, 5, 6\}$.

If the die is perfectly symmetrical about all its axes of symmetry then there will be no tendency for one side of the die to land face up more often than any other side. This is a physical property of the die which translates into a probability statement about the outcomes being equally likely. If, for example, one side of the die was slightly heavier than the opposite side then the die would have a tendency for the heavier side to land face down. Thus the die would be biased and the six elementary outcomes would not be equally likely.

In the example it is relatively easy to evaluate the probabilities of the events by writing down the list of possible outcomes and counting. It is not difficult to see that there will be practical difficulties even with relatively simple probability experiments as the sample space can become very large. With one die the sample space had 6 outcomes, with two there will be $6 \times 6 = 36$, and so with 3 dice there will be $6 \times 6 \times 6 = 216$ possible outcomes. There are laws of probability which control the calculation of the probabilities of complex events, knowing the probability of the elementary outcomes, and these will be discussed in Section 8.5.

The UK National Lottery is another example where the use of equally likely outcomes is used to determine the probabilities of events. There are 49 numbers and each individual entry is a choice of 6 of these numbers. Thus the sample space for all possible choices has $49 \times 48 \times 47 \times 46 \times 45 \times 44 = 1.0068 \times 10^{10}$ outcomes, i.e. over 10 billion choices. There are 49 possible numbers to choose for the first number. Having selected the first number, there are then only 48 numbers left for the second choice, 47 for the third, and so on. There is no replacement at each choice, and the selection is made **without replacement**.

Actually, the order in which the numbers are chosen is unimportant. Suppose a specific choice, in order of choice is 23, 43, 10, 25, 37, 11. This is exactly the same as a choice in a different order such as 10, 43, 37, 25, 11, 23 or 25, 11, 23, 43, 10, 37. In order to work out the total number of *different* choices, ignoring the order in which the numbers were chosen, it is necessary to work out the number of permutations of any specific 6 numbers. There are 6 choices for the first number, 5 for the second, 4 for the third, and so on. Thus there are $6 \times 5 \times 4 \times 3 \times 2 \times 1 = 720$ different orders (permutations) for 6 numbers.

Putting the two permutations together will give the total number of different combinations of 6 numbers from the 49 numbers. There are $49 \times 48 \times 47 \times 46 \times 45 \times 44$ ways of choosing 6 numbers from 49, where you retain knowledge of the order in which the numbers are selected. When it comes to winning the lottery, the order in which the numbers were selected is not important and as there are 720 different orderings of 6 numbers there must be only

$$\frac{49 \times 48 \times 47 \times 46 \times 45 \times 44}{720} = 13\,983\,816$$

different combinations of number choices. As there is only one jackpot prize, obtained when your choice of 6 numbers matches the 6 numbers drawn, the probability of getting the jackpot prize is $1/13\,983\,816 = 0.000\,000\,071\,511\,2$, which is extremely small but not totally impossible.

The size of the sample space for the problem with three dice was calculated by multiplying the sample size for one die by itself three times, i.e. $6 \times 6 \times 6$. In contrast, for the lottery the multiplier was the sample size for one number, decreased by one each time, i.e. $49 \times 48 \times 47 \times 46 \times 45 \times 44$. The difference occurs because the numbers in the lottery are chosen **without replacement**, whereas with the dice there is effectively a choice **with replacement**. The difference between the two schemes is important. When the first die is thrown there are 6 possible outcomes and when the second is thrown the same 6 outcomes are still available, giving 6×6 outcomes. The situation is identical if two separate dice are thrown, or if one die was thrown two separate times. In the lottery, having selected a number means that number is no longer available for selection at the next choice.

8.3.2 Long-run relative frequency

There are many probability experiments where there are not a fixed number of elementary outcomes which are equally likely. Examples include the number of traffic accidents per month at a road junction. Here, the elementary outcomes are 0, 1, 2, 3, . . . so the sample space does not have a fixed number of outcomes in it. Also, they are not going to be equally likely, as ten accidents per month will not have the same probability as no accidents per month. Another example is the probability that the price of a particular share will rise over the course of a day. On a particular day, the share price can either increase, decrease or stay the same. It would be a rash financial investor who assumed that these three outcomes were equally likely and invested on that basis.

If a probability experiment can be repeated a large number of times under completely identical conditions then the probability of an event can be calculated as the relative frequency of the event. Suppose that the experiment is repeated n times and the event E occurs on r of them. The probability of the event is

$$P(E) = \frac{r}{n}.$$

In samples where n is small, this is not a good way of calculating the probability of an event as there is a great deal of variation in the responses in small samples. It is a valid method in very large samples. Probability experiments with games of chance can be repeated many times and the equivalence of the equally likely outcomes definition and the long-run relative frequency definition of probability can be demonstrated.

Example 8.3: Tossing a coin
A simple experiment involves tossing a coin 20 times and recording the number of heads. This experiment was carried out 10 times.

Trial	1	2	3	4	5	6	7	8	9	10
Number of heads	13	9	11	10	13	10	11	11	13	10
Proportion	0.65	0.45	0.55	0.50	0.65	0.50	0.55	0.55	0.65	0.50

If the coin was unbiased the equally likely outcomes method gives $P(\text{Head}) = 0.5$. From the 10 samples of size 20 you can see that there is a great deal of variation in the estimated proportion of heads. Even when all 10 samples are combined to give a single sample of 200 the proportion of heads is 0.555.

The results of two computer simulations to estimate the probability of heads through the long-run relative frequency are displayed in Figure 8.1. The graph was created by using a computer program to simulate the probability experiment of tossing a coin. This technique is known as computer simulation, and is a very powerful mechanism for studying probabilities. It is also an excellent method of studying models which are difficult to investigate by more traditional mathematical means. Computer simulation relies on having a program with a reliable random number generator. This is a computer program which will generate numbers by a mathematical process in such a way that the numbers appear to be completely random. If the generator is not of a very high quality then the random number generator may be biased and the results of the simulation misleading.

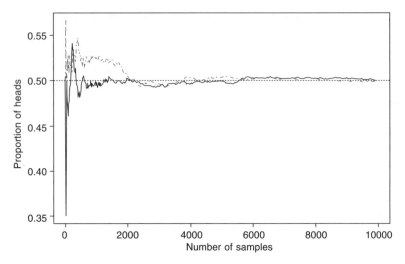

Figure 8.1 The long-run frequency of heads with increasing sample sizes.

In the graph a dotted horizontal line is drawn at 0.5 and the lines representing the convergence of the long-run relative frequency calculations of the probability of heads approach this as the sample size gets large. There are two simulations, each represented by a separate line. The important point is that the sample sizes need to be extremely large and in the example it is only when the sample size is in excess of 2000 that both lines remain close to 0.5. The second point is that in small samples the relative frequency estimate from samples can be extremely variable.

Example 8.3 showed that the long-run relative frequency definition of probability coincides with the equally likely outcomes definition when both are applicable. In the next example the more common case, where the equally likely outcomes method is not applicable, is considered.

Example 8.4: Scottish Young Persons Survey: calculating probabilities

The data in Table 8.1 represent a random sample of all secondary school pupils in S5 in Scotland in session 1986–87. The probability that a pupil chosen at random obtained 3+ Highers can be estimated from this table. There are 1979 pupils in the sample and 706 obtained 3+ Highers. Thus, to 4 decimal places,

$$P(3 + \text{Highers}) = \frac{706}{1979} = 0.3567.$$

As there are 1057 female pupils in the sample the probability that a pupil in S5 is Female is estimated as

$$P(F) = \frac{1057}{1979} = 0.5341.$$

There are a total of 922 + 377 = 1299 pupils who are male or have 3+ Highers. Thus the probability that a pupil in S5 is male or has 3+ Highers is

$$P(M \text{ or } 3+ \text{ Highers}) = \frac{1299}{1979} = 0.6564.$$

In the final illustration, there are $376 + 164 = 540$ pupils who are female and who obtained less than 2 Highers in S5. The probability of selecting such a pupil is

$$P(F \text{ and less than 2 Highers}) = \frac{540}{1979} = 0.2729.$$

In practice, the above method is certainly a more common way of estimating probabilities than equally likely outcomes. It is worth pointing out that probabilities calculated using the long-run relative frequency method are estimates of the true value, which is unknown. With equally likely outcomes the true value can be calculated based upon the assumption that the outcomes are equally likely. This assumption can be checked from knowledge of the physical properties of the probability experiment.

Within Example 8.4 there is a subtle but important point about the meaning of probability. If we consider just the 1979 pupils in the sample and nothing else. An individual is to be selected from this sample at random. This can be achieved by assigning a unique number between 1 and 1979 to each individual in the sample and selecting a number at random. The probability of selecting a male pupil from this group of pupils is then $922/1979 = 0.4659$. This is the equally likely outcomes definition of probability as there are 1979 outcomes which are equally likely if the selection is made at random.

If we ask what is the probability that a pupil in S5 in session 1986–87 was male, then we are referring to the whole population of pupils in S5 at that time. We do not have any information on all of them, only on the random sample of 1979. In this case we can estimate the probability as $922/1979 = 0.4659$ This is the long-run frequency estimate of the probability of selecting a male pupil from all pupils in S5 in 1986–87.

The same numerical value is used in two situations with two different meanings. In the first case it is the exact probability for the selection of one pupil from a fixed and known number of pupils in the sample. In the second it is an estimate of the probability for the selection of one pupil from all pupils in the population with a larger but unknown total number of pupils. The key feature which makes the last estimation valid is the fact that the original sample was selected at random from the population.

8.3.3 Subjective probability

There are many situations in which it is not feasible to estimate a probability by repeating an experiment many times or by selecting a large sample, which is effectively the same thing. This can arise because the process is completely new and there are no previous records. It may be that the event is possible but is so rare that it has never previously occurred and so there is no information on the probability. In such cases, a numerical value can be assigned to the probability to represent the degree of belief that an event will occur. **Subjective probability** and **subjective degree of belief** are interchangeable names for the same idea.

The financial manager of a company has to set a budget for contingency repairs to the manufacturing equipment in the company's factory. He can use past records on the reliability of the equipment to estimate the probability that a machine will have to be repaired during the budget period and what type of repairs are required. He will then be able to use this information, together with information on the cost of the repairs, to estimate how much money he will have to allocate for the repair budget. This is the long-run frequency approach.

If the manufacturing machines are replaced by new ones, then there will be no previous data to help the financial manager. He cannot therefore rely on long-run relative frequency, and the equally likely outcomes approach will not do. In this case he may call on the assistance of engineers to help him to make a subjective assessment of the probability of a machine breaking down. Here the engineer and financial manager are using their experience to make an 'estimate' or 'educated guess' of the unknown probability because that is really all that they can do under the circumstances.

All of us are familiar, in some sense, with the idea of making judgements about probabilities. The decision to step outside for a walk with or without a raincoat is often based upon a quick glance at the sky and an assessment of the likelihood of rain. Two different people in the same situation may easily attribute different subjective probabilities of rain.

An investment analyst may decide to invest in one company rather than another on the basis of a 'gut feeling' about the companies, when all other financial considerations about these companies are similar. Here, she is judging that the probability of a successful investment with one company is greater than the probability of success with another. Again, two investors considering the same two companies may attribute different subjective probabilities of success to each company.

Subjective probabilities are particularly associated with the Bayesian school of statistical thinking. The long-run frequency definition of probability is known as the frequentist or classical school. The laws of probability are common and are the same in both schools, but the method of assigning numerical values to these probabilities is different.

8.3.4 Interpreting probability

A probability is a numerical measure of the likelihood of an event occurring, but the interpretation of the meaning of probability is often tricky. In ordinary everyday speech we are often very loose in our use of language about probabilities, and this can confuse the precise understanding of the meaning of probability. The trouble is that the precise meaning requires precise language and our discussion of probability can become convoluted.

When tossing a coin we may also say that there is a 50% chance of a head appearing. Probabilities are defined, as proportions, to lie between 0 and 1 inclusive, but generally they are expressed, in normal conversation and writing, as percentages. This does not cause any difficulties provided you remember to work with them as proportions in all calculations, expressing the final result in percentage terms if you wish.

In the National Lottery the overall chance of winning a prize is quoted as being approximately 1 in 54. What does this mean and how are we to interpret it? Let us suppose for the sake of argument (and so that we can work with exact fractions whenever we talk about winning the lottery) that the probability is exactly 1/54. Then there is one chance in 54 of winning, so the probability of winning is $1/54 = 0.0185$. We might also say that there is just under a 2% chance of winning any prize.

Another way of looking at the interpretation is to say that if you bought a single lottery ticket each week for 54 weeks, then you would expect to win a prize on one of these weeks. This does not mean that you are *guaranteed* to win a prize on 1 week out of the 54, merely that this is what is *expected*. If you do the experiment it will cost you £54 and you may win on 0 weeks, 1 week, 2 weeks, 3, weeks, and so on.

If you have gone for 53 weeks without winning a prize with your one ticket per week, do not be convinced that you will win on the 54th week. The probability of winning on the 54th week is still the same as on the first week, namely 1/54, because each draw of numbers is completely separate and independent of previous draws. You can see that this must be the case because of the way in which the numbered balls are selected each Saturday night.

One of the more common failings in the interpretation of the meaning of probability is to inflate your expectations by effectively modifying the probability by a subjective degree of belief. This is illustrated by the statement that 'I have bought one ticket for the lottery every week for 53 weeks and, as the chance of winning is 1/54, I must win next week'. Here the correct interpretation that you expect to win one week out of 54 weeks is modified into a feeling that eventually you must win because one win is expected every 54 weeks. This is fallacious because an expectation of 1 win in 54 weeks is not a guarantee of 1 win.

Another misconception about probability is the notion that there is something regular to the occurrence of events. When you roll a die, the probability of a 6 is 1/6, so every 6 throws you expect to get one 6. This does not mean that you expect every sixth throw to result in a 6, or that in any batch of six throws there will be one 6.

The use of probabilities in simple games of chance, such as throwing a die, tossing a coin, or drawing cards is a good way of putting a probability on to a scale of reference. One is drawn in Figure 8.2 and most of the events listed there have probabilities which are towards the zero (impossible) end of the scale. One way to interpret probabilities is to refer them to events with which you are relatively familiar, and this is the purpose of

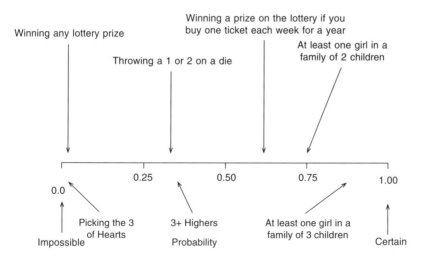

Figure 8.2 Probability scale.

the illustration. The probability of winning any prize on the lottery with one ticket (1/54) is much the same as the probability of drawing a specified card from a pack of cards (1/ 52). The probability that a randomly selected pupil from S5 in Scotland obtained 3+ Highers is about the same as the probability of throwing a 1 or a 2 with a die.

8.4 Mutually exclusive and independent events

The probabilities for simple events can often be obtained by counting. More complex events are obtained by combining simple events together, and the laws of probability which are presented in Section 8.5 enable us to work out the probability of these combined events. There are two special type of events which make these laws much simpler in special cases, and they are discussed here.

8.4.1 Mutually exclusive events

Two events are said to be **mutually exclusive** if they have no elementary outcomes in common. This means that there is absolutely no overlap between these events and they are totally separate from each other. It also means that if one of the events occurs then the other one cannot. So the second event is totally impossible given that the first one has occurred.

Example 8.5: Mutually exclusive events
Three events taken from Example 8.1 are repeated below:

Event	Description	Outcomes
E_1	Highers in S5	{1, 2, 3, 4, 5, 6+}
E_2	No Highers in S5	{0}
E_3	3+ Highers in S5	{3, 4, 5, 6+}

Event E_2 is mutually exclusive of the other two events E_1 and E_3. If a pupil has no Highers in S5 then he or she cannot have at least one Higher at the same time. If a pupil has three or more Highers then he or she cannot simultaneously have no Highers.

On the other hand, the events E_1 and E_3 are not mutually exclusive as both can easily occur at the same time. The elementary outcomes 3, 4, 5, 6+ Highers are in common to both events.

8.4.2 Conditional probability

The second special type of events are independent events. Two events are said to be **independent** if the occurrence of one of them does not influence the probability that the other one will occur. Here we have a situation in which we wish to discuss the probability of one event given that another event has occurred. In order to do this it is necessary to introduce a new concept of **conditional probability**, and a new notation for it.

Given two events, E_1 and E_2, and the conditional probability of event E_2 given that E_1 is known to have occurred is written as $P(E_2 \mid E_1)$. With these two events there is a second conditional probability, $P(E_1 \mid E_2)$, which is the probability of E_1 given that E_2 is known to have occurred. This is not the same probability as $P(E_2 \mid E_1)$. Occasionally you may find that they take the same value, but this is not true in general.

The calculation of the numerical value for the conditional probability follows on in the same fashion as for the calculation of the unconditional probabilities, $P(E_1)$ and $P(E_2)$. The only difference is that the sample space is restricted. To calculate $P(E_2 \mid E_1)$, the denominator is the number of outcomes in which E_1 is known to have occurred and the numerator is the number of times E_2 occurs at the same time as E_1 does. Thus,

$$P(E_2 \mid E_1) = \frac{\text{Number of outcomes in which both } E_2 \text{ and } E_1 \text{ occur}}{\text{Number of outcomes in which } E_1 \text{ occurs}}.$$

Example 8.6: Scottish Young Persons Survey: conditional probability

In Table 8.1 let E be the event that a pupil has 0 Highers in S5. The unconditional probability is

$$P(E) = \frac{729}{1979} = 0.3684.$$

The conditional probability of getting 0 Highers given that the pupil is male is

$$P(E \mid M) = \frac{353}{922} = 0.3829.$$

There are 922 pupils in the sample who are male and this gives the denominator. Out of these 922 pupils, 353 did not get any Highers in S5 and this is the numerator in the calculation. Similarly,

$$P(E \mid F) = \frac{376}{1057} = 0.3557.$$

For a randomly selected pupil from S5 the chance that a pupil will get no Highers is 37%, given that the pupil is male this increases slightly to 38% and given that the pupil is female it drops slightly to 36%.

Conditional probabilities can also be calculated by conditioning on the number of Higher passes. For example,

$$P(F \mid E) = \frac{376}{729} = 0.5158.$$

There are 729 pupils with 0 Higher passes in S5 and 376 of them are Female.

In this example the two orderings of the conditional events have reasonable interpretations. What is the probability of getting 0 Higher passes in S5 given that you are a male pupil, and what is the probability that the pupil is male given that he or she got 0 Higher passes in S5? In some instances there is a natural ordering to the events, and it usually makes more sense to condition on the event which occurs first.

8.4.3 Tree diagram

Conditional events and conditional probabilities can be pictured in a **tree diagram**. This is often a simpler way of seeing the problem than a two-way table and can help to simplify the calculation of the probabilities of more complex events. Figure 8.3 is the tree diagram for the data in Example 8.6.

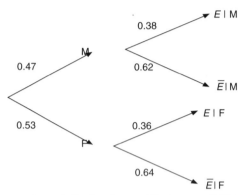

M denotes a male pupil, F a female one, E is the event '0 Highers in S5' and $\overline{E}$ (not E) the event '1 or more Highers in S5'.

Figure 8.3 Tree diagram.

The tree diagram has information about the unconditional events, in the branch at the first level, and their probabilities. The unconditional events are the genders of the pupils in S5 – male with probability 0.47 and female with probability 0.53. The conditional events are in the second branch. There are two sets of branches at the second level as the branch is repeated for every unconditional event. In the first we have the events no Highers in S5 conditional on a male pupil, with probability 0.38, and Highers in S5 conditional on a male pupil, with probability 0.62. In the second of the branches at this level we condition on the pupil being female, and the events are no Highers in S5 conditional on a female pupil, with probability 0.36, and Highers in S5 conditional on a female pupil, with probability 0.64.

You should note that in the tree diagram the sum of the probabilities within each of the two branches at the second level is 1, as it is for the branch at the first level. This must occur because at each stage you write down all the possible outcomes, and the probabilities of all the elementary outcomes must add up to 1.

8.4.4 Independent events

The conditional probability gives you the probability that one event occurs, given that another is known to have occurred, and this is related to independent events. If two events are independent then knowing that one of them has occurred does not change the probability that the other one will occur. This is simply the equality of the unconditional and conditional probability: if they are equal then the two events are independent. If they are unequal then they are not independent events.

If two events, E_1 and E_2 are independent then the probability that E_2 occurs is not changed by knowledge that E_1 has occurred. If you know that E_1 has occurred then the probability the E_2 occurs is $P(E_2 \mid E_1)$. If you do not know if E_1 has occurred then the probability of E_2 is the unconditional probability $P(E_2)$. If E_1 and E_2 are independent then

$$P(E_2 \mid E_1) = P(E_2).$$

Conversely, if you know that $P(E_2 \mid E_1) = P(E_2)$ then this implies that the two events are independent. It does not matter if you condition on E_1 or E_2 as, if the events are independent, then both $P(E_2 \mid E_1) = P(E_2)$ and $P(E_1 \mid E_2) = P(E_1)$.

Example 8.7: Throwing two dice
Suppose we throw two dice. Let E_1 denote the event '6 on the first die', with E_2 denoting the event '6 on the second'. These two events are completely independent of each other as the outcome on the first die has no bearing on the outcome on the second. In terms of the conditional probabilities

$$P(E_1) = \frac{1}{6}, \quad P(E_2) = \frac{1}{6}, \quad P(E_2 \mid E_2) = \frac{1}{6}.$$

Given that E_1 is known to have occurred has no influence whatsoever on the probability of E_2 occurring.

Example 8.8: Scottish referendum
On 11 September 1997 people in Scotland voted in a referendum on two questions. One was about setting up a parliament in Scotland again, and the other about giving the parliament the powers to vary the income tax rates in Scotland compared to England. The number of people voting was 2.4 million. The information in Table 8.1 gives the percentages voting yes of those who voted.

The percentage voting yes to a parliament is taken as an estimate of the probability that a randomly selected voter in Scotland is in favour of a parliament. This is 0.74, and it can also be seen that this probability is not the same over all the four main cities. Knowing that the voter lives in Glasgow changes the estimate of the probability to 0.84. It is not the same probability as the latter is a conditional probability, conditioning on where the voter lives:

Table 8.2 Selected percentages of voters voting 'Yes'

	Yes to a parliament	Yes to tax varying powers
All Scotland	74	63
Four main cities		
Aberdeen	72	60
Dundee	76	65
Edinburgh	72	75
Glasgow	84	62

$$P(\text{voter in Scotland voted yes to a parliament}) = 0.74,$$

$$P(\text{voter in Scotland voted yes to a parliament} \mid \text{voter in Glasgow}) = 0.84.$$

As knowledge that the person is from Glasgow changes the probability of the event that the voter is in favour of a parliament the events 'voter lives in Glasgow' and 'yes to a parliament' are not independent.

8.4.5 Relationship between mutually exclusive events and independent events

Independence and mutually exclusive events are two different properties of some events. Deciding whether or not two events are mutually exclusive is generally easier than deciding if two events are independent. This occurs because independence is defined in terms of changes in probabilities given that one event has occurred, whereas all you have to do to decide if two events are mutually exclusive is to work out if they can possibly occur at the same time. If they cannot both occur together then they must be mutually exclusive. If they can both occur at the same time then they are not mutually exclusive.

To decide if two events are independent you have to work out if the occurrence of one event can possibly change the probability that the other one will occur. If the answer is yes then the events are not independent; if it is no then the events are independent. In most cases the answer will be 'I don't know' or 'I think so, but I'm not sure' or some other uncertain response. Independence of events is a physical property of the probability experiment and you need to know a great deal about the mechanism by which the outcomes arise in order to say if two events are definitely independent. This is normally easier for the probability experiments which fall into the equally likely outcomes type than it is for more general probability experiments.

It is easy to see that if you throw two fair dice then the second dice is physically unconnected to the first and so the outcome of the first die will be completely independent of that of the second. If you have five cards labelled A, B, C, D, E and two cards are drawn at random without replacement, the outcome on the card drawn second cannot be independent of the outcome on the card drawn first. The probability of drawing card B at the first draw is 1/5. If card C is selected at the first draw then the probability of drawing card B at the second draw is 1/4, and if card B is drawn first the probability of drawing B second is 0. Thus when you have selection without replacement then you cannot have complete independence of events.

If two events, E_1 and E_2, are mutually exclusive then knowing that one has occurred means that the other one cannot possibly occur:

$$P(E_2 \mid E_1) = 0 \quad \text{and} \quad P(E_1 \mid E_2) = 0.$$

This means that mutually exclusive events cannot be independent. Indeed, such mutually exclusive events represent one of the most extreme forms of dependence.

Conversely, independent events cannot be mutually exclusive events. If E_1 and E_2 are independent then

$$P(E_2 \mid E_1) = P(E_2).$$

This is not going to be equal to zero unless $P(E_2) = 0$, which can only occur if the event E_2 is impossible.

8.5 Addition law, multiplication law and complementary events

These laws give the rules for adding, subtracting, multiplying and dividing probabilities. They perform the same function as adding and multiplying numbers but are more complicated as the answer, at all times, needs to lie between 0 and 1. This has particular implications for adding, subtracting and dividing probabilities. If you multiply two numbers which lie between 0 and 1 the answer will always lie between 0 and 1, so multiplying probabilities is straightforward.

8.5.1 Complementary events

The law of complementary events is one instance of subtracting probabilities. The probability of an event E is derived from the set of all elementary outcomes to the experiment in which E occurs. The **complementary event** to E is denoted $\overline{E}$, and comprises all elementary outcomes in which E does not occur.

Over a period of 24 hours the price of a share can increase, decrease or remain the same. These are the three elementary outcomes of the experiment. If E is the event that the price increases, then $\overline{E}$ is the event that the price does not increase and the elementary outcomes associated with this event are {price decreases, price stays the same}.

In a probability experiment the sample space, S, is the set of all possible outcomes to the experiment. This allows us to define an event S with associated outcomes {price increase, price decrease, price stays the same}. The probability of this must be 1 as this contains all the possible outcomes to the experiment: E = {price increase} and $\overline{E}$ = {price decrease, price stays the same}.

Together, E and $\overline{E}$ make up the whole sample space, and furthermore they are mutually exclusive as they do not have any outcomes in common. As the probability of the sample space S is 1, we must have

$$P(E) + P(\overline{E}) = 1,$$

and so we have the law for complementary events

$$P(\overline{E}) = 1 - P(E).$$

Example 8.9: Winning on the lottery
If the probability of winning any prize of the lottery is 1/54, then the probability of not winning is simply

$$P(\text{not winning}) = 1 - P(\text{winning}) = 1 - \frac{1}{54} = \frac{53}{54} = 0.9815.$$

8.5.2 Addition law

The addition law deals with the calculation of the probability of combined events linked with 'or', $P(E_1$ or $E_2)$ and its extension to more than two events. In the discussion of the law for complementary events above there was the addition of the probabilities of two complementary events. Complementary events must also be mutually exclusive and so we have the addition law for two mutually exclusive events, E_1 and E_2:

$$P(E_1 \text{ or } E_2) = P(E_1) + P(E_2).$$

Considering again the illustration of the share price, let E_1 be the event 'a price decrease', and E_2 the event 'no change in price'. The two outcomes which comprise the combined event E_1 or E_2 are {a price decrease, no change in price}. The probability of 'a price decrease or no change in price' is simply the sum of the probabilities of these two mutually exclusive events.

If all the events are mutually exclusive then there is absolutely no problem about adding up the probabilities. There are, at most, three mutually exclusive events in the share price movements, namely the three elementary outcomes. The sum of the probabilities of these outcomes is 1 as the combined event is the whole sample space. For our share price, let E_3 be the event 'a price increase'. Then

$$P(E_1) + P(E_2) + P(E_3) = 1.$$

This illustrates that if you add up the probabilities of all the mutually exclusive events in an experiment however many there are, the largest value you can obtain for a probability is 1. In fact the sum of the probabilities of all the elementary outcomes in an experiment will always be exactly equal to one. This is sometimes referred to as the **law of total probability**.

The difficulty with adding up probabilities comes when dealing with events which are not mutually exclusive, which applies to most events in general. If the probabilities of the individual events are just added up then it is quite possible to obtain an answer greater than 1, which is not feasible.

Example 8.10: Scottish Young Persons Survey: addition law

We define two events based upon Table 8.1: E_1 is the event '2 or more Highers in S5' and E_2 is the event 'at most 3 Highers in S5'. The elementary outcomes and probabilities are listed below:

Event	Outcomes	Probability
E_1	{2, 3, 4, 5, 6+}	929/1979 = 0.4694
E_2	{0, 1, 2, 3}	1496/1979 = 0.7559

These two events are not mutually exclusive as they have outcomes {2, 3} in common. Also it is clear that the probability of $(E_1$ or $E_2)$ cannot be calculated simply as $P(E_1$ or $E_2) = P(E_1) + P(E_2)$ because the result is clearly greater than 1. Thus a correction needs to be undertaken to account for the overlap, or intersection, between these two events. This is E_1 and E_2, and contains the outcomes {2, 3}.

Considering the sum of the probabilities of the two events,

$$P(E_1) + P(E_2) = P(\{2, 3, 4, 5, 6+\}) + P(\{0, 1, 2, 3\}).$$

Each of the outcomes are mutually exclusive, which means that we can write

$$P(E_1) + P(E_2) = \underline{P(2) + P(3) + P(4) + P(5) + P(6)} + \underline{P(0) + P(1) + P(2) + P(3)}. \quad (*)$$

By considering the outcomes in the union of the two events E_1 or E_2 we can calculate the probability by an alternative route. The outcomes in E_1 or E_2 are $\{0, 1, 2, 3, 4, 5, 6+\}$. This is the whole sample space and so $P(E_1 \text{ or } E_2) = 1$. This can also be written as

$$P(E_1 \text{ or } E_2) = P(0) + P(1) + P(2) + P(3) + P(4) + P(5) + P(6). \qquad (**)$$

By comparison of the two sums (*) and (**) it can be seen that the intersection of the two events, E_1 and E_2, with probability $P(\{2, 3\}) = P(2) + P(3)$, is present twice in the sum (*). One of the occurrences must be removed to obtain the correct probability:

$$P(\{2, 3, 4, 5, 6+\} \text{ or } \{0, 1, 2, 3\}) = P(\{2, 3, 4, 5, 6+\}) + P(\{0, 1, 2, 3\})$$

$$- P(\{2, 3, 4, 5, 6+\} \text{ and } \{0, 1, 2, 3\}).$$

This leads to the addition law for probabilities:

$$P(E_1 \text{ or } E_2) = P(E_1) + P(E_2) - P(E_1 \text{ and } E_2).$$

The subtraction of the term for the probability of the intersection between the two events corrects the sum $P(E_1) + P(E_2)$ for the double occurrence of outcomes which are common to both events.

It can also be seen that the addition law for mutually exclusive events is a special case of this general law. If events are mutually exclusive then they cannot both occur at the same time. This means that $P(E_1 \text{ and } E_2) = 0$, and for mutually exclusive events all you need to do is to add up the probabilities of the individual events.

Example 8.11: Share price rises

A stock market analyst has studied the behaviour of the share prices of two companies. She has estimated that over a 24-hour period the probability that the share price of company A increases is 0.45, and for company B it is 0.35. She has also observed that the prices of both shares increase on the same day with a probability of 0.30. We can use this information to calculate the probability that one or other of the share prices will increase. We have:

$$P(A) = 0.45; \qquad P(B) = 0.35; \qquad P(A \text{ and } B) = 0.30.$$

Using the addition law,

$$P(A \text{ or } B) = P(A) + P(B) - P(A \text{ and } B) = 0.45 + 0.35 - 0.30 = 0.50.$$

We would interpret this by saying that there was a 50% chance that the price of either share (possibly both) will increase over a 24-hour period.

8.5.3 Multiplication law

The multiplication law makes use of conditional probabilities to give a rule for evaluating the probability of the intersection of events. The probability can be simplified if the events are independent of each other.

Example 8.12: Selecting cards

By way of introduction, let us consider the simple probability experiment of selecting two cards from five, labelled A, B, C, D, E, and calculate the probability that the cards are drawn in the sequence C followed by E. We evaluate the probability $P(C$ and $E)$. In terms of the sequence of the experiment, first of all card C is selected. As there are five cards and as the selection of each card is carried out at random, $P(C) = 1/5$. Once the first card has been selected, there are only four cards left, and at the second draw we need to calculate the conditional probability of selecting card E given that card C has already been selected. This is $P(E|C) = 1/4$. Putting the two parts of the experiment together gives

$$P(C \text{ and } E) = P(C) \times P(E|C) = \frac{1}{5} \times \frac{1}{4} = \frac{1}{20}.$$

This can be verified by considering all the possible outcomes for the experiment where the order in which the cards are selected is important. There are $5 \times 4 = 20$ possible arrangements:

AB	AC	AD	AE
BA	BC	BD	BE
CA	CB	CD	CE
DA	DB	DC	DE
EA	EB	EC	ED

There is only one arrangement with the order CE, and so the probability of this arrangement occurring is 1/20. This is exactly the same value as calculated before and thus verifies the original calculation.

The multiplication law was used in the calculation of the probability in the above example. If there are two events E_1 and E_2 then the probability that they both occur together is

$$P(E_1 \text{ and } E_2) = P(E_1) P(E_2|E_1).$$

This can also be written as

$$P(E_1 \text{ and } E_2) = P(E_2) P(E_1|E_2).$$

If the two events are independent of each other then the conditional probabilities are exactly the same as the unconditional probabilities, $P(E_1|E_2) = P(E_1)$ and $P(E_2|E_1) = P(E_2)$, and so the multiplication law simplifies to

$$P(E_1 \text{ and } E_2) = P(E_1)P(E_2).$$

In order to evaluate the probabilities for the intersection of two events the conditional and unconditional probabilities are multiplied together. The only time that the unconditional probabilities are multiplied together is when the two events are independent of each other. Extending this law to more than two events is straightforward in the case of independent events. If E_1, E_2 and E_3 are independent events then

$$P(E_1 \text{ and } E_2 \text{ and } E_3) = P(E_1)P(E_2)P(E_3).$$

If E_1, E_2 and E_3 are dependent events then it is a little more complex:

$$P(E_1 \text{ and } E_2 \text{ and } E_3) = P(E_1)P(E_2|E_1) \, P(E_3|E_1 \text{ and } E_2).$$

Example 8.13: Selecting three cards

If three cards are to be selected then the probability of obtaining, for example, the sequence CEA is

$$P(\text{CEA}) = P(\text{C})P(\text{E}|\text{C}) \, P(\text{A}|\text{CE}) = \frac{1}{5} \times \frac{1}{4} \times \frac{1}{3} = \frac{1}{60}.$$

Once C and E have been selected there are only three cards left, A, B, D, and the probability of selecting A is $P(\text{A}|\text{CE}) = 1/3$.

8.6 Bayes' theorem

This is an important theorem in elementary probability as it permits you to reverse the order of the conditioning using the multiplication law. If we have two events A and B, then the multiplication law tells us that

$$P(A \text{ and } B) = P(A)P(B|A) = P(B)P(A|B).$$

By taking $P(A)$ over to the other side of the equation, we end up with

$$P(B|A) = \frac{P(A|B)P(B)}{P(A)}.$$

This is **Bayes' theorem** in its simplest form, and we use it to calculate the probability of B given A, when we already know the probability of A given B and the two unconditional probabilities. The use of this theorem is best seen from an example.

Example 8.14: Home working

In 1997 in the United Kingdom just under 27 million men and women were in employment. Of these 55% were men and 45% women. Of those women in employment, 4.0% worked from home, compared to 1.4% of men in employment (Office for National Statistics 1998a, Tables 4.11 and 4.20). If a home worker is selected at random, what is the probability that she is a woman?

   We have unconditional probabilities about people in employment, for example $P(\text{Man})$ = 0.55, and conditional probabilities about home working conditional on gender, for example $P(\text{Home working}|\text{Man}) = 0.014$. We need to reverse the order of conditioning from that presented to calculate $P(\text{Woman}|\text{Home working})$. Let M stand for 'man', W for 'woman, and H the event that a person works at home. The calculations are best laid out in a tree diagram:

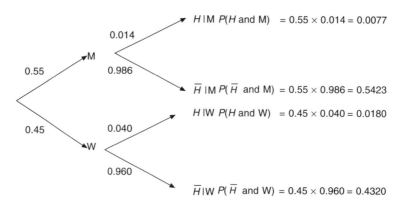

The two unconditional events are in the branch at the first level of the tree diagram, while the conditional events are at the branches at the second level. Multiplying the probabilities down the branches of the first level with those of the second level uses the multiplication law to get the probabilities of the combined events. These four probabilities at the right-hand side of the diagram all add up to 1 as they cover the whole sample space.

We want to calculate

$$P(W \mid H) = \frac{P(H \mid W)P(W)}{P(H)}$$

and from the tree diagram you can see that we know the values of the two probabilities on the numerator, $P(H \mid W) = 0.040$ and $P(W) = 0.45$. The probability on the denominator is not directly available but it can be calculated from the tree diagram using a version of the law of total probability. Among the elementary outcomes on the right-hand side of the tree diagram there are only two which are associated with the event H. These are $(H$ and $M)$ and $(H$ and $W)$. Thus the event H corresponds to the union of these two events, and we can write this out as

$$P(H) = P((H \text{ and } M) \text{ or } (H \text{ and } W))$$

$$= P(H \text{ and } M) + P(H \text{ and } W) = 0.0077 + 0.0180 = 0.0257$$

as these two elementary outcomes are mutually exclusive. Substituting in the equation for $P(W \mid H)$ gives

$$P(W \mid H) = \frac{P(H \mid W)\, P(W)}{P(H)} = \frac{0.040 \times 0.45}{0.0257} = \frac{0.0180}{0.0257} = 0.70.$$

This implies that if a person in employment is chosen at random then the probability that a woman is picked is 0.45. If, however, you know that the person is engaged in home working then the conditional probability that you will pick a woman changes to 0.70.

Bayes' theorem is used to reverse the conditioning in probability calculations. This theorem has great importance in statistical inference, though we will not cover its use in statistical inference in this book. If we write Bayes' theorem in a slightly different way as

$$P(B \mid A) = \frac{P(A \mid B)}{P(A)} \times P(B),$$

then we see that it gives a way of transforming the unconditional probability of B into a conditional probability given that the event A is known to have occurred. Within the Bayesian framework the unconditional probability is known as the **prior probability**, and this is modified by the term $P(A \mid B)/P(A)$ to give the **posterior probability** conditional on the information in event A.

8.7 Summary

The material in this chapter forms the basis of the probability calculations in the next chapter, which in turn form the basis for the application of statistical models, tests and estimation procedures in all the subsequent chapters. The principles of probability link the descriptive applications of statistics in the preceding chapters with the more formal inferential statistical procedures of the later chapters. Probability and the laws of probability are studied so that we can go on to discover how we can describe the outcomes of experiments which have a large random component and where the final outcome is not predictable exactly.

The main points of importance from this chapter are: the definition of probability and what it means; the difference between mutually exclusive events and independent events; the laws of probability and their application. The concept of independence is one which will keep being referred to throughout this book. It is a concept which is also closely linked with random samples, and we have already seen how important they are in statistical procedures.

9

Probability distributions as a model for populations

- Key ideas of random variables, expectation and variability
- Knowing the difference between a discrete and continuous random variable
- Knowing what a probability distribution is and how to calculate the expectation and variance of a discrete random variable
- Knowing some results about combining expectations and variances
- Being familiar with the binomial distribution and the assumptions on which it is based
- Knowing how to calculate probabilities using the binomial distribution
- Knowing the expected value and variance of a discrete random variable following a binomial distribution
- Being familiar with the normal distribution and type of random variables it can be used for
- Knowing how to calculate probabilities using tables of the normal distribution
- Knowing about normal ranges
- Knowing how to use a normal probability plot to check that a set of data are consistent with the normal distribution.

9.1 Key ideas on random variables, expectation and variability

9.1.1 Random variables

In the study of probability and events we look at the outcomes of an experiment and evaluate the probability of these outcomes using the laws of probability. In many statistical investigations there is a variable under consideration, and we now turn to investigating probability models for variables which are measured in an experiment or survey. **Random variables** are derived from the elementary outcomes of a probability experiment but generally focus on restricted aspects of the experiment.

In the context of this book, random variables will be numerical quantities whose value in any one experiment depends upon chance. These numerical quantities can either be **discrete**, in that they can only take distinct vales, or **continuous**, where they can take any value within a given range. This terminology is related to the distinction between different types of data variables discussed previously (see Chapter 2). The basic ideas will be developed for discrete random variables, taking only a fixed number of possible values. Then they will be extended to cover discrete random variables with an unknown number of possible values (see book website). Finally, we look at continuous random variables.

In a study of the reliability of mobile telephone components we can use long-run relative frequency methods to estimate the probability that the keypad is defective. If a sample of 8 keypads are tested out of every batch, we can use the laws of probability to work out the probability of a specified number of defective keypads among the sample. The variable 'the number of defective keypads' is known as a **random variable**. This is a counted variable and its value can change according to probability rules from one sample of 8 keypads to another. When 8 keypads are selected, the full sample space for the probability experiment has $2^8 = 256$ outcomes, taking into account the state of each specific keypad and ranging from all eight defective to none of the eight defective. The random variable 'the number of defective keypads' can take only one of 9 values {0, 1, 2, 3, 4, 5, 6, 7, 8}. This is clearly a simplification, and we can work out the probabilities of each of these values using the laws of probability. This is an example of a discrete random variable taking only a small number of distinct values.

The set of possible values of a random variable and the associated probabilities of each of these values form the **probability distribution** of the random variable. This is the probability model equivalent of the frequency distribution of a sample of data (Chapter 3). The frequency distribution gives the observed values and the relative frequency of each of these observed values. There can be two main differences between the frequency distribution and the probability distribution. Firstly, there may be fewer values in the frequency distribution than in the probability distribution. This will occur when a value does not occur in the repetitions of the experiment. Secondly, the relative frequencies are very unlikely to be the same as the probabilities. In terms of the long-run relative frequency definition of probability (Chapter 8) the frequency distribution will become more and more similar to the probability distribution as the number of repetitions of the experiment increases.

A probability model is a mathematical description of the set of possible values of the random variable, and the frequency distribution is the observed values of the random variable. One of the goals of developing probability models is to find a model which is a good description of the random variable and then to use the model to make predictions about future values of the random variable and hence make rational decisions and judgements.

The technical discussion of random variables is motivated by the use of an example which arose in a marketing and advertising company. The latter was involved in a promotional campaign by a manufacturer and distributor of sweets to try to increase orders for a new variety. The company sold sweets to small retailers by sending salesmen to call on the retailers. To promote a new variety, each retailer was to be given a prize if he or she placed an order to the value of £10 for the new variety. If the retailer ordered £30 then he or she got three prizes.

The prize was a scratch card, illustrated in Figure 9.1. The retailer was given a card and instructed to scratch out only one of the panels, and would get a cash prize to the value uncovered. As you can see from the card, there is no possibility of failing to win some money and it is only a question of how much.

This problem came to our attention as part of statistical consulting work at the University

£1	£5	£1	£5
£2	£1	£1	£2
£10	£1	£2	£1

Figure 9.1 Scratch card.

of Strathclyde. The marketing company contacted staff in the Department of Statistics and Modelling Science as one of their marketing executives had gone to the university and had previously attended a course in statistics for business students. A total of 10 000 scratch cards were going to be prepared and distributed to salesmen and the company wanted to know how much this publicity campaign was going to cost.

It was fairly easy for the marketing people to work out that it was going to cost a minimum of £10 000 and a maximum of £100 000, but they knew that both of these were extremely unlikely. Furthermore, they knew that the sweet manufacturer would not undertake a campaign that was going to cost more than £30 000.

The full solution to this problem requires (1) the study of the probability distribution of the random variable 'the amount won on each card', (2) working out the average or expected value of the amount won per card and (3) working out the variation in the amount won per card. Then it is necessary to use this information to work out what will happen with the 10 000 cards. The first three requirements are now considered.

9.1.2 Probability distribution of the amount won per card

The random variable is the amount won per card. It is a discrete random variable with four possible values, {£1, £2, £5, £10}. In general terms we denote the random variable by the capital letter X, and specific values of the random variable by the small letter x. The probability that the random variable, X, takes the value x is written as $P(X = x)$, or $P(x)$.

It is possible to work out the values for $P(x)$, for x equal to £1, £2, £5, or £10 on the assumption that the retailer chooses completely at random which panel on the card to scratch out. In practice this means that the retailer has no idea of the value hidden under each panel; for example, it is not possible to see the value written by holding the card up to a light. Also the salesman must have no idea of where the different prize values are located throughout the card, otherwise he can tell the retailer and they can split the maximum prize money. If the retailer has no idea of the location of the prizes of different value then all that he or she can do is to select a panel to scratch out at random.

As there are 12 panels the probability of choosing any one of them is 1/12. Six of the squares have a prize of only £1, so the probability of getting £1 is $P(1) = 6/12 = 1/2$. By a similar counting mechanism we can see that $P(2) = 3/12$, $P(5) = 2/12$ and $P(10) = 1/12$. In this case, and many others, it is convenient to arrange the probability distribution in a table:

x, (£)	1	2	5	10
$P(x)$	$\frac{6}{12}$	$\frac{3}{12}$	$\frac{2}{12}$	$\frac{1}{12}$

A probability distribution of a discrete random variable has two important characteristics. The first is that for all possible values of the random variable, x, the probability is greater than or equal to zero. Secondly, the sum of the probabilities of all distinct possible values must be equal to 1. This is the same as saying that the sum of the probabilities of all possible elementary outcomes in the sample space must be 1. These requirements can be written mathematically as

$$P(x) \geq 0, \text{ for all } x, \text{ and } \sum_{x} P(x) = 1.$$

9.1.3 Expectation

The **expectation** of a random variable gives the 'average value' or 'expected value'. Its interpretation is similar to the interpretation of the mean of a sample of observations. It tells you something about the location of the distribution. The expected value of the random variable X is denoted E[X] or by the Greek letter μ, which is the Greek 'm' pronounced 'mu'. The expected value is defined, in an analogous fashion to the mean of a sample, as

$$\mu = \text{E}[X] = \sum_{x} xP(x),$$

and is calculated by multiplying each possible value by its associated probability and adding all of these products up. The calculations are best laid out in a table such as that below:

x, (£)	1	2	5	10	Total
$P(x)$	$\frac{6}{12}$	$\frac{3}{12}$	$\frac{2}{12}$	$\frac{1}{12}$	$\frac{12}{12} = 1$
$xP(x)$	$\frac{6}{12}$	$\frac{6}{12}$	$\frac{10}{12}$	$\frac{10}{12}$	$\frac{32}{12}$

The expected value of the prize on one card is than

$$\mu = \text{E}[X] = \sum_{x} xP(x) = \frac{32}{12} = £2.67.$$

This means that the expected amount of money won per card is £2.67. Note that this is not an amount that can be won on any one card. Thus although the expected value of a random variable is a measure of the average value, it need not correspond to one of the possible values.

9.1.4 Variance

The **variance** measures the spread of the probability distribution by evaluating the expected value of the square of the distance of each possible value from the expected value. It is analogous to the sample variance. It is usually denoted σ^2, or $E[(X - \mu)^2]$, or $Var[X]$. For a discrete random variable it is defined as

$$\sigma^2 = E[(X - \mu)^2] = \sum_x (x - \mu)^2 P(x).$$

This is not the easiest method of calculating it, by hand with a calculator, and a more reliable method is

$$\sigma^2 = \sum_x x^2 P(x) - \mu^2.$$

Again the calculations are best set out in a table:

x, (£)	1	2	5	10	Total
$P(x)$	$\frac{6}{12}$	$\frac{3}{12}$	$\frac{2}{12}$	$\frac{1}{12}$	$\frac{12}{12} = 1$
$xP(x)$	$\frac{6}{12}$	$\frac{6}{12}$	$\frac{10}{12}$	$\frac{10}{12}$	$\frac{32}{12}$
$x^2P(x)$	$\frac{6}{12}$	$\frac{12}{12}$	$\frac{50}{12}$	$\frac{100}{12}$	$\frac{168}{12}$

Thus,

$$\sigma^2 = \sum_x x^2 P(x) - \mu^2 = \frac{168}{12} - \left(\frac{32}{12}\right)^2 = 6.8889 \ (\pounds)^2.$$

The units of the variance of a random variable are in squared units, which is not the same as the units for the variable or for its expected value. Consequently, it is more usual to use the **standard deviation** of the probability distribution, σ, the square root of the variance. This has the same units as the variable and its expected value.

The standard deviation of the value won with each card is

$$\sigma = \sqrt{6.8888} = \pounds2.62.$$

This gives a numerical measure of the variability of the prizes for each card.

In order to evaluate the impact of the campaign it is necessary to know how to manipulate expectations and variance and also to make use of the normal distribution, an important continuous probability distribution. The first two will not be covered in detail (see Section 9.1.5) and the latter will not be discussed until a little later (see Section 9.4). However, details are now provided to complete the example. We know that 10 000 cards are to be distributed. The expected value of the prize on each card is £32/12. Thus the expected value of the prize for all 10 000 cards will be $10\,000 \times \pounds32/12 = \pounds26\,667$. As the cards are scratched independently of each other the variance for the total amount of prize money is $10\,000 \times 6.8889 = 68\,889 \ (\pounds)^2$, which corresponds to a standard deviation of £262. This allows us to say that we expect the total costs of the campaign to lie between £26 152 and £27 204 with a probability of 0.95, and between £25 991 and £27 343 with a probability of 0.99. Thus, while we cannot completely discount the possibility that the total costs will exceed £30 000 they are very unlikely to do so and the company should expect the total cost to lie between £26 000 and £27 500.

9.1.5 Combining expectations and variances

If X and Y are two independent random variables with means, μ_X and μ_Y, and standard deviations σ_X and σ_Y, and a and b are two constants, then

$$E[aX + bY] = a\mu_X + b\mu_Y,$$

$$Var[aX + bY] = a^2\sigma_X^2 + b^2\sigma_Y^2.$$

This means that the expectation and variance of the sum of two independent random variables are

$$E[X + Y] = \mu_X + \mu_Y,$$

$$Var[X + Y] = \sigma_X^2 + \sigma_Y^2.$$

The difference of two independent random variables has the same variance as the sum of the two variables but the expected value is the difference of the expected values:

$$E[X - Y] = \mu_X - \mu_Y,$$

$$Var[X - Y] = \sigma_X^2 - \sigma_Y^2.$$

These results have lots of implications. If $b = 0$ then we have

$$E[aX] = a\mu_X,$$

$$Var[aX] = a^2\sigma_X^2.$$

This corresponds to changing the scale of the variable X, say from pounds sterling to euros, where a represents the rate of exchange.

If Y is not a random variable at all and is just a constant, then $\sigma_Y = 0$ and we specify that $\mu_Y = c$. If $b = 1$ and $a = 0$ then we have

$$E[X + c] = \mu_X + c,$$

$$Var[X + c] = \sigma_X^2.$$

This corresponds to changing the origin of the random variable by adding or subtracting a constant.

The results can be extended to more than two independent random variables, and in general we have

$$E[X_1 + X_2 + \ldots + X_n] = \mu_1 + \mu_2 + \ldots + \mu_n,$$

$$Var[X_1 + X_2 + \ldots + X_n] = \sigma_1^2 + \sigma_2^2 + \ldots + \sigma_n^2.$$

If the random variables are not independent of each other the calculation of the variance of the sum and difference is more complex. The expected value is the same as for independent random variables but the variance has an additional term to take into account the non-independence or association between X and Y. In Chapter 6 the association between two variables was measured by the covariance or correlation between them. These terms are used in the calculation of the variance of a sum or difference of two dependent random variables:

$$E[X + Y] = \mu_X + \mu_Y,$$

$$\mathrm{Var}[X + Y] = \mathrm{Var}[X] + \mathrm{Var}[Y] + 2\mathrm{Cov}[X, Y] = \sigma_X^2 + \sigma_Y^2 + 2\rho\sigma_X\sigma_Y,$$

$$E[X - Y] = \mu_X - \mu_Y,$$

$$\mathrm{Var}[X - Y] = \mathrm{Var}[X] + \mathrm{Var}[Y] - 2\mathrm{Cov}[X, Y] = \sigma_X^2 + \sigma_Y^2 - 2\rho\sigma_X\sigma_Y.$$

The covariance is defined in terms of the joint probability distribution of X and Y together, and is give by

$$\mathrm{Cov}[X, Y] = \sum(x - \mu_X)(y - \mu_Y)P(x, y);$$

this is related to the correlation between X and Y through

$$\rho = \frac{\mathrm{Cov}[X, Y]}{\sigma_X\sigma_Y}.$$

We do not use these results here but will make use of them in Chapters 10–12.

9.1.6 Cumulative distribution function

The probability distribution of the random variable X is the set of possible values of the random variable and the associated probabilities, $P(x)$. This is the population, or model, equivalent of the frequency distribution in the sample. The cumulative distribution function for the random variable is the probability that the random variable takes a value less than or equal to a specified value, x, i.e. $P(X \le x)$. This is used for finding the quantiles, or percentiles, of the distribution, and for finding probabilities of certain events. Quantiles and percentiles are often most easily obtained from a graph of the cumulative probability function against the possible values of the distribution.

The cumulative distribution function is presented in the table below for the scratch-card example, and is plotted in Figure 9.2. The values of the random variable are plotted on the x axis and the cumulative distribution function on the y axis. The random variable is discrete and so the plot is a step function. The changes in the value of $P(X \le x)$ only take place at the discrete values of x. For as long as the winnings are less than £1 the probability is zero; the instant the winnings become £1 the probability changes to 0.5, and it remains at this level until the next change takes place at £2, where there is a jump to 0.75.

x, (£)	1	2	5	10
$P(x)$	$\frac{6}{12}$	$\frac{3}{12}$	$\frac{2}{12}$	$\frac{1}{12}$
$P(X \le x)$	$\frac{6}{12}$	$\frac{9}{12}$	$\frac{11}{12}$	$\frac{12}{12}$

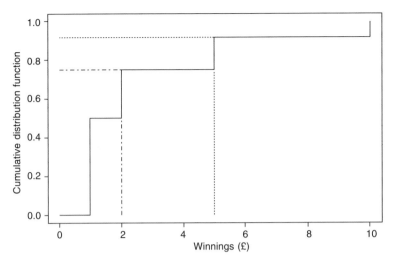

Figure 9.2 Cumulative distribution function for the amount won on each scratch card.

There are two main uses for graphs such as Figure 9.2. The first is to read off the probability associated with specified values of the random variable. This is illustrated by the dot-dashed line. You can find the probability that the winnings are less than or equal to 2 by following the dot-dashed line vertically up from 2, until you reach the line representing the cumulative distribution function; then you go horizontally until you reach the y axis. Reading off the value gives $P(X \leq 2) = 0.75$. As the cumulative distribution function makes its jump exactly at 2, we have $P(X < 2) = 0.5$.

The second use is in reading off percentiles based upon specified probabilities. This is illustrated by the dotted line. From the y axis, corresponding to a probability of $^{11}/_{12} = 0.92$, follow the dotted line horizontally until it reaches the cumulative distribution line, then go down vertically until reaching the x axis at 5. Thus, $P(X \leq 5) = 0.92$, i.e. there is an 92% chance that the card will have winnings of £5 or less.

The median and quartiles can be obtained using this method, but in this example we can see that they are not all unique. Following across from 0.5 we see that any point from £1 to £2 (including £1 but not including £2), corresponds to 0.5. Thus there is no unique median and we average the two end points to give a median of £1.50. There is a similar situation with the upper quartile. Following across from 0.75, we see that any value of the winnings from £2 up to, but not including, £5 corresponds to 0.75; thus we say the upper quartile is £3.50. There is a unique lower quartile as the line from 0.25 cuts the cumulative distribution function in only one place, at £1.

9.2 Binomial distribution

9.2.1 Derivation from binary trials

In the previous section the probability distribution, cumulative distribution, expectation, standard deviation and quartiles of a general discrete random variable were discussed and illustrated with a simple example. There are many types of discrete random variables, and some of them are used in quite a wide variety of situations. In this section we move on to look at the most important of these specific distributions, namely the **binomial** distribution.

The simplest type of discrete random variable is a binary trial in which an event either occurs or does not. This is very common as many events can be thought of in this way. For example, the sex of a child is either female or it is not (i.e. it is male); when a die is thrown either a 6 appears or it does not; the price of a share on the stock market increases or it does not increase; an electrical component either functions properly or it does not; a customer either buys a product or does not. The characteristic of these events is that there are only two outcomes. The binomial distribution arises when we consider a sequence of these binary events such as the number of females in a group of 20 children, the number of shares with a price increase out of the 100 shares in the FT index, or the number of customers purchasing a product out of 30 making enquiries.

As a first step in this section we will derive the probability function for the binomial distribution in a simple situation where it can be enumerated. This will enable us to focus on the assumptions on which the distribution is based. Then we will move on to looking at the distribution in general.

Example 9.1: The distribution of the number of girls in families of size three

If we consider all the possible arrangements of the sex of children in families of three children we have the following table, where the order of the births is taken into account:

	BBB	BBG	BGB	BGG	GBB	GBG	GGB	GGG
Number of girls	0	1	1	2	1	2	2	3

For example, BBG corresponds to the eldest two children being boys and the youngest a girl.

Although there are eight different combinations, taking into account the birth order, there are only four distinct values for the number of girls, $\{0, 1, 2, 3\}$. In order to evaluate the probability distribution for this random variable all we have to do is to work out the probabilities of each of the separate arrangements and add them up. To do this we have to know the probability that any particular child will be a boy or a girl and we need to know how to work out the probability of the combined event 'B and B and B'. The former can be done by considering the two events to be equally likely and the latter by arguing that the sex of the second child does not depend upon the sex of the first one, that is, that they are independent.

There are two outcomes, a boy or a girl, and if we assume that they are equally likely then we have $P(B) = P(G) = \frac{1}{2}$. Using the multiplication law for independent events we can see that $P(BBB) = \frac{1}{2} \times \frac{1}{2} \times \frac{1}{2} = \frac{1}{8}$. This calculation can be repeated for all the other seven arrangements, and each arrangement has the same probability of $\frac{1}{8}$. Inserting these probabilities in the table gives

	BBB	BBG	BGB	BGG	GBB	GBG	GGB	GGG
Number of girls	0	1	1	2	1	2	2	3
Probability	$\frac{1}{8}$	$\frac{1}{8}$	$\frac{1}{8}$	$\frac{1}{8}$	$\frac{1}{8}$	$\frac{1}{8}$	$\frac{1}{8}$	$\frac{1}{8}$

The probability distribution for the number of girls is obtained by combining the above probabilities

Number of girls	0	1	2	3
Probability	$\frac{1}{8}$	$\frac{3}{8}$	$\frac{3}{8}$	$\frac{1}{8}$

9.2.2 Assumptions of the binomial distribution

If the following assumptions are satisfied the binomial distribution may be used:

1. Each trial has only two possible outcomes which, in general, we call a positive response or a negative response.
2. The trials are independent of each other.
3. Each trial has the same probability of a positive response.
4. There are a fixed number of trials.
5. The random variable is the number of positive responses.

Let us check the distribution of the number of girls:

1. Each child is either a boy or a girl.
2. We assume that the fact that the first child is a boy does not change the probability that the second child is a boy.
3. Using an equally likely outcomes argument we concluded that $P(B) = P(G) = \frac{1}{2}$.
4. Each family had three children only.
5. The random variable is the number of girls and a girl corresponds to a positive response.

In any specific example criteria 1, 4, and 5 are usually fairly easy to check. The independence criterion and the constant probability assumption are less easy to check. Often independence is an assumption which is made but there is not sufficient information to completely validate it. In the example we can be reasonably sure that our assumptions are fair, however we need to be careful that families with twins and triplets are excluded. Also, there is some evidence that the probability that a child is a boy is slightly greater than 0.5. Figures published by the General Registrar for Scotland stated that in 1996 there were 1061 male live births to every 1000 female live births. Over the years this ratio tends to fluctuate around 1050, giving the probability of a boy of about 0.51, rather than 0.5.

Returning to the list of examples at the beginning of the section, it is reasonable to assume that a binomial distribution is valid for the number of sixes in 10 throws of a die. This is so because each throw has only two outcomes, 'a 6' and 'not a 6', there are 10 throws, each throw is completely independent of the others, the probability of a 6 is known, 1/6, and is the same over all throws. Thus the five criteria can be checked. The binomial distribution is not likely to be valid for the number of shares whose price increases, although criteria 1, 2 and 5 are satisfied. The shares are not likely to behave independently – if one rises the others may be more likely to rise in price also – and they are unlikely to have the same probability of increasing in price. It is reasonable to believe that the number of customers buying an article out of 30 making enquiries would follow a binomial distribution provided we could be sure that the enquiries are unrelated to each other and that one person buying the article does not change the probability that the others will. There need to be enough articles for all 30 customers to be able to buy one; clearly the binomial distribution will not be valid if there is only one article for sale.

9.2.3 Evaluating binomial probabilities

In order to evaluate the probabilities associated with the binomial distribution in the above example we wrote down all the possible events, worked out their probabilities and then added them up to get the distribution. In principle this can always be done, but the process can very quickly become quite unwieldy. When there were three trials, there were $2^3 = 8$ outcomes, with four trials there will be $2^4 = 16$ outcomes, and with 10 trials there will be $2^{10} = 1024$ outcomes. Fortunately there are some mathematical formulae which can be used for counting up the numbers of combinations, and these mean that we can use a formula to work out the probabilities in the binomial distribution.

If there are n trials, each with a probability, p, of a positive response occurring then the formula for the binomial probability distribution is

$$P(x) = \binom{n}{x} p^x (1 - p)^{(n-x)}.$$

To calculate the probability for any particular number of positive responses all we need to have is the total number of trials or events, n, and the probability of a positive response on any one of them, p. The value of p will either come from sample data where we use the observed proportion to estimate p (see Example 9.3), or from detailed knowedge of the probability experiment to work out the value of p using, for example, equally likely outcomes, (see Example 9.4). The binomial coefficient $\binom{n}{x}$ gives the number of ways of choosing x objects from n, and is evaluated as follows:

$$\binom{n}{x} = \frac{n!}{x!(n-x)!},$$

where $x! = x \times (x - 1) \times (x - 2) \times \ldots \times 2 \times 1$, with 0! defined to be equal to 1.

Example 9.2: Binomial calculations

$$5! = 5 \times 4 \times 3 \times 2 \times 1 = 120, \ 3! = 3 \times 2 \times 1 = 6, \ 1! = 1, \ 0! = 1,$$

$$\binom{5}{3} = \frac{5!}{3!2!} = \frac{5 \times 4 \times 3 \times 2 \times 1}{(3 \times 2 \times 1) \times (2 \times 1)} = \frac{5 \times 4}{2 \times 1} = \frac{20}{2} = 10,$$

$$\binom{10}{2} = \frac{10!}{2!8!} = \frac{10 \times 9 \times 8 \times 7 \times 6 \times 5 \times 4 \times 3 \times 2 \times 1}{(2 \times 1) \times (8 \times 7 \times 6 \times 5 \times 4 \times 3 \times 2 \times 1)} = \frac{10 \times 9}{(2 \times 1)} = \frac{90}{2} = 45,$$

$$\binom{6}{1} = \frac{6!}{1!5!} = \frac{6 \times 5 \times 4 \times 3 \times 2 \times 1}{(1) \times (5 \times 4 \times 3 \times 2 \times 1)} = \frac{6}{(1)} = 6,$$

$$\binom{5}{0} = \frac{5!}{0!5!} = \frac{1}{0!} = \frac{1}{1} = 1, \quad \binom{5}{5} = \frac{5!}{0!5!} = \frac{1}{0!} = \frac{1}{1} = 1.$$

These binomial coefficients give the number of ways of selecting x objects from n. There are 6 ways of selecting 1 object from 6, 45 ways of selecting 2 from 10, and only one way of selecting 5 objects from 5.

Example 9.3: Truancy in Scottish schools

In 1994, an article was published in the *Scotland on Sunday* newspaper on the question of truancy. The following information was presented on the percentage of fourth-year pupils who never played truant (*source:* Scottish Young Persons Survey: School Levens Series)

	1983	1985	1987	1989	1991	1993
Never truant	45%	44%	42%	47%	39%	40%
Sample size	1370	6206	5591	4567	3513	NA

In a random sample of 10 fourth-year pupils in Scotland in 1993 what are the following probabilities.

(a) The probability that exactly 3 truanted?
(b) The probability than more than 8 truanted?
(c) The probability that at least 7 never truanted.

This is the type of problem for which the binomial distribution is likely to be applicable, and we will use it here. There are 10 trials (pupils), each one with two outcomes – truant or never truant – and the random variable is the number of pupils who truant. From the table we see that in 1993 40% of pupils never truanted and so 60% did at some time. Thus, we estimate that the probability that a pupil chosen at random will truant as 0.6. That just leaves us with the question of independence, and this is likely to be valid if the pupils are selected at random. Obviously, it is less likely to be valid if all the pupils were in the same class as peer pressure is likely to play a part.

The random variable, X, is the number of pupils who truant, $n = 10$ and $p = 0.6$.

(a) The probability is calculated using the binomial distribution:

$$P(X = 3) = \binom{10}{3}(0.6)^3(0.4)^7 = 120 \times 0.216 \times 0.001\,638\,4 = 0.042\,467.$$

If 10 pupils are selected at random then the probability that exactly three of them truanted is 0.042.

(b) In this part it is necessary to calculate

$$P(X > 8) = P(X \geq 9) = P(9) + P(10)$$

$$= \binom{10}{9}(0.6)^9 (0.4)^1 + \binom{10}{10}(0.6)^{10} (0.4)^0$$

$$= 0.040\,310 + 0.006\,047 = 0.0464.$$

The key features of this calculation are recognizing that we are looking for more than 8, not more than or equal to 8, and that for a discrete random variable $P(X > 8) = P(X \geq 9)$. Secondly, $P(X \geq 9)$ is calculated by adding up the probabilities of the individual outcomes in the event $(X \geq 9)$, essentially using the addition law for mutually exclusive events. Thirdly, note that any real number to the power 0 is 1.

(c) Here we are asked about the number of pupils who never truant. There are two ways to go about this. One is to change the definition of the random variable to the number of pupils who never truant and the other is to phrase the question in terms of truanting. We will do the latter.

If 7 or more pupils never truant, this means that 3 or less truant. Thus

$P(7$ or more pupils never truant$)$

$$= P(3 \text{ or fewer pupils truant})$$

$$= P(X \leq 3) = P(0) + P(1) + P(2) + P(3)$$

$$= \binom{10}{3}0.6^3 0.4^7 + \binom{10}{2}0.6^2 0.4^8 + \binom{10}{1}0.6^1 0.4^9 + \binom{10}{0}0.6^0 0.4^{10}$$

$$= 0.042\,467 + 0.010\,617 + 0.001\,573 + 0.000\,105$$

$$= 0.0443.$$

The calculation of the probabilities is fine as an exercise, but it is the interpretation of them which is important in any given situation. If we believe that the model is correct, and from a consideration of the applicability of the criteria this would appear to be reasonable, then we can interpret the probabilities as measuring the likelihood of the events occurring. Thus if we have a group of 10 fourth-year pupils in 1993 then there is only 4.4% chance that seven or more of then never truanted. This event is slightly more common than the chance of getting five heads in a row when you toss a coin five times $(1/32 = 0.03125)$ but not as common as getting four heads in a row when you toss a coin four times $(1/16 = 0.0625)$.

Having calculated a probability based upon a model, we have a means of assessing how likely a given event is. Sometimes you will find that an event has occurred which is extremely rare, and this is possible. The model provides information on how frequently you can expect this to occur. A key issue, which we will come across in more detail in the section on statistical inference, is that if the model is true then we can use it to make predictions. If events which, according to the model, are rare occur too frequently in practice then this will cast doubt on the validity of the model. This is a good reason for checking the criteria on which the model is based, and for the binomial distribution this can be done in many cases.

9.2.4 Expected value and standard deviation

With the binomial distribution it is possible to derive simple equations for the expected value and standard deviation of the random variable. We will not do so here but will go through the calculations based upon a small example and show that the general result is valid.

Example 9.4: Expected number of wins on the National Lottery

From figures published by the UK National Lottery the probability of winning any prize in the Saturday draw is approximately 1/54. The event includes the £10 prize for guessing three numbers correctly as well as the jackpot for getting all six correct and all the other prizes. If a person plays the lottery on four consecutive occasions how many times does he expect to win?

We know how to do these calculations based upon the equations for the expected value of a discrete random variable, i.e. using $\mu = E[X] = \sum_x x P(x)$ for the mean and $\sigma^2 = \sum_x x^2 P(x) - \mu^2$ for the variance. The calculations are set out as follows:

x	$P(x)$	$P(x)$	$xP(x)$	$x^2P(x)$
0	$\binom{4}{0}\left(\frac{1}{54}\right)^0\left(\frac{53}{54}\right)^4$	0.927 958 3	0	0
1	$\binom{4}{1}\left(\frac{1}{54}\right)^1\left(\frac{53}{54}\right)^3$	0.070 034 59	0.070 034 59	0.070 034 59
2	$\binom{4}{2}\left(\frac{1}{54}\right)^2\left(\frac{53}{54}\right)^2$	0.001 982 111	0.003 964 222	0.007 928 444
3	$\binom{4}{3}\left(\frac{1}{54}\right)^3\left(\frac{53}{54}\right)^1$	0.000 024 932 2	0.000 074 796 64	0.000 224 389 9
4	$\binom{4}{4}\left(\frac{1}{54}\right)^4\left(\frac{53}{54}\right)^0$	0.000 000 117 6	0.000 000 470 41	0.000 001 881 67
Total	1	1.000 000 0	0.074 074 07	0.078 189 30

So

$$\mu = E[X] = \sum_x x P(x) = 0.074\,074,$$

$$\sigma^2 = \sum_x x^2 P(x) - \mu^2 = 0.078\,189\,30 - 0.074\,074\,07^2 = 0.072\,702\,3.$$

Thus the standard deviation is $\sigma = 0.269\,634$.

This example has two points. The calculations are tedious, particularly because of the small number involved. The small probabilities mean that you need to retain a large number of significant digits to make sure that the answer is correct.

The equations for the expected value and variance of a random variable following a binomial distribution are:

$$\mu = np,$$

$$\sigma^2 = np(1 - p).$$

With these two equations you can see that the expected value and variance of the binomial distribution are related to each other. This means that we cannot discuss the variability of the binomial distribution completely separately from the expected value. The equation for the variance is a quadratic function of p. If, $p = 0$, or $p = 1$, then the variance is exactly equal to zero. This is quite natural as if $p = 1$ ($p = 0$) then every trial will result in a positive (negative) response and so there will be no random variation. The variance will increase as p moves away from 0 or 1 towards 0.5, where the variance will take its maximum value.

In Example 9.4, $n = 4$ and $p = 1/54$. This means that the expected value and variance using the simple formulae are

$$\mu = 4 \times \frac{1}{54} = \frac{4}{54} = 0.074\,074,$$

$$\sigma^2 = 4 \times \frac{1}{54} \times \left(\frac{53}{54}\right) = 0.072\,702\,3.$$

These values are exactly the same as those obtained by going through the calculation in the long way. If we had used slightly fewer significant figures in Example 9.4 then we would not have obtained an exact agreement because of rounding errors in the calculations.

Continuing with Example 9.4, if you now played for 54 consecutive weeks then you would expect to have one win, $\mu = 54 \times \frac{1}{54} = 1$, with a variance of $\sigma^2 = 54 \times \frac{1}{54} \times \frac{53}{54} = \frac{53}{54}$ and so a standard deviation of $\sigma = \sqrt{\frac{53}{54}} = 0.9907$ wins. One way of looking at this is to say that after playing the lottery for 54 weeks you would expect to have one win. It does not guarantee one win. If you did not win any prize then you have performed worse than expected and if you won twice then you have done better than expected.

Incidentally

$$P(0) = \left(\frac{53}{54}\right)^{54} = 0.3644, \qquad P(1) = \binom{54}{1}\left(\frac{1}{54}\right)\left(\frac{53}{54}\right)^{53} = 0.3713,$$

$$P(2) = \binom{54}{2}\left(\frac{1}{54}\right)^2\left(\frac{53}{54}\right)^{52} = 0.1857$$

and the probability of getting 0, 1 or 2 wins is 0.92. The most likely single value is 1 win as the probability is the greatest. This is known as the **modal value**. The second most likely value is no wins and this is predicted to happen with a chance of 36%, which is much the same as throwing a dice and getting a 5 or a 6 (2/6 = 0.33).

9.3 Continuous models

The simplest probability functions to use are those associated with the discrete distributions such as the binomial. For these distributions the random variables take discrete values and the number of occurrences can be counted. Continuous distributions are required for random variables which can take any value within a given range. They are known as **continuous random variables**, and examples include the usual measurements such as height, weight, distance, time. Also, money variables, such as revenues, profits, sales and various financial ratios all tend to be considered as continuous random variables. The most commonly used continuous probability model is the normal distribution (Section 9.4). Another common distribution is the exponential distribution (see book website).

When dealing with discrete distributions we had the probability function, $P(x)$, which gave the probability that the random variable X, took the value x. There was also the cumulative distribution function $P(X \leq x)$, which gave the probability that X took values less than or equal to x. There are corresponding quantities for continuous distributions, but they have slightly different names and meanings as a result of the differences between discrete and continuous random variables.

The main difference between a continuous random variable and a discrete one is that a continuous random variable can take any value in a specified range while a discrete one can only take fixed values. This means that the probability function, $P(X = x)$, has a meaning for a discrete random variable, namely the probability that X takes the value x, which can be observed. If X is continuous then a specific value can never be observed exactly because there will always be some measurement error. When we say that a man is 180 centimetres tall what we really mean is that his height is between 179.5 and 180.5 centimetres. Thus the probability that he is 180 centimetres would be calculated as $P(179.5 < X < 180.5)$, to within 0.5 centimetres. So the probabilities for individual observed events are calculated using the cumulative distribution as

$$P(179.5 < X < 180.5) = P(X < 180.5) - P(X < 179.5).$$

Note that for a continuous random variable there is no distinction between, for example, $P(179.5 < X)$ and $P(179.5 \leq X)$. 'Less than' and 'less than or equal to' are interchangeable for continuous random variables, as are 'greater than' and 'greater than or equal to'. This is definitely not the case for discrete random variables where there is a great distinction between 'less than' and 'less than or equal to' which occurs because specific values can occur. Suppose we have a discrete random variable taking the values 0, 1, 2, 3 only, each with a probability of 1/4. $P(X \leq 1) = P(0) + P(1) = 1/2$, which is not equal to $P(X < 1) = P(0) = 1/4$.

With a continuous random variable the quantity corresponding to the probability function for a discrete random variable is the **probability density function**, denoted $f(x)$. In a discrete distribution the cumulative probability function is obtained from the probability function by addition. In a continuous distribution the **cumulative density function** is obtained from the probability density function by mathematical integration. We will not go into details of this here, and it is not needed for the normal distribution as there are tables of probabilities which can be looked up.

9.4 Normal distribution

9.4.1 Parameters and probability density function

The normal distribution is the most commonly used distribution in statistics. It has widespread use as a model for many quantities which are measured as the distributions of these quantities tend to be symmetric about the mean and tail off towards very high and very low values. A few examples are heights and weights, daily changes in share prices, rates of returns on investments and amounts spent by customers at supermarkets.

Unlike the binomial (see Section 9.2), Poisson and exponential distributions (see the book's website), where it is possible to assess if the distribution is likely to be valid on theoretical grounds by consideration of the validity of a number of assumptions, this is not possible with the normal distribution. The validity of the use of the distribution as a model in any specific circumstance has to be established by comparison of the model with observed data. This is most commonly carried out by looking at a histogram and seeing if it has the same shape as a normal distribution, i.e. symmetric and bell-shaped as in Figure 9.3. Other plots can be used as well, and one that we will use later (Section 9.4.5) is the normal probability plot.

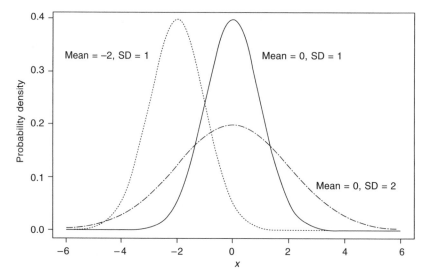

Figure 9.3 Examples of normal distributions.

There are two parameters in the normal distribution, the mean μ and the variance, σ^2 (or the standard deviation σ). The probability density function involves the exponential function and the constant, π, and is complicated:

$$f(x) = \frac{1}{\sigma\sqrt{2\pi}}\, e^{-\frac{1}{2}\left(\frac{x-\mu}{\sigma}\right)^2}.$$

Furthermore the cumulative distribution function cannot be evaluated explicitly, though it can be calculated and is tabulated for a special normal distribution, known as the *standard normal distribution* where $\mu = 0$ and $\sigma = 1$. Usually we use Z to denote a random variable with a standard normal distribution and X for any other normal distribution. Examples of the probability density functions for a number of normal distributions with different means are shown in Figure 9.3. These illustrate that the mean, μ, controls the location of the distribution along the x axis, while the standard deviation, σ, controls the shape of the distribution. The larger the standard deviation, the greater the variability and the more spread out the distribution. The smaller the standard deviation, the more compact the distribution.

9.4.2 Using tables of the standard normal distribution

The table of the standard normal distribution is/are given in Appendix A. The easiest way to use these tables is in conjunction with diagrams of the standard normal distribution such as those in Figure 9.4. The tables give $P(Z < z)$, for positive values of z, where z is known to 2 decimal places.

We will go through the use of the tables and the calculation of normal probabilities in conjunction with the graphs in Figure 9.4. The shaded area in Figure 9.4a represents the

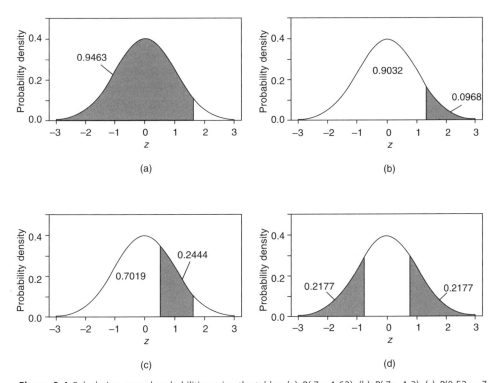

Figure 9.4 Calculating normal probabilities using the tables: (a) $P(Z < 1.62)$; (b) $P(Z > 1.3)$; (c) $P(0.53 < Z < 1.62)$; (d) $P(Z < -0.78)$.

probability that Z is less than 1.62. This is read off directly from the table – we see that $P(Z < 1.62) = 0.9463$.

To evaluate the probability that Z is greater than 1.3 (Figure 9.4b) it is necessary to use the fact that the total probability, which is the total area under the curve, is equal to 1. This means that $P(Z > 1.3) = 1 - P(Z < 1.3)$. We already know that $P(Z < 1.3)$ can be looked up in the tables and is equal to 0.9032. Thus

$$P(Z > 1.3) = 1 - P(Z < 1.3) = 1 - 0.9032 = 0.0968.$$

Suppose we wish to evaluate the probability that Z lies between two limits, 0.53 and 1.62 (Figure 9.4c). We already know how to evaluate $P(Z < 1.62)$ and $P(Z < 0.53)$ by reading the values off from the tables. By studying Figures 9.4a and 9.4c it can be seen that the area under the curve between 0.53 and 1.62 is equal to the area under the curve to the left of 1.62 minus the area under the curve to the left of 0.53. This is, in fact, the addition law for mutually exclusive events.

$$P(0.53 < Z < 1.62) = P(Z < 1.62) - P(Z < 0.53) = 0.9463 - 0.7019 = 0.2444.$$

The last type of probability calculation is illustrated in Figure 9.4d. This is to calculate the probability that Z is less than -0.78. Negative numbers do not appear in the tables of the probabilities of the normal distribution and it is necessary to use the symmetry property of the distribution to work out the probability. As the standard normal distribution is symmetric about its mean of 0, $P(Z < -0.78) = P(Z > 0.78)$. This can be seen from the shaded areas in Figure 9.4d. We already know how to evaluate $P(Z > 0.78)$ as it is the same type of calculation as in Figure 9.4b:

$$P(Z < -0.78) = P(Z > 0.78) = 1 - P(Z < 0.78) = 1 - 0.7823 = 0.2177.$$

These examples illustrate the four type of probability calculations that you need to do with the normal distribution. There are other types of calculations, such as $P(-1.6 < Z < 2.4)$, but all of then can be reduced to one of the four illustrated in Figure 9.4. The main concepts that you need to use are the law of total probability (the area under the curve is equal to one), symmetry ($P(Z < -z) = P(Z > z)$) and the addition law for mutually exclusive events.

When faced with probability calculations for normal distributions other than the standard normal all that you need to do is to convert the normal distribution into the standard normal distribution by standardizing it. This is achieved by using the formula

$$Z = \frac{X - \mu}{\sigma},$$

where Z represents the standard normal distribution and X is the general normal distribution with mean μ and standard deviation σ. If the population mean and standard deviation are unknown then we estimate them using the sample mean and sample standard deviation (see the book's website).

Example 9.5: Duration of court cases

On 3 February 1994 the *Guardian* published an article on the duration of child abuse cases. The mean length of time from arrival of a case at court to the completion of the case was 142 days. If a case is completed within 12 weeks, i.e. in less than 85 days, is this a particularly unusual event? In this instance we can assume that the duration of the cases follows a normal distribution, approximately, with a standard deviation of 52 days.

We have to calculate $P(X < 85)$, when $\mu = 142$ and $\sigma = 52$. This is accomplished by converting to the standard normal distribution using

$$P(X < 85) = P\left(Z < \frac{85 - 142}{52}\right) = P(Z < -1.10) = 0.1357.$$

This is not particularly small, 14%, and occurs with about the same probability as throwing a 6 on a die.

Note that we have used the normal distribution even though the random variable was a duration, which can never take a value less than zero. We often expect durations to follow an exponential distribution (see the book's website). However, for an exponential distribution to be valid the mean and standard deviation would have to be about the same, and this is not the case here.

The normal distribution is only an approximation here, and this is best seen by consideration of the probability of having a case take less than zero days. This is of course impossible, but the model we are using predicts

$$P(X < 0) = P\left(Z < \frac{0 - 142}{52}\right) = P(Z < -2.73) = 0.0032.$$

Although this is a very small probability, it is not zero and so the model predicts that an impossible event will happen with a very small probability. This often happens with statistical models in that they can produce impossible predictions. This does not invalidate the use of the model in areas in which it gives consistent results but does mean that you have to recognize that many models are only approximations and do not mimic reality exactly.

9.4.3 Looking up values of z in the standard normal tables

In the above examples we have looked at finding probabilities associated with the normal distribution given a value of z. Now we turn our attention to the situation where the probability is known and it is necessary to find the corresponding value of z. This situation arises when we ask questions such as 'above what value do 10% of child abuse cases lie?' and 'between what two limits do 95% of child abuse cases last?' with reference to Example 9.5.

Let us first find the value of z corresponding to $P(Z < z) = 0.95$. This is done by scanning through the central portion of the table of the normal probabilities looking for 0.95. It is found in the row labelled 1.6:

	0.00	0.01	0.02	0.03	0.04	0.05	0.06	0.07	0.08	0.09
1.6	0.9452	0.9463	0.9474	0.9484	0.9495	0.9505	0.9515	0.9525	0.9535	0.9545

From this row it can be seen that $P(Z < 1.64) = 0.9495$ and $P(Z < 1.65) = 0.9505$. From this you can see that we can estimate $P(Z < 1.645) = 0.95$, as 0.95 is mid-way between 0.9495 and 0.9505.

Now suppose we need to find the value of z corresponding to $P(Z > z) = 0.025$. Before looking a value up in the table it may be necessary to do some manipulation; the probabilities in the table are all greater that 0.5, and our probability is 0.025. Furthermore, we have to find $P(Z > z)$ while the tables only give $P(Z < z)$. This manipulation is easily accomplished with the help of a diagram (Figure 9.5a). We are looking for the value of z such that the probability in the shaded area is 0.025. As the total probability under the curve is 1, this means that $P(Z < z) = 0.975$. This is now a value that can be looked up in the tables. It is to be found in the row headed 1.9, giving $P(Z < 1.96) = 0.975$. This means that $P(Z > 1.96) = 0.025$.

	0.00	0.01	0.02	0.03	0.04	0.05	0.06	0.07	0.08	0.09
1.9	0.9713	0.9719	0.9726	0.9732	0.9738	0.9744	0.9750	0.9756	0.9761	0.9767

What is the value of z corresponding to $P(Z < -z) = 0.10$? As the probability is 0.10, we know that z must be less than 0. This can easily be seen from the tables of the normal distribution as $P(Z < 0) = 0.5$. So in Figure 9.5b, the value we are looking for is denoted '$-z$' with a probability of 0.1 below it in the shaded area. Using the symmetry rule we move to z also with a probability of 0.1 above it. So we know that $P(Z > z) = 0.10$ and hence $P(Z < z) = 0.90$. This can be found in the tables, under the row headed 1.2.

	0.00	0.01	0.02	0.03	0.04	0.05	0.06	0.07	0.08	0.09
1.2	0.8849	0.8869	0.8888	0.8907	0.8925	0.8944	0.8962	0.8980	0.8997	0.9015

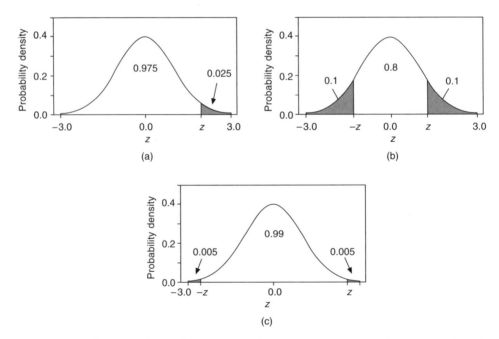

Figure 9.5 Finding percentage points of the standard normal distribution: (a) $P(Z > z) = 0.025$; (b) $P(Z < -z) = 0.10$; (c) $P(-z < Z < z) = 0.99$.

We find that $P(Z < 1.28) = 0.8997$ and $P(Z < 1.29) = 0.9015$ and so we take the closest value to 0.90. This gives $P(Z < 1.28) = 0.90$ and so $P(Z < -1.28) = 0.10$.

If we were to be more accurate we would interpolate between 1.28 and 1.29, but this is not really necessary. If you are using a statistical program or spreadsheet which has a function in it for calculating the cumulative distribution of the normal distribution then you will get the exact value to many decimal places.

Finally, how do we obtain the value of z such that $P(-z < Z < z) = 0.99$? This problem is tackled in much the same way as the previous one. From Figure 9.5c, we can see that $P(Z < z)$ must be equal to 0.995. This value is to be found in the row headed 2.5 of the tables of the standard normal distribution, where it is mid-way between $P(Z < 2.57) = 0.9949$ and $P(Z < 2.58) = 0.9951$. Thus we would use $P(Z < 2.575) = 0.995$ and conclude that $P(-2.575 < Z < 2.575) = 0.99$.

	0.00	0.01	0.02	0.03	0.04	0.05	0.06	0.07	0.08	0.09
2.5	0.9938	0.9940	0.9941	0.9943	0.9945	0.9946	0.9948	0.9949	0.9951	0.9952

When the probability that you are looking for is half-way between two values in the tables then you can interpolate to the midpoint. If it is closer to one value than the other then just choose the value it is closer to.

The points that we have been calculating here are known as **percentage points of the standard normal distribution**. They are usually denoted z_α, so that $P(Z > z_\alpha) = \alpha$. The common ones that will be used are $z_{0.10} = 1.28$, $z_{0.05} = 1.645$, $z_{0.025} = 1.96$ and $z_{0.01} = 2.33$.

Example 9.6: Normal ranges for cholesterol

Individuals who have a high cholesterol intake are at a greater risk of heart disease. From a survey of healthy women aged 50–59 we find that the distribution of total body cholesterol is normally distributed with a mean of 230 mg/dl and a standard deviation of 35 mg/dl. What cholesterol limits contain the middle 95% of women in this age group?

We have to find x_1 and x_2 such that $P(x_1 < X < x_2) = 0.95$. Earlier on in this section we saw how to find z such that $P(-z < Z < z) = 0.95$, and this is the first step in the solution. Once we have z then we will be able to get the values of x using the relationship between the standard normal distribution and the normal distribution:

$$Z = \frac{X - \mu}{\sigma},$$

where $\mu = 230$ and $\sigma = 35$.

From the tables of the normal distribution we find that $P(-1.96 < Z < 1.96) = 0.95$. This means that

$$1.96 = \frac{x_2 - 230}{35}$$

which implies that $x_2 = 230 + 1.96 \times 35 = 298.6$. The corresponding calculations for x_1 are

$$-1.96 = \frac{x_1 - 230}{35},$$

leading to $x_1 = 230 - 1.96 \times 35 = 161.4$.

We would interpret this as saying that 95% of women in the age group 50–59 are expected to have a total body cholesterol level within between 161 and 299 mg/dl. These are often referred to as 'normal ranges' or 'reference ranges', in that they give limits within which 95% of the population is expected to lie. A woman with a cholesterol level of 320 mg/dl is among the top 2.5% of cholesterol values, and a woman with a cholesterol level of 150 mg/dl is in the bottom 2.5% of values.

These limits are often used by doctors as guidelines for deciding if an individual patient has abnormally high or low values. This, in conjunction with other information about the patient, may help the doctor to form a diagnosis about the patient and lead to a suggested treatment, if necessary. A normal range is just a range of values within which 95% of the population are expected to lie. If you have a value of cholesterol which is outside the normal ranges this does not necessarily mean that you are ill and need to go on a reducing or cholesterol increasing diet.

Similar calculations are used in many businesses for setting tolerance limits for products. In precision engineering these tolerance limits will be very small, but the basic principle is the same. The length of a certain item is expected to lie within two limits. These limits will be set based upon the standard deviation of the lengths. As with all measurements, the smaller the standard deviation the greater the precision as the two limits will be closer to each other. Generally the limits are set such that 95% or 99% of all items are expected to have lengths between these two limits.

9.4.4 Combining two or more normal distributions

If X and Y are two independent random variables following normal distributions with means, μ_X and μ_Y, and standard deviations σ_X and σ_Y, and a and b are two constants, then $aX + bY$ is also normally distributed. The expected value and variance are

$$E[aX + bY] = a\mu_X + b\mu_Y,$$

$$\text{Var}[aX + bY] = a^2\sigma_X^2 + b^2\sigma_Y^2.$$

This means that the sum of two independent normally distributed random variables is also normally distributed with

$$E[X + Y] = \mu_X + \mu_Y,$$

$$\text{Var}[X + Y] = \sigma_X^2 + \sigma_Y^2;$$

and the difference between two independent normally distributed random variables is also normally distributed with

$$E[X - Y] = \mu_X - \mu_Y,$$

$$\text{Var}[X - Y] = \sigma_X^2 + \sigma_Y^2.$$

We will not illustrate these results with an example here. They will be used in Chapters 10–12.

9.4.5 Quantile plots and normal probability plots

A normal probability plot is a plot designed to illustrate whether data come from a normal distribution. With this plot if the data all lie along a straight line then we conclude that the data come from a normal distribution. If there is any deviation from the straight line, such as a pronounced curve, then this is evidence that the data do not come from a normal distribution. The curve is best described through an example, using the data in Table 9.1.

Table 9.1 Normal probability plot data

Data	−1.69	−1.38	−0.46	−0.29	−0.26	−0.2	−0.04	0.17	0.3	0.43	1.03	1.29
Proportion	0.042	0.125	0.208	0.292	0.375	0.458	0.542	0.625	0.708	0.792	0.875	0.958
Quantile	−1.73	−1.15	−0.81	−0.55	−0.32	−0.11	0.11	0.32	0.55	0.81	1.15	1.73

1. Sort the observations in ascending order of magnitude, as in the first row of Table 9.1.
2. Calculate the cumulative proportion less than each point using the expression $(i - 0.5)/n$, where i goes from 1 to n and indexes the rank of the observation. These proportions are in the second row, where $n = 12$.
3. Calculate the percentiles of the standard normal distribution corresponding to these proportions. In the first instance this means finding z such that $P(Z < z) = 0.042$. Using the procedures illustrated earlier in this section we find that $z = -1.73$. These percentiles are also known as the quantiles of the normal distribution. We do not use i/n for the proportion as this would mean finding $P(Z < z) = 1$ and this implies that z would be equal to infinity and cannot be plotted.
4. Plot the sorted data in row 1 against the percentiles of the normal distribution in row 3, Table 1.1. If the data come from a normal distribution then all the points should lie approximately on a straight line, with no major curvature, as shown in Figure 9.6.

The basis behind this plot is to look at the relationship between the observed data, on the y axis, and the data that you would expect to get from the normal distribution, on the x axis. Any deviation from a straight line suggests that the observed data do not mimic the data that you would expect to get from a normal distribution.

9.5 Summary

In this chapter we have looked in detail at discrete probability distributions and have studied the most important one, the binomial distribution. We have also investigated probability distributions for continuous random variables and have worked with the most important of these, the normal distribution. These two distributions are not the only ones that can be used and there are many others.

Other important discrete distributions are the Poisson, geometric and negative binomial distributions. The Poisson distribution is used as a model for events which occur randomly in space or time, such as the number of machine breakdowns per week and the number

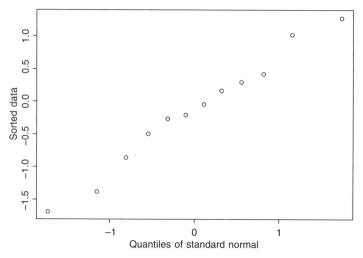

Figure 9.6 Normal plot of the data in Table 9.1.

of purchases of a product per day by an individual. It is closely linked to the binomial distribtuion and shares many of its underlying assumptions. Details are presented on the book website.

In later chapters of the book in the context of estimation and testing (Chapters 11–14) we will use two other continuous distributions, the t distribution and the χ^2 distribution. The normal distribution is symmetric and consequently cannot be used for skew distributions. In some cases a random variable with a skew distribution can be transformed into one with a normal distribution. The common transformations are the same as those for transforming a curved relationship into a straight-line relationship, namely by using the square root or the logarithm (see Chapter 6).

The exponential distribution is a model for a skew random variable which can take only positive values. It is also the probability model for the time between events which follow a Poisson distribution. So if the number of accidents per month on a road follows a Poisson distribution the time between successive accidents will follow an exponential distribution. Details of this distribution are also given in the book website.

Case studies with applications of probability distributions are provided on the book website.

10

Sampling distributions

- Understanding the relationship between random samples, repeated samples and sampling distributions
- Appreciating that the act of sampling induces a sampling distribution
- Knowing about the standard error of an estimate
- Appreciating what bias is and what an unbiased estimate means
- Understanding the effect of sample size on the sampling distribution of the sample estimate.
- Knowing about the importance of the central limit theorem, which states that in large samples the sample mean follows a normal distribution with mean μ and standard error of $\sigma/\sqrt{n}$.
- Knowing how to use the normal distribution to work out probabilities regarding the sample mean.
- Knowing about the sampling distribution of the sample proportion
- Understanding the importance of precision and accuracy and knowing the difference between them
- Appreciating that the results about sampling distributions mean that you can think of deciding how large a survey to carry out on the basis of sample size calculations for standard errors.

10.1 Introduction

One of the main uses of statistical techniques is concerned with the appropriate ways of collecting and displaying information. This places emphasis on good graphical displays, tabulations and descriptive interpretation based upon making sure that the data you have collected will provide appropriate information on the quantities that you wish to measure. Examples of this work include the calculation and presentation of price indices and other indicators of social and economic trends. A second use concerns the application of probability models in order to make predictions about the likely behaviour of random variables or systems of random variables. These have more use in the engineering and investment sectors of business and are concerned with reliability, machine failure and predicting share price changes. This is based upon having a probability model which is valid for the process under study.

   In both of these applications sample information is used. In the first it is usual to collect information only on a sample of individuals rather than on all individuals in the population. For example, the retail price index is based upon the expenditure information collected from a random sample of 7000 households per year in the Family Expenditure Survey. In the second, we saw in Chapter (9) and on the book's website how to use sample information to estimate unknown parameters of the probability models. The exponential distribution is often used as a model for the length of time that you have to wait for something to occur. From the historical records of machine breakdown you can estimate the mean time between failures, and combining this with an appropriate probability distribution such as the exponential allows you to estimate the probability that the machine will breakdown within certain periods. This information is used in planning the servicing and production schedules for a manufacturing company.

The parameters of the probability models were derived from the sample mean, sample standard deviation or sample proportion. When describing the characteristics of a population the same sample quantities are used. In any particular situation we will generally have one sample from a population. If the sample were selected at a slightly different time, or if a second sample were selected, then the sample values would not necessarily be the same. In this chapter we look at the behaviour of these sample quantities and see how we can use this information to help us in our interpretation of the sample information. This information can also help us decide on how large a sample is needed in the first place.

10.2 Random samples, repeated samples and sampling distributions

The data presented in Figures 10.1 and 10.2 were collected during a survey of first-year students attending Strathclyde University in session 1991–92. This was a not particularly well-designed survey based, in terms of the types of samples discussed in Chapter 2, on a convenience sample. The respondents were all in a statistics class at 9.00 a.m on a

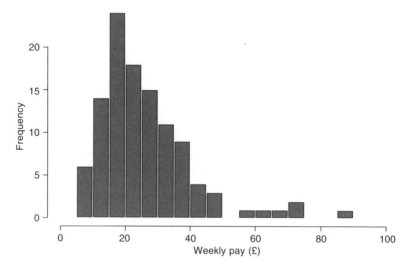

Figure 10.1 Distribution of total pay for students in part-time employment.

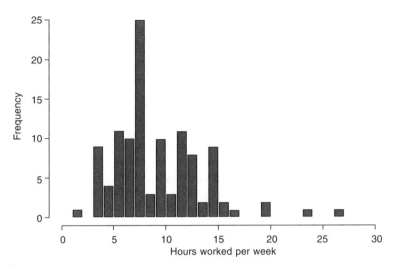

Figure 10.2 Distribution of hours worked by students in part-time employment.

Wednesday. A total of 317 students responded and the data presented in the histograms refer to the 115 students who had a part-time job.

These data will be used to illustrate the process of taking random samples, and this will then provide information on the sampling distributions of the statistics calculated from samples. For the purposes of this investigation we will pretend that the population consists of only 115 individuals, and we will select small samples of sizes 5, 10 and 20 from it. The aim of the investigation is to see what distribution of the sample mean looks like and to see how good an estimate of the population mean it is.

This investigation will be carried out for both of the variables illustrated in Figures 10.1 and 10.2. One feature of both of these distributions is that neither the normal distribution nor any of the other distributions in Chapter 9 is valid. Both distributions are skew to a greater or lesser extent. While the distribution of weekly pay in Figure 10.1 is skew, it looks relatively smooth. This contrasts with the distribution of weekly hours worked in Figure 10.2, which has local modes at discrete values. The most common number of hours worked is 8 but there are also local peaks at 3, 10, 12 and 15 hours.

The mean number of hours worked per week is 9.6 hours and the median is 8 hours. The mean amount earned is £27.50 with a median of £25. The mean and median are both measures of the central tendency of a distribution, and they are the main parameters that we shall focus on in this investigation.

When a sample is selected from a population this is selected at random and without replacement from all individuals in the population. The random sampling mechanism is such that each individual in the population has exactly the same chance of being sampled. Having obtained a sample, the statistician then attempts to describe the population in terms of statistics calculated from the data in the sample. Possibly the most important piece of information in the population is the central value – the mean or median – and the sample mean or sample median is used to estimate the corresponding population quantity.

If the sample selected is representative, then the distribution of the values in the sample will be representative of the distribution of all of the values in the population. This means that we do not always need to look at the whole population and that a subset – a random sample of it – will be sufficient. Random sampling should, on average, result in representative samples.

In the first part of the investigation we will select samples of size n from the population of 115 values. The sample sizes will be $n = 5$ or $n = 10$ or $n = 20$, corresponding to samples which are approximately 5%, 10% and 20% of the total population sizes. (Normally sample sizes will be a much smaller percentage of the population than this.) For each sample we will calculate the sample mean. This whole process will be repeated 1000 times. We use such a large number of samples because we wish to see the distribution of the sample mean.

The rows in Table 10.1 correspond to the first four samples selected from the population of weekly wages from part-time employment. In the first sample there are the observations, £45, £40, £25 £34 and £15. The sample mean of these numbers is £31.8 and the sample median is £34. The second sample yields completely different values with a mean of £32.62 and a median of £25. Looking through the other rows of Table 10.1 reveals that there is a great variation in the observations which are sampled and this in turn leads to different sample means and different sample medians. Thus random sampling induces a distribution of the statistics calculated from the samples. This distribution is known as the **sampling distribution**.

Table 10.1 First four samples of size 5

Sample	Observations					Sample mean	Sample median
1	45	40	25	34	15	31.8	34
2	23.1	25	65	25	25	32.62	25
3	20	14	40	25	17	23.2	20
4	30	33	21	30	28	28.4	30

The sampling distribution is always associated with a specific quantity. Thus, for example, you can have the sampling distribution of the (sample) mean, the sampling distribution of the (sample) median, and the sampling distribution of the (sample) proportion. The word sample is in brackets in the previous sentence because it is often left out when talking about sampling distributions and we often just talk about the sampling distribution of the mean and so on.

The sampling distributions of samples of size 5, 10 and 20 from the population of the 115 individuals with a weekly wage are presented in Figure 10.3. Summary information on these sampling distributions is presented in Table 10.2. From these you can see a number of things, particularly in relation to the population in Figure 10.1. Firstly, the original population distribution is skew with a long tail towards the right. This skewness is also clearly visible in the sampling distribution of the mean in samples of size 5. As the sample size is increased to 10 and 20 the skewness becomes less and the sampling distribution approaches a symmetric distribution.

The curves which are superimposed on the histograms are the probability curves of normal distributions. There is quite a good match between the sampling distribution of the mean based on samples of size 20 and the corresponding normal distribution. It is not

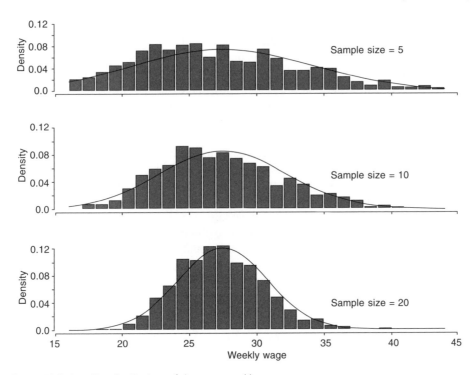

Figure 10.3 Sampling distributions of the mean weekly wage.

Table 10.2 Summary statistics for the sampling distribution of the mean weekly wage based on 1000 samples

Sample	Minimum	First quartile	Median	Mean	Third quartile	Maximum	Standard deviation
Original	5.50	18.00	25.00	27.49	33.94	90.0	14.84
Size 5	13.00	22.50	26.60	27.38	31.43	59.0	6.74
Size 10	17.10	24.11	27.09	27.49	30.38	46.4	4.69
Size 20	17.55	24.95	27.15	27.30	29.43	39.7	3.27

a perfect match as the sampling distribution is a little skew. In smaller samples the match is not so good, especially when the samples are as small as 5. The original distribution is so skew that it does not resemble a normal distribution at all.

The range of the original distribution is from £5.50 to £90 per week. As the sample size increases the range of the sampling distribution of the mean decreases. When the sample size is 5 the range of the sample mean is from £13 to £59 but when the sample size is 20 the range of the sample mean is from £17.55 to £39.70. Thus as the sample size increases the sampling distribution of the mean becomes more compact around its central value.

The mean of the original 115 wages is £27.49. The means of the sampling distributions are all very close to this value. In the investigation the mean of the 1000 sample means from samples of size 5 is £27.38, from samples of size 10 it is £27.49 and from samples of size 20 it is £27.30.

The median also measures the central tendency of a distribution and there is a difference between the median and the mean in the original population of £2.49. This is a result of skewness in the distribution over the population. In the other three rows of Table 10.2 the median and mean refer to the distribution of sample means. The difference between the median and mean decreases as the sample size increases. By the time we have the sampling distribution of samples of size 20 the median is only £0.15 different from the mean. This illustrates that the sampling distribution becomes more symmetric as the sample size increases.

In symmetric distributions the standard deviation is the most appropriate measure of variability of the distribution. It is not so appropriate if the distribution is very skew. The standard deviation of the original distribution is £14.84. The standard deviations in the sampling distributions are all smaller than this. As the sample size increases the standard deviation of the sampling distribution decreases.

In the above investigation we have made some observations based upon looking at the distribution of 1000 samples of size 5 from the population of 115 individuals. In fact, there are $\binom{115}{5}$ possible samples of size 5 from 115 which is in excess of 150 million, and the numerical results in Table 10.2 and the graphical results in Figure 10.3 will not be exactly the same if a different set of 1000 samples is selected. The results from a second set of 1000 samples are presented in Table 10.3. In this table you can see that the means of the sampling distributions are all close to the mean of the original distribution. The median of the sampling distribution is closer to the mean of the sampling distribution in larger samples than in smaller samples, indicating that the sampling distribution is more symmetric in larger samples. The range and standard deviation of the sampling distributions get smaller as the sample size increases, indicating that the sampling distribution gets closer to its central value in larger samples. Thus exactly the same conclusions are drawn from the second illustration as from the first.

Table 10.3 Summary statistics for the sampling distribution of the mean weekly wage based on a second set of 1000 samples

Sample	Minimum	First quartile	Median	Mean	Third quartile	Maximum	Standard deviation
Original	5.50	18.00	25.00	27.49	33.94	90.00	14.84
Size 5	14.22	22.80	26.40	27.45	31.30	53.40	6.55
Size 10	14.80	24.37	27.00	27.58	31.50	48.50	4.65
Size 20	18.80	25.17	27.19	27.45	29.50	39.68	3.32

We now turn our attention to the sampling distribution of the mean in the other population chosen for illustration. Compared to the distribution in the population of the weekly wage (Figure 10.1), the population distribution of the number of hours worked per week had more local peaks but was a little bit more symmetric (Figure 10.2). In the histograms of the sampling distributions (Figure 10.4), the local peaks are still visible in the samples of size 5, but have largely disappeared in samples of size 10, and have completely disappeared in samples of size 20. As before, the larger the sample size the more symmetric is the sampling distribution. Indeed, the histogram of the sampling

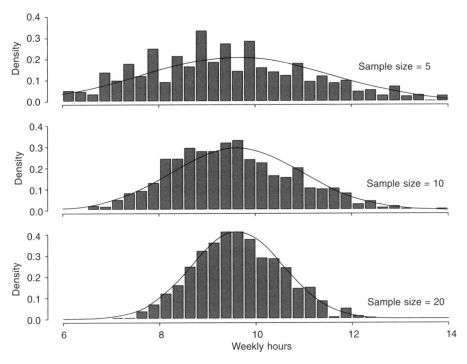

Figure 10.4 Sampling distributions of the mean number of hours worked per week.

Table 10.4 Summary statistics for the sampling distributions of the mean weekly hours worked based on 1000 samples

Sample	Minimum	First quartile	Median	Mean	Third quartile	Maximum	Standard deviation
Original	2.0	7.0	8.00	9.615	12.00	27.00	4.27
Size 5	5.0	8.2	9.30	9.488	10.60	17.80	1.82
Size 10	6.2	8.6	9.45	9.556	10.35	14.45	1.32
Size 20	6.3	9.0	9.60	9.646	10.28	12.85	0.93

distribution of the mean based on samples of size 20 matches up well with the normal distribution. The summary statistics in Table 10.4 show that the mean of the sampling distributions is the same as the mean of the original distribution over the population, irrespective of the sample size. As the sample size increases the variability in the sampling distribution decreases, the range and the standard deviation become smaller.

The two illustrations have all told the same tale:

- The average value of the sampling distribution of the mean is close to the mean of the population in all three sample sizes illustrated.
- The variability in the sampling distribution decreases as the sample size increases.
- The sampling distribution of the sample mean becomes more symmetric as the sample size increases, even if the original distribution is very skew and has multiple peaks.

- In large samples the normal distribution looks to be a reasonable distribution to describe the sampling distribution.

We will now go on to investigate each of these points in turn and then see how we can use them in a statistical analysis.

10.3 Bias

The first point is concerned with bias. This has already been discussed in Chapters 2 and 5 in connection with survey bias caused by non-response or poor sample selection resulting in a biased sample, that is, one that is not representative of the population. When talking about sampling distributions, bias has a special meaning which is more specific than the meanings used previously.

Looking at the sampling distribution of the sample means, the observation in the previous section that the mean value of the means of samples of size 5, 10 and 20 was more or less the same as the average value in the population, whatever the variable under study, can be expressed more formally in terms of bias. If the population mean is denoted μ and the sample mean is denoted $\overline{X}$, with a specific value in any one sample denoted $\bar{x}$, then the sample mean is an **unbiased** estimate of the population mean. This means that the expected value of the sample mean $\overline{X}$ over the sampling distribution of the sample mean, is equal to the population mean. In symbols, this is written

$$E[\overline{X}] = \mu.$$

If an estimate is unbiased, this means that there is no tendency for the estimate to be larger or smaller than the population value that it is supposed to estimate. This is a good property for an estimate to have. But what does it mean if we use an estimator which is biased?

As an illustration let us try to estimate the population mean by using $n\overline{X}/(n + 1)$, rather than $\overline{X}$. We know that this is not a particularly bright thing to do in reality, but it will serve as an illustration of bias. If $\overline{X}$ is unbiased then we would expect $n\overline{X}/(n + 1)$ to be biased because it will always be smaller than $\overline{X}$. The sampling distributions are illustrated in Figure 10.5. The same normal distribution curves are superimposed as were superimposed in Figure 10.3, and you can see that the bias in the estimate means that the sampling distribution is located towards lower values. This is the case even in the largest sample. In Figure 10.3, there was a reasonably good agreement between the sampling distribution and the superimposed normal probability curve, but in Figure 10.5 the bias towards lower values is evident.

In the summary statistics in Table 10.5 you can see that the mean of the sampling distribution is consistently below the population mean. The extent of the bias in this case decreases as the sample size increases, but does not disappear. The other features of the sampling distribution which were noted in Tables 10.2–10.4 are also present here. The median and mean of the sampling distribution get closer to each other as the sample size increases, indicating that the sampling distribution gets more symmetric as the sample size increases. The range and standard deviation decrease as the sample size increases, indicating that the sampling distribution is becoming more compact around its central value.

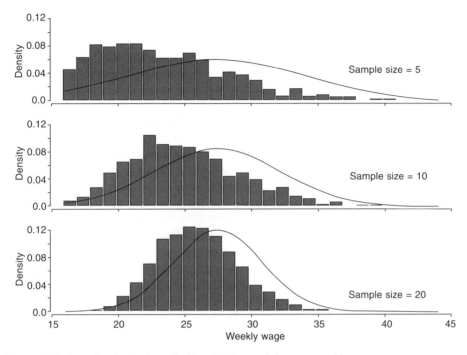

Figure 10.5 Sampling distributions of a biased estimate of the mean weekly wage.

Table 10.5 Summary statistics for a biased estimate of the mean weekly wage

Sample	Minimum	First quartile	Median	Mean	Third quartile	Maximum	Standard deviation
Original	5.50	18.00	25.00	27.49	33.94	90.00	14.84
Size 5	10.83	18.75	22.17	22.82	26.19	49.17	5.62
Size 10	15.55	21.92	24.62	24.99	27.61	42.18	4.27
Size 20	16.71	23.76	25.86	26.00	28.03	37.81	3.12

In terms of estimating any features of a population, bias is the tendency for the sampling distribution of any sample estimate to be located above or below the true value in the population. In any individual sample, the sample estimate may be above or below the true value, as can be seen in the histograms. In large samples, where a normal distribution is appropriate, if the estimate is unbiased then the probability that the sample estimate will be above the true population value is the same as the probability it will be below. If the estimate is biased downwards then the probability that the estimate from any one sample will be below the true value is greater then the probability that it will be above. If the estimate is biased upwards then the probability that the estimate from any one sample will be below the true value is less than the probability that it will be above.

10.4 Standard error

The second of the points noted at the end of the investigation in Section 10.2 was concerned with the variability in the sampling distribution. In all cases the variation decreases as the sample size increases. As the sampling distributions also tend to get more symmetric as the sample size increases, the best way of measuring the variability in the sampling distribution is by the standard deviation of the sampling distribution. This has a special name. The standard deviation of the sampling distribution of the sample mean is known as the **standard error** of the (sample) mean. Again, the term in brackets is often omitted.

You can have the standard deviation of a random variable, the standard deviation of a population and the standard deviation of a sample of observations, but the standard error is not the same as the standard deviation. The standard error always refers to some quantity calculated from the sample. So you can have the standard error of the sample mean and the standard error of the sample proportion. These are the two most important. You can also have the standard error of other quantities such as the sample median or the sample standard deviation.

The variation of the distribution of the sample mean is less than the variation in the distribution over the population. If the standard deviation in the population is denoted σ, and the sample size is n, then the standard error of the sample mean is given by

$$SE(\overline{X}) = \sigma_{\overline{X}} = \frac{\sigma}{\sqrt{n}}.$$

As the sample size decreases, the standard error of the sample mean decreases. A comparison of the standard deviations obtained from the sampling distributions in Tables 10.2 and 10.4 with the value given by the formula is presented in Table 10.6. These show a good agreement between the observed standard errors in the investigation and those given by the formula. This illustrates the validity of the formula.

Table 10.6 A comparison of standard errors using the formula and those observed in the sampling distributions

| Sample | Weekly wage | | Hours | |
	Standard diviation from sampling distributions	Standard error using $\sigma/\sqrt{n}$	Standard deviation from sampling distributions	Standard error using $\sigma/\sqrt{n}$
Population standard deviation	14.84		4.27	
5	6.64	6.74	1.91	1.82
10	4.69	4.69	1.35	1.32
20	3.32	3.27	0.95	0.93

The important point of this section is that there is a direct link between the standard error of the sample mean, the population standard deviation and the sample size. The standard error of the mean is a measure of variability in the distribution of the sample mean. It is thus a measure of sampling variability. By selecting larger samples we are able to make sure that this sampling variability is reduced. It can never be eliminated as there will always be variation in the population, σ, and the sample size cannot increase indefinitely. However, it is possible to control sampling variability to some extent, and we will investigate this in Section 10.9.

10.5 Central limit theorem

The third and fourth points at the end of the investigation in Section 10.2 are both concerned with the probability distribution of the sample mean. As the sample size gets larger the distribution of the sample mean tends to become more symmetric, eventually following a normal distribution. This pattern was observed in both examples, even though the original distributions over the population were not normal themselves. This is an important mathematical result and is known as the **central limit theorem**. It is also one of the main reasons for the importance of the normal distribution in statistical practice and theory.

The result is usually expressed by saying that in large random samples the sampling distribution of the sample mean follows a normal distribution approximately, or by saying that as the sample size increases the sampling distribution of the sample mean tends to a normal distribution. The implication of this result is that we can use the normal distribution to carry out probability calculations about the sample mean provided the sample is large enough. In order to use the normal distribution we need to know the expected value and its standard deviation. These have already been established for the sample mean in the preceding two sections. The expected value of the sample mean is the population mean, μ, and the standard deviation of the sample mean is the standard error of the sample mean, given by $\sigma/\sqrt{n}$.

The practical application of this result depends upon when it is reasonable to use the normal distribution as an approximation to the sampling distribution of the sample mean. If the variable under study itself has a normal distribution over the population, then the sample mean will have a normal distribution whatever the sample size. Most variables do not have normal distributions, as the examples in Section 10.2 have illustrated. However, the normal distribution will be applicable provided that sample size is large enough – and samples of size 30 or more are usually considered to be large enough.

These guidelines should not be slavishly adhered to. In smaller samples we can still use the normal distribution for the sample mean but here we need to be careful about the original distribution. If it is close to normal – for example, symmetric with a single peak – then a normal distribution for the sample mean will be valid even in samples as small as 5. If it is a flat distribution then sample sizes of as small as 10 will be sufficiently large, as illustrated in Figure 10.6. If it is very skew, as in Figure 10.7, where the population is the number of hours per week of part-time work among all students where those with no part-time work have 0 hours, then the normal approximation will require larger samples. Even in samples of size 30 the sampling distribution is still a little skew compared to the normal distribution.

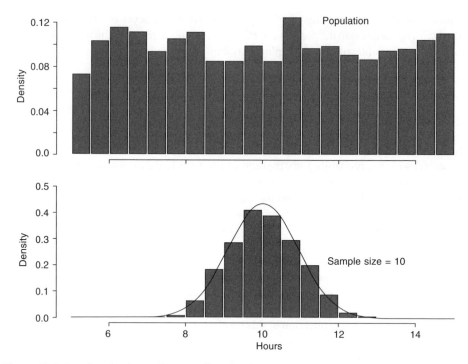

Figure 10.6 Sampling distribution from a uniform distribution.

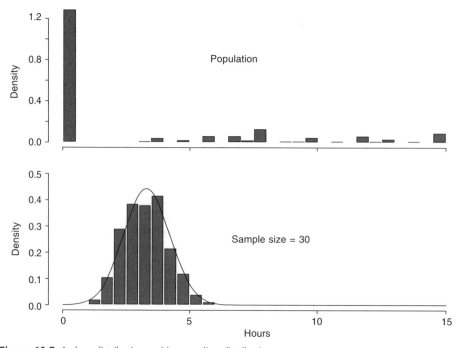

Figure 10.7 A skew distribution and its sampling distribution.

10.6 Sampling distribution of the sample mean

In this section we will look at the practical use of the result about the sampling distribution of the sample mean through two examples.

Example 10.1: Graduate pay

The *Guardian* of 18 November 1998 published an article reporting on a survey of graduate expectations with regard to employment and pay. The survey was carried out by NOP and included a questionnaire survey of 2000 final-year students who were due to graduate from university in the summer of 1998. The following paragraph is of interest to us here.

> The survey found that women graduates had substantially lower expectations when it came to starting salaries. Only 15 per cent expected to earn £16 000 or more, compared to 31 per cent of men. Only 9 per cent expected to earn more than £17 000, compared to 21 per cent of men. And 2 per cent expected more than £20 000, compared to 5 per cent of men.
>
> On average graduates expected their starting salary to be £14 097, rising to £22 157 after five years. Women's expectations were £1500 below men's at the start of their careers.

The article states that there is a £1500 difference in the anticipated starting salary between men and women graduates. This is obviously a big difference in financial terms, but in terms of the variation in anticipated salaries is it an important difference, or just a difference that we might expect by chance alone? From our investigation of the sampling distribution of the mean we know that the sample mean is a random quantity and that a different random sample can produce different results. Before making claims about differences in expectations between men and women graduates we need to establish the sampling error in these results. Such claims can be tested using the methods to be discussed in Chapter 12.

We know that in large samples the sample mean will follow a normal distribution with a mean of the true population mean, μ, and a standard error of $\sigma/\sqrt{n}$, where n is the sample size and σ is the standard deviation in the population. From information provided in the article, we calculate this to be £2561.

If we use the mean and standard deviation from the sample of 2000 students as the best estimates of the population values then we can make some probability calculations about the behaviour of the sample mean in a future sample of students. Questions which are of interest include: (a) the probability that the sample mean will be in excess of £14 500; (b) the probability that the sample mean will lie between £14 000 and £15 000; (c) the limits between which you would expect 90% of the sample means to lie.

The probability calculations are all about a future sample of students, say with 50 men and 50 women. We cannot make probability calculations about the observed sample as we already know the result. It is pointless to try and work out the probability that the sample mean is in excess of £15 000 when we know that it was calculated as £14 097. In the next chapter we will see how to incorporate the information in this chapter about sampling distributions and apply it in the context of one sample from a population. For the minute, we are using the information in the original published sample of 2000 students to provide information about the population mean and standard deviation. Then we make probability statements about the sample mean in a future sample. These calculations all make use of normal probability calculations from Chapter 9.

We assume that $\mu = £14\,097$ and that $\sigma = £2561$. There are to be 50 men and 50 women in the second sample, giving $n = 100$. Thus the standard error of the sample mean is going to be

$$\sigma_{\overline{X}} = \frac{\sigma}{\sqrt{n}} = \frac{2561}{\sqrt{100}} = 256.1.$$

For the sample of 50 men the standard error will be

$$\sigma_{\overline{X}} = \frac{\sigma}{\sqrt{n}} = \frac{2561}{\sqrt{50}} = 362.2.$$

The sample mean $\overline{X}$ follows a normal distribution.

(a) What is the probability that the sample mean is in excess of £14 500? Here, we need to calculate $P(\overline{X} > 14\,500)$, where $\overline{X}$ follows a normal distribution. In order to make the calculation we need to convert $\overline{X}$ into a standard normal variable and this is achieved, as before, by subtracting the mean and dividing by the standard deviation of $\overline{X}$. Note that the standard deviation of $\overline{X}$ is the same as the standard error of $\overline{X}$. Thus

$$Z = \frac{\overline{X} - \mu}{\sigma_{\overline{X}}} = \frac{\overline{X} - \mu}{\sigma/\sqrt{n}}.$$

This means that

$$P(\overline{X} > 14\,500) = P\left(Z > \frac{14500 - 14097}{256.1}\right) = P(Z > 1.57) = 0.0582.$$

We interpret this probability by saying that in 5.8% of samples of size 100 we would expect the sample mean to be in excess of £14 500, given that the true population mean is £14 097 with a true population standard deviation of £2561. Thus obtaining a sample mean in excess of £14 500 is a relatively rare event which is expected to occur about once every 17 times.

(b) What is the probability that the sample mean is between £14 000 and £15 000? This is calculated as

$$P(14\,000 < \overline{X} < 15\,000) = P\left(\frac{14\,000 - 14\,097}{256.1} < Z < \frac{15\,000 - 14\,097}{256.1}\right)$$

$$= P(-0.38 < Z < 3.53) = P(Z < 3.53) - P(Z < -0.38)$$

$$= 0.9998 - 0.3520 = 0.6478.$$

So we would expect the sample mean to lie within the range £14 000 to £15 000 with a probability of 0.65. This indicates that this would be a relatively common event.

(c) Between what two limits would you expect 90% of the sample means to lie? This is the same type of problem as discussed in Section 9.4.3. We seek two limits, 14 097 − k and 14 097 + k, such that the probability that the sample mean lies between these two limits is 0.90. This can be written as

$$P(14\,097 - k < \overline{X} < 14\,097 + k) = P\left(\frac{-k}{256.1} < Z < \frac{k}{256.1}\right) = 0.90.$$

From the tables of the normal distribution we know that $P(-1.645 < Z < 1.645) = 0.90$. Equating the terms means that we can write

$$\frac{k}{256.1} = 1.645$$

$\therefore$
$$k = 1.645 \times 256.1 = 421.3.$$

Thus 90% of the sample means from samples of size 100 would be expected to lie between £$(14\,097 - 421) = £13\,676$ and £$(14\,097 + 421) = £14\,518$.

Example 10.2: Payment of invoices

Late payment of invoices is a major problem for European businesses. A leading European market research company, NOP, carried out research on this for Intrum Justitia. This research was supported by the European Commission and detailed results are available from the survey archive of the NOP website (http://www.nop.co.uk). The main results are:

> The average payment time on domestic trade for all European firms is 53 days. With an average contractual credit period of 39 days, this means all payments are on average 14 days overdue.
> Overall in Europe, 61% of invoices are paid on time; 21% are settled 1–30 days after due date; 12% 31–90 days after due date; 4% 91–180 days late; and 1% over 180 days late. One per cent of bills are never paid. Most countries follow a broadly similar pattern, except Ireland, Portugal and UK where only around 40% of invoices are paid on time. A comparison of different business sectors shows that across Europe banking/insurance companies are the best payers, while government and construction firms are the worst payers.

Suppose a survey is to be carried out using a sample of 500 invoices. (a) What is the probability that the sample mean payment time is greater than 60 days? (b) What is the probability that the sample mean is between 50 and 55 days? (c) Between what two limits, $\mu \pm k$, would you expect 95% of the sample means to lie?

In order to answer these questions, we need to have information about the standard deviation of the payment times. This was not available in the publication but we know that the payment time must be greater than zero and if we assume that the distribution for the payment time follows an exponential model (see book website) with a mean of $\mu = 53$ days then the standard deviation is the same as the mean and so is $\sigma = 53$ days. With a sample size of 500 the standard error of the mean is

$$\sigma_{\overline{X}} = \frac{\sigma}{\sqrt{n}} = \frac{53}{\sqrt{500}} = 2.37.$$

The sample size of 500 is so large that we do not need to worry about the original distribution of the payment times, which we assumed to be exponential, and can use a normal distribution for the probability calculations about the sample mean.

(a) What is the probability that the sample mean payment time is greater than 60 days? We calculate

$$P(\overline{X} > 60) = P\left(Z > \frac{60 - 53}{2.37}\right) = P(Z > 2.95) = 0.0016.$$

In a sample of 500 invoices the chance that the sample mean is larger than 60 days is 0.16% which is a very rare event.

(b) What is the probability that the sample mean is between 50 and 55 days? Here the calculation is

$$P(50 < \overline{X} < 55) = P\left(\frac{50 - 53}{2.37} < Z < \frac{55 - 53}{2.37}\right) = P(-1.27 < Z < 0.84) = 0.6975.$$

There is almost a 70% chance that the sample mean will lie between 50 and 55 days.

(c) Between what two limits, $\mu \pm k$, would you expect 95% of the sample means to lie? The population mean is $\mu = 53$, and we are required to find k to give limits either side of the population mean within which 95% of sample means are expected to lie. Thus we have to find k such that,

$$P(53 - k < \overline{X} < 53 + k) = P\left(\frac{-k}{2.37} < Z < \frac{k}{2.37}\right) = 0.95.$$

From the table of the normal distribution (Appendix A) we know that $P(-1.96 < Z < 1.96) = 0.95$. Equating the terms means that we can write

$$\frac{k}{2.37} = 1.96$$

$$k = 1.96 \times 2.37 = 4.65.$$

This means that the limits are $53 - 4.65 = 48.35$ days and $53 + 4.65 = 57.65$ days. The probability that the sample mean payment time will lie within these limits is 0.95. This means that there is a 5% chance that the sample mean will lie outside these limits.

There are two further points to make about this example. The first is that it is based upon the use of published information which does not give all the details that are required for the statistical analysis. Specifically, the information about the standard deviation is missing. In a serious problem the best way forward is to contact the original suppliers of the data and to try to obtain the information directly. In this example we have made an assumption which, in terms of the data supplied, cannot be tested. Specifically, we assumed that an exponential distribution would be a good model for the payment times. As the mean and standard deviation are the same for the exponential distribution, this gave us a value for the standard deviation. The use of this model is based upon prior experience which suggests that the exponential model may be useful for random variables which measure the length of time for something to occur.

Here, we do not know if the assumed standard deviation is appropriate or not. One way to investigate the sensitivity of the conclusions to this assumption is to carry out what is known as a **sensitivity analysis**. In this case this is achieved by taking different values of the standard deviation and seeing the effect that these different values have on the calculated probabilities. The results are illustrated in Table 10.7. They show that large changes in the assumed standard deviation do not have such a great effect on the standard error of the mean. This is due to the effect of the sample size. The probability calculations in the centre of the distribution, (b) and (c), are influenced by the standard deviation but the probabilities in the tails, (a), are not influenced as much.

Table 10.7 Payment time sensitivity analysis

σ	33	53	73
$\sigma/\sqrt{n}$	1.48	2.37	3.26
(a) $p =$	0.0000	0.0016	0.0159
(b) $p =$	0.9333	0.6975	0.5515
(c) $k =$	2.90	4.65	6.39

The second point is to draw your attention to the fact that the interpretation in the first sentence of the quote is not completely accurate. With the misinterpretation italicized, the original interpretation is:.

The average payment time on domestic trade for all European firms is 53 days. With an average contractual credit period of 39 days, *this means all payments are on average 14 days overdue.*

The average payment time is the time from the date of the invoice to the payment of the invoice. Each invoice has a contractual credit period within which the bill should be paid. The payment is overdue if the payment is made after the end of the contractual credit period. The average payment time is 53 days and the average credit period is 39 days. The difference between the two is 14 days, and this is the *average length of the overdue period.* It does not mean that *all* payments are overdue by an average of 14 days.

Table 10.8 Overdue period on European invoices: distribution of the number of days overdue

Interval	0	1–30	31–90	91–180	181+	Never
Midpoint	0	15	60	135	270	
Per cent	61	21	12	4	1	1

In fact, the distribution of the overdue period is given (see Table 10.8). A better way of expressing the information about the overdue period is to focus on the invoices which were overdue. Of the 38% of payments which were overdue and eventually paid, the average overdue period was 48.6 days, calculated as.

$$\frac{15 \times 21 + 60 \times 12 + 135 \times 4 + 270 \times 1}{21 + 12 + 4 + 1}.$$

Of all invoices which were paid, including those paid on time, the average overdue period was 18.7 days. This was calculated including the 61% of payments made on time as having no overdue period.

The reason why there is a difference in the two estimates of the average of the overdue periods is that, in the first (14 days), any invoices paid before the end of the credit period will have a negative overdue period. This will have the effect of decreasing the average length of the overdue period. In the second (18.7 days) any invoices paid before the end of the credit period are just treated as being paid on time and have an overdue period of 0 days.

10.7 Sampling distribution of the sample proportion

All of the foregoing illustrations of sampling distributions have focused on the sampling distribution of the sample mean. The sample mean is an important quantity for many quantitative variables, but is not appropriate for qualitative variables. Many surveys deal largely with qualitative variables and the sample proportion, or sample percentage, is the appropriate statistic. In this section we will review the sampling distribution of the sample proportion

10.7.1 Studies reporting percentages and the effects of sampling variability

Graduate survey

The NOP survey of 1998 graduates (*Guardian*, 18 November 1998) reported that 49% of men had job offers by the time they had left university, compared to 42% of women. We know that these figures are based upon a sample of 2000 students but until we have some information about the sampling distribution of a sample proportion we do not really know whether this 7% difference is likely to be something which could have arisen simply by chance or whether it reflects a real systematic difference in the ability of male and female graduates to get jobs.

Family expenditure survey and leisure goods

Results from the Family Expenditure Survey were published in the *Guardian* (28 November 1998). These were based upon a sample of 6400 families. The share of household expenditure devoted to leisure goods and services was 17% in 1997. The corresponding percentage in 1990 was 14%. Is this difference an indication of a real increase, or can it just be attributed to the effects of chance? After all, a difference of 3 percentage points is not particularly staggering.

Household formation by young people

In a study of the social context of spending in youth, two sociologists, Gill Jones and Chris Martin, from the Centre for Educational Sociology at Edinburgh University, investigated changes in young people's households. They investigated 16–25-year-olds who had been included in the Family Expenditure Surveys of 1982, 1987 and 1992. The changes in the percentages of young people in the different household types are presented in Table 10.9. Over the 10-year period the percentage of 16–25-year-olds living with their parents changes from 61% to 58%. Is this difference indicative of any social trend or just something which can occur by chance? The percentage living with partners and children has decreased by 4 percentage points from 14% to 10%. Is this again the sort of change that might be expected by chance? In a study using the 1997 FES, what range would be expected for the percentage of young people who are lone parents on the basis of the 1992 figure of 3%?

Table 10.9 Household types, from FES

Household type (%)	1982	1987	1992
Parental	61	58	58
Kin	2	2	2
Peer	5	5	7
One Person	4	6	6
Lone Parent	1	3	3
Partners	14	14	14
Partners + Children	14	13	10
N	2925	2704	2248

Source: Jones and Martin (1997).

10.7.2 Sampling distribution investigation

The investigation of the sampling distribution will use the student survey which was used in Section 10.2 to investigate the sampling distribution of the sample mean. A total of 317 students responded in the survey, of whom 115 had a part-time job. We will investigate the sampling distribution of the proportion of students who had part-time jobs. Assuming that the population is the 317 students, then the population proportion is $p = 0.36 = 115/317$. We will select samples of size 10, 30 and 50 from the population, and will look at the sampling distribution of the sample proportions from each of these samples.

The sampling distributions are presented in Figure 10.8, and you can see that as the sample size increases the sampling distribution of the sample proportion tends to a symmetric distribution. The central value of the sampling distribution is the same as the population proportion, and the variation in the sampling distribution decreases as the sample size increases. The mean of the sampling distribution is close to the population proportion of 0.36, irrespective of the sample size, as can be seen from Table 10.10. The standard deviation and range of the sampling distribution decrease as the sample size increases. These are the same type of observations as were made about the sampling distribution of the sample mean.

It is important to remember that the sample proportion must be a discrete random variable. It must lie between 0 and 1, but it can only take a finite number of possible values. This is most easily seen from the first graph in Figure 10.8, which is based upon samples of size 10. In a sample of 10 students the number with a part time job can only be 0, 1, 2, . . . , 10 and so the proportion can only be 0.0, 0.1, 0.2, . . ., 1.0 and so is discrete. As the sample size increases the discrete nature of the sampling distribution of the sample proportion becomes much less noticeable. In the samples of size 50 there are 51 possible values for the sample proportion. The curves which are superimposed on the histograms in Figure 10.8 are from a normal distribution, and the use of this distribution

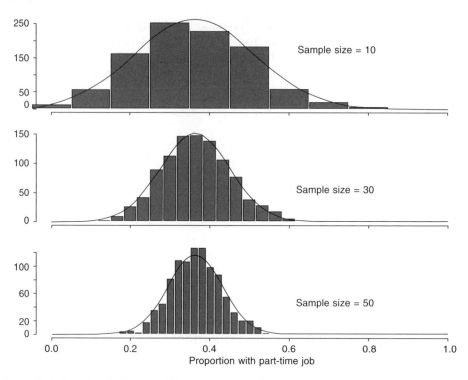

Figure 10.8 Sampling distributions of the sample proportion.

Table 10.10 Summary statistics for the sampling distribution of the proportion of students with a part-time job

Sample size	Minimum	First quartile	Median	Mean	Third quartile	Maximum	Standard deviation
10	0.00	0.30	0.40	0.3590	0.50	0.90	0.1512
30	0.13	0.30	0.37	0.3682	0.43	0.70	0.0886
50	0.16	0.32	0.36	0.3646	0.40	0.66	0.0654

will not be valid in small samples where the discrete nature of the sampling distribution will be more noticeable.

According to Table 10.9, 3% of young people aged 16–25 lived in lone-parent households. The histograms in Figure 10.9 represent the sampling distributions of the sample proportion based upon samples of size 10, 30 and 50, while the summary statistics are presented in Table 10.11. The summary statistics exhibit the same type of behaviour as those in Table 10.10, in that the mean of the sampling distributions is the same as the original population proportion of 0.03. The standard deviation of the sampling distribution decreases as the sample size increases. However, the histograms are not at all symmetric, even in the largest sample size, and the normal curve is not a good description of the sampling distribution.

This small investigation has shown that, although the sampling distribution of the sample proportion is discrete, in large samples we can effectively ignore the discrete

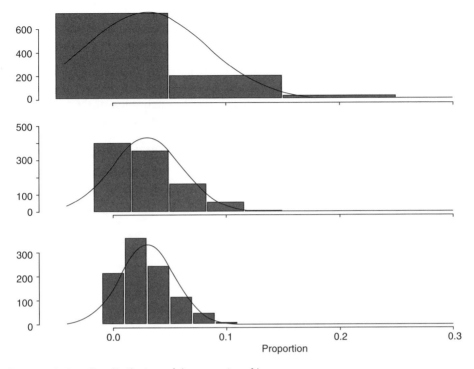

Figure 10.9 Sampling distributions of the proportion of lone parents.

Table 10.11 Summary statistics for the sampling distributions of the proportion of lone parents

Sample size	Minimum	First quartile	Median	Mean	Third quartile	Maximum	Standard deviation
10	0	0.00	0.00	0.0303	0.10	0.20	0.0545
30	0	0.00	0.03	0.0307	0.03	0.17	0.0320
50	0	0.02	0.02	0.0293	0.04	0.12	0.0235

nature and use a continuous normal distribution to calculate the probabilities. The use of the normal distribution depends not only on the sample size but also on the population proportion. If the population proportion is close to 0 or 1 then the sampling distribution of the sample proportion will be skew, as in Figure 10.9, and this means that the normal distribution, which is symmetric, will not be valid. Irrespective of the sample size, the mean of the sample proportion is the same as the true population proportion and so the sample proportion is an unbiased estimate of the population proportion. As the sample size increases the standard deviation of the sampling distribution of the sample proportion decreases. This means that the sampling distribution becomes more and more tightly distributed around the true population proportion as the sample size increases.

10.7.3 Sampling distribution result

The sampling distribution of the sample proportion is used in situations where a random sample of size n is selected from a population where the true proportion is p. The random variable representing the sample proportion in repeated random samples is denoted $\hat{P}$, and a specific value in one sample is denoted $\hat{p}$, The application of the central limit theorem to the sampling distribution of the sample proportion yields the following results:

- The expected value of the sample proportion is the same as the population proportion:

$$E[\hat{P}] = \mu_{\hat{p}} = p.$$

- The variance of the sampling distribution of the sample proportion is given by

$$\sigma_{\hat{p}}^2 = \frac{p(1-p)}{n}.$$

Thus the standard error of the sample proportion is

$$\sigma_{\hat{p}} = \sqrt{\frac{p(1-p)}{n}}.$$

- In large samples the sampling distribution of the sample proportion follows a normal distribution. The rule used to decide if the sample size is large enough is based upon a combination of the sample size and the population proportion. The normal approximation is valid provided that

$$np > 5 \quad \text{and} \quad n(1-p) > 5.$$

In Figure 10.8, where the normal approximation looked to be reasonable, $p = 0.36$. The sample would be large enough to use the normal approximation whenever $n \times 0.36 > 5$ and $n \times 0.64 > 5$. This is certainly true for samples of size 20 or more. By comparison, in Figure 10.9 where the population proportion was 0.03, none of the illustrated sample sizes was large enough for the normal approximation as $50 \times 0.03 = 1.5$ which is below the cut-off value of 5. In fact you would need a sample of at least 167 to use the normal approximation when the population proportion is as small as 0.03.

Example 10.3: One-tenth of workers test positive for drugs

The *Guardian* of 10 November 1998 contained an article based upon the results of data supplied by the Forensic Science Service.

Drug abuse is so prevalent among British workers that one in 10, including over-60s and senior executives, are testing positive for illicit substances. The Forensic Science Service (FSS), which carries out 1 million drug tests on employees a year, says that in some companies up to 15 per cent of the workforce is taking drugs. . . . Drugs detected in random tests on workers and applicants include cannabis, amphetamines, cocaine, ecstasy and heroin. . . . Drug testing has been pioneered in Britain by transport and construction companies, whose workers are considered to run a high risk of suffering accidents. Catering companies and the financial industries have also followed suit.

If we assume that the percentage of workers testing positive for drugs is 10% and a random sample of 500 workers in a factory are to be tested for drugs, what is the probability that (a) more than 15% of the sample will test positive, (b) less than 7% will test positive, and (c) between 5% and 9% will test positive? (d) Between what two limits, $p \pm k$, does the sample proportion have an 80% chance of lying?

In terms of our notation, $n = 500$ and $p = 0.10$ so $np = 50$ and $n(1 - p) = 450$. This means that the normal approximation should be valid. The standard error of the sample proportion is

$$\sigma_{\hat{p}} = \sqrt{\frac{p(1 - p)}{n}} = \sqrt{\frac{0.1 \times 0.9}{500}} = 0.0134.$$

(a) What is the probability that more than 15% of the sample will test positive?

We calculate

$$P(\hat{P} > 0.15) = P\left(Z > \frac{0.15 - 0.10}{0.0134}\right) = P(Z > 3.73) = 0.0001.$$

If the true proportion testing positive is 0.10 then the probability of obtaining a sample in which the sample proportion is in excess of 0.15 is very small. It is only to be expected once in every 10 000 samples.

(b) What is the probability that less than 7% will test positive?

$$P(\hat{P} < 0.07) = P\left(Z < \frac{0.07 - 0.10}{0.0134}\right) = P(Z < -2.24) = 0.0125.$$

Again, this is an unlikely event, expected once in every 80 samples.

(c) What is the probability that between 5% and 9% will test positive?

$$P(0.05 < \hat{P} < 0.09) = \left(\frac{0.05 - 0.10}{0.0134} < Z < \frac{0.09 - 0.10}{0.0134}\right)$$

$$= P(-3.73 < Z < -0.75) = 0.2265.$$

We would expect this event in just under a quarter of all samples.

(d) Between what two limits, $p \pm k$, does the sample proportion have a probability of 0.8 of lying?

$$P(p - k < \hat{P} < p + k) = P\left(\frac{-k}{0.0134} < Z < \frac{k}{0.0134}\right) = 0.80.$$

From tables of the normal distribution (Appendix A) we find that

$$P(-1.28 < \hat{P} < 1.28) = 0.80.$$

Equating the corresponding terms gives

$$\frac{k}{0.0134} = 1.28$$

$$k = 1.28 \times 0.0134 = 0.0172,$$

which means that the two limits are 0.0828 and 0.1172 in terms of proportions, or 8.28% and 11.72% in terms of percentages.

In a random sample of 500 independent tests there is an 80% chance that the sample percentage of positive tests will lie between 8.3 and 11.7 per cent. This is quite a narrow band and illustrates that with a large sample you should not expect a great deal of sampling variability.

10.8 Precision and accuracy

Precision and accuracy are two terms which are often confused with each other. In terms of the estimation of population quantities, precision is concerned solely with the sampling variability of the estimate, while accuracy is concerned both with the precision and the bias. A precise estimate may have a large bias, but an accurate estimate should have a low variability and a low bias.

We will briefly look at the distinction between the two terms through some diagrams of sampling distributions. The population in Figure 10.10 is that of weekly wages for part-time work. The vertical line is at the population mean, and it is this quantity that is to be estimated. The lines represent the sampling distributions of three estimates, A, B and C. All of the sampling distributions are symmetric. The sampling distributions of two of these estimates, A and B, are centred on the population mean and so are unbiased. The sampling distribution of C is located at a higher value than the population mean and so is biased.

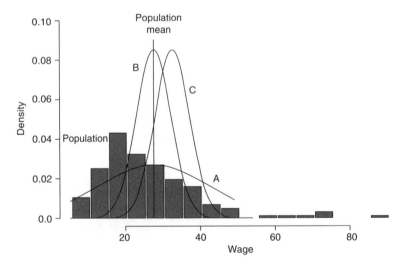

Figure 10.10 Precision and accuracy.

Comparing the distributions of A and B we can see that the distribution of A is spread over a wider range than that of B. Thus A is a more variable estimate of the population mean than B. This means that B has a higher precision than A, or B is a more precise estimate than A. The shapes of the sampling distributions of B and C are identical. In particular, they both have the same variation about the means of the sampling distributions. Thus they both have the same precision. However, B is more accurate as it has no bias whereas C does have some bias. Thus of the three estimates, B is the best one as it has the higher precision and is unbiased.

The precision of an estimate is the inverse of the variance of the estimate, and so the precision of the sample mean is

$$\text{Precision}(\overline{X}) = \frac{1}{\text{Var}[\overline{X}]} = \frac{n}{\sigma^2}.$$

The accuracy of an estimate is measured by the mean squared error which is a combination of variance and bias:

$$\text{MSE} = (\text{Bias})^2 + \text{Variance}.$$

An estimate is accurate if it has a low bias and a low sampling variance. In Figure 10.10 the bias of estimate C is £5, the population standard deviation is £14.84. Estimate A is a mean of three observations, which means that its standard error is $\sigma_A = 14.84/\sqrt{3} = 8.57$, and estimate C is a mean of 10 observations and so has a standard error of $\sigma_C = 14.84/\sqrt{10} = 4.69$. The mean square errors of the two estimates are

$$\text{MSE(A)} = 0^2 + 8.57^2 = 73.41,$$

$$\text{MSE(C)} = 5^2 + 4.69^2 = 47.02.$$

Thus, compared to A, estimate C is more accurate as it has a lower mean square error.

The important point from this section is to note that precision and accuracy are not the same thing. Precision is concerned with the ability of the sampling distribution to be close to its expected value and so is only concerned with the variance, or equivalently the standard error, of the sampling distribution. A precise estimate has a sampling distribution which has a low variability. Accuracy is concerned with how close the estimate is to the true value, and so has a component associated with bias. Generally, unbiased estimates which are precise are to be preferred. However, it is possible that a slightly biased estimate may be preferable to an unbiased estimate with a low precision. In practical terms it may be preferable to use a slightly biased estimate from a large sample, where the standard error will be small, instead of an unbiased estimate from a smaller sample, where the standard error may be larger.

10.9 Sample size calculations for standard errors

For both the sample mean and the sample proportion the standard error of the sampling distribution decreases as the sample size increases. This will generally be the case for all sample estimates. The larger the sample, the more information there is and the greater the precision of the estimate.

In Section 10.7.3 we calculated that when the population percentage was 10% there was a probability of 0.80 that the sample percentage would lie between 8.3% and 11.7%, based upon a sample size of 500. The 0.80 probability limits are shown below for a number of different sample sizes.

Sample size	Lower limit (%)	Upper limit (%)
100	6.16	13.84
300	7.78	12.22
500	8.28	11.72
700	8.55	11.45
900	8.72	11.28

As the sample size increases the width of the 0.80 probability interval decreases and so the precision of the estimate increases. With information such as that in the above table we can decide how large a sample we need in order to achieve a specified precision. If we wanted the 0.80 probability limits to go from 8.7% to 11.3% then we would need a sample of 900 tests. If we only wanted the limits to go from 7.8% to 12.2% then a sample of only 300 would be sufficient. This brings us on to the use of statistical arguments to help decide how large a sample to select.

The only quantity which is changing in the probability calculations is the sample size, and this only comes into the calculations through the standard error, $\sigma_{\hat{p}} = \sqrt{p(1-p/n}$ in the case of the sample proportion here, and $\sigma_{\bar{x}} = \sigma/\sqrt{n}$, for the sample mean. The precision of the sample estimate is related to its standard error. By specifying the value of the standard error we wish to achieve we can then calculate the sample size required.

Example 10.4: City of London housing costs

The *Evening Standard* of 26 November 1997 published a statistical report of the local authority performance indicators for 1997 for the Corporation of London. The average weekly management costs per dwelling were £34.61, and the average weekly repair costs were £26.85. In planning a survey of housing costs, how large a sample of houses should be selected so that the standard error of the mean is less than £0.50, when the standard deviation of the weekly management costs per house is £6.34 and the standard deviation of the average weekly repair costs per house is £4.76?

Weekly management costs. The standard deviation is £6.34 and we want to find the sample size, n, such that $\sigma_{\bar{x}} = 0.50$. This means choosing n so that $\sigma/\sqrt{n} = 0.50$, that is,

$$\frac{6.34}{\sqrt{n}} = 0.50$$

$$\sqrt{n} = \frac{6.34}{0.50} = 12.68$$

$$\therefore \qquad n = 161.$$

Thus a sample size of 161 is sufficiently high to ensure that the standard error of the mean management cost per week is less than £0.50.

Weekly repair costs. The standard deviation is £4.76 and we want to find the sample size, n, such that $\sigma_{\bar{x}} = 0.50$. This again means choosing n so that $\sigma/\sqrt{n} = 0.50$, that is

$$\frac{4.76}{\sqrt{n}} = 0.50$$

∴ $n = 91$.

In this case a sample of 91 is sufficiently high to ensure that the standard error of the mean repair cost per week is less than £0.50.

Putting the two calculations together allows you to conclude that a sample size of 161 would be sufficient for both purposes.

Example 10.5: City of London Corporation rent arrears

In 1997 the City of London reported that 3.80% of their council house tenants were significantly behind with their rent. This meant that this percentage of tenants were owing over 13 weeks' rent but excluded those who owed a small amount of money, less than £250. If a survey of rent arrears were to be carried out, how large a sample of tenants would you need to include in the study to ensure that the standard error of the sample percentage was less than 0.5%?

You will find that in common speech proportions and percentages are both used. It is probably best if you do all the calculations in terms of proportions, and this is what we shall do here. We need to find the sample size, n, such that $\sigma_{\hat{p}} = \sqrt{p(1-p)/n} = 0.005$, when the population proportion is equal to 0.038:

$$\sqrt{\frac{0.038(1-0.038)}{n}} = 0.005$$

$$\sqrt{\frac{0.0366}{n}} = 0.005$$

$$\frac{0.0366}{n} = 0.005^2$$

$$n = \frac{0.0366}{0.005^2} = 1463.$$

A sample size of at least 1463 tenants will be required in order for the standard error of the percentage of tenants who are significantly in arrears to be less than 0.5%.

There are two points that you should note from these two examples. The first is that in the estimation of the sample size for a mean you do not need to know the population mean, only the population standard deviation. For the proportion it is necessary to know in advance the population proportion. Secondly, the sample size for the proportion is much larger than for the mean. Although the calculation of the sample size depends upon the population standard deviation and the specified standard errors, you generally find that for acceptable standard errors you require samples of the order of thousands when you are interested in a proportion, but only hundreds when you are dealing with the sample mean.

11

Estimation and confidence intervals

- Knowing about the relationship between confidence intervals and precision of an estimate
- Knowing how to calculate the confidence interval for the mean, the difference of two means, the proportion and difference of two proportions
- Knowing how to interpret a confidence interval – what it is and what is not
- Knowing about the *t* distribution and why it has to be used
- Knowing the difference between having two independent samples and two dependent, or paired, samples and the effect that this has on the choice of the appropriate confidence interval
- Understanding the use of confidence intervals with regard to opinion polls
- Being familiar with the use of specifying the width of the confidence interval (i.e. specifying precision) as a means of deciding how large a sample needs to be selected

11.1 Introduction

In October and November 1998 a series of investigations took place into the private lives of UK Cabinet members. This involved the *Sun* newspaper in particular, and was generally concerned with the sexual proclivities of the politicians. A public opinion poll was carried out by ICM who interviewed a random sample of 1222 adults aged 18 or over on 6 and 7 November by telephone. One of the estimates quoted in an article about this survey was that 49% of the public think that politicians' private lives should not enjoy legal protection. In a similar survey one year previously 53% of the public thought that politicians' private lives should not enjoy legal protection (*Guardian*, 10 November 1998).

In the previous chapter on sampling distributions we saw that one of the effects of using samples was that a sampling distribution of the quantity of interest was induced. This distribution describes the variation in the sample estimates of the population parameter. Consequently, you will be aware that the true percentage of the public who think that politicians' private lives should not enjoy legal protection is unknown and that the figure of 49% is an estimate based upon a large random sample of 1222 individuals. If we knew the true value of the percentage in the population then we would be able to evaluate probabilities such as the probability that the sample percentage was (a) less than 49%, (b) more than 52%, or (c) between 49% and 52%. With percentages such as these we would be able to take into account the effects of sampling variability and hence inform our judgement about whether or not there has been a change in public opinion.

We know that the sample size was 1222, and that if the true proportion in the population were 0.53 then the standard error of the proportion would be

$$\sigma_{\hat{P}} = \sqrt{\frac{0.53 \times 0.47}{1222}} = 0.0143.$$

We can also say that the standard error of the percentage would be 1.4%. Based upon the simulations and probability calculations we made in Chapter 10, we would judge that if the true percentage were 53%, leading to a standard error of 1.4%, then a percentage of 49% is a reasonable figure to expect on the basis of sampling variability. This is a judgement only and is not based upon any probability calculations; as we know, our judgement of probabilities can be wrong.

The material to be covered in this chapter relates to the interpretation of samples, surveys and experiments. As in the example above, we have a sample statistic which is an estimate of a population quantity. We know that sampling variability means that there is a sampling distribution of the estimate and we need to use this information when reporting the results of the study. The solution to reporting an estimate when you know that there is a distribution of values for the estimate is to report not only the single point estimate but also a range of values or an interval estimate. This is known as a **confidence interval**.

Confidence intervals are one of the most difficult quantities to interpret in elementary statistics. They should be used when reporting the results of any experiment or survey as they give some indication of the variation in the experiment in relation to the size of the estimated effect. Only this way will you be able to fully interpret your results and, more importantly, let others evaluate your survey results relative to theirs. A confidence interval is a range of values, and this range of values tells you something about the sampling variation in the results and also about the population quantity you are trying to estimate. If the confidence interval is wide then you know that you do not have very precise information about the population quantity, but if it is narrow then you know that you have precise information.

This is illustrated in Figure 11.1 with respect to the estimation, using a large sample, of the population proportion who think that the private lives of politicians should not enjoy legal protection. We envisage that the true value of the population proportion is located at the vertical line. This has a unique single value, which is unknown. If a complete census were taken then it would be possible to find out this value. This seldom happens, and sample information is used to provide an estimate of the population value.

The curves represent the sampling distribution of the sample proportion from two separate large samples. Sample 1 is larger than sample 2, and so the sampling distribution is more tightly clustered around the true but unknown population proportion. The confidence interval from sample 1 is narrower than the confidence interval from sample 2. As it covers a narrower range of values for the population proportion, sample 1 will provide a more precise estimate of the unknown value of the population proportion.

From this illustration we can deduce that the formation of confidence intervals will be very closely related to the results about the sampling distributions. As in Chapter 10, we will only focus on the confidence intervals for the population mean and for the population proportion, though it is possible to calculate confidence intervals for any population quantity such as the population standard deviation or the population median or other

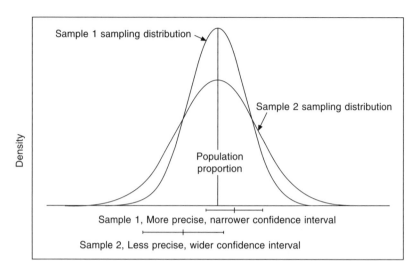

Figure 11.1 Confidence intervals and precision.

quantiles. While the formulae that you need to use will be different, the concepts and general techniques are similar.

The key ideas on which the formation of confidence intervals is based are that there exists a true value of the population parameter and that this value is unknown. The sampling distribution results about the sampling distribution of the sample mean or sample proportion tell us the likely values of the estimated quantities in any sample. Confidence intervals are based upon an inversion of these results to provide a range of values for the population parameter based upon the sampling distribution results.

In Figure 11.2 we envisage a population and from this population we randomly select a sample of a specified size. The observed sample is depicted in the diagram, as are some

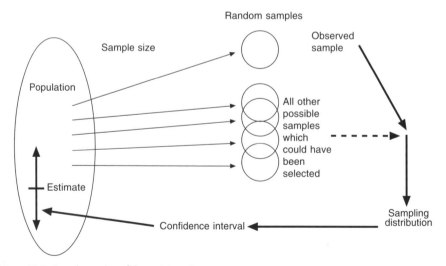

Figure 11.2 Samples and confidence intervals.

of the other random samples which could have been selected. The sample estimate based upon these samples gives rise to the sampling distribution of the sample estimate. The confidence interval is calculated from the data in the observed sample, using the sampling distribution results to provide a range of values back in the population.

11.2 Confidence interval for the mean of a population

The population mean is the most important parameter for many quantitative variables. The discussion of the properties of confidence intervals and their calculation will be couched in terms of a situation in which the standard deviation in the population is known but the mean in the population is not known. This is obviously not a situation which is very common in practice, as you need to know the population mean in order to calculate the population standard deviation. The reason why we start off with this situation is that it only involves the normal distribution. When we consider the more realistic setting of having no knowledge of the standard deviation in the population when trying to estimate the population mean, we need to use another probability distribution known as the t distribution.

11.2.1 Population standard deviation known

Table 4.16 of *Social Trends 28* Office for National Statistics 1998a provides information on the number of hours worked per week by all men and women in employment in the United Kingdom for 1992 and 1997. This information is available from the Labour Force Surveys. The mean weekly number of hours worked by men in 1992 was 42.16 hours, with a standard deviation of 11.24 hours. The corresponding figures for women in 1992 were a mean of 29.43 hours and a standard deviation of 13.53 hours. The figures for 1997 for women were a mean of 29.60 hours and a standard deviation of 13.44 hours. These are relatively unchanged from the 1992 figures. For men the mean in 1997 was 41.79 hours, but the standard deviation had increased to 12.40 hours. As an aside, this increase in the standard deviation comes about because more men were undertaking part-time work with low hours worked per week in 1997 compared to 1992, while others were working longer hours. Both of these factors led to an increase in the variability in the number of hours worked per week.

Suppose we have a sample of 150 female workers from a broad spectrum of industries, collected in 1998, where the mean number of hours worked was 31.07. How do we go about forming a confidence interval for the mean number of hours worked by women in employment in the United Kingdom in 1998? We start off by setting up some notation to help with the algebra. As before, we denote the unknown population mean by μ and use σ for the population standard deviation. The sample size is n and the observed value of the sample mean is $\bar{x}$. The result that we have established for the sampling distribution of the sample mean states that, in large samples, the probability distribution of the sample mean from random samples is a normal distribution with a mean of μ and a standard error of

$$\sigma_{\overline{X}} = \frac{6}{\sqrt{n}}.$$

Using this information, we can establish from probability calculations that

$$P(\mu - 1.96\sigma_{\overline{X}} < \overline{X} < \mu + 1.96\sigma_{\overline{X}}) = 0.95.$$

This probability statement says that there is a 95% chance that the sample mean will lie within an interval of

$$(\mu - 1.96\sigma_{\overline{X}}, \quad \mu + 1.96\sigma_{\overline{X}}).$$

When we are trying to estimate the population mean, μ is unknown, and if we replace it by its sample estimate of $\overline{x}$ then we get the 95% confidence interval for the population mean:

$$(\overline{x} - 1.96\sigma_{\overline{X}}, \quad \overline{x} + 1.96\sigma_{\overline{X}}).$$

In the example the sample mean was 31.07 hours, the sample size was 150 and we use the 1997 population standard deviation as the 1998 value. This gives $\sigma = 13.44$. Carrying out the calculations gives the 95% confidence interval:

$$\sigma_{\overline{X}} = \frac{\sigma}{\sqrt{n}} = \frac{13.44}{150} = 1.90,$$

$$\overline{x} - 1.96\sigma_{\overline{X}} = 31.07 - 1.96 \times 1.90 = 31.07 - 3.72 = 27.35,$$

$$\overline{x} + 1.96\sigma_{\overline{X}} = 31.07 + 1.96 \times 1.90 = 31.07 + 3.72 = 34.79.$$

The 95% confidence interval for the mean weekly number of hours worked in the population of all women in employment in the United Kingdom in 1998 is from 27.4 hours to 34.8 hours, with a mean of 31.1 hours. This confidence interval provides information on the precision of the estimate as well as the location of the value.

11.2.2 Interpretation of a confidence interval

The 95% confidence interval for the population mean gives a range of values. The sample mean is at the centre of this range. The two limits are derived from the probability that the sample mean lies between two limits equidistant from the population mean. We will now investigate how to interpret the confidence interval correctly. This investigation will be based upon a simulation study where we will randomly select samples from a population, calculate the confidence intervals and look at them. This is very similar to the studies we did to establish the sampling distribution of the sample mean.

The investigation is based upon the weekly number of hours worked by men in 1997. The observed distribution is shown in Figure 11.3, where you can see that the distribution is skew, with almost 70% of men working 30–50 hours per week. The mean is 41.8 hours, with a standard deviation of 12.4 hours. The data are not normally distributed, but we will be using samples of size 150 from this distribution in Figure 11.3, so a normal distribution is valid for the sampling distribution of the sample mean as the sample is large.

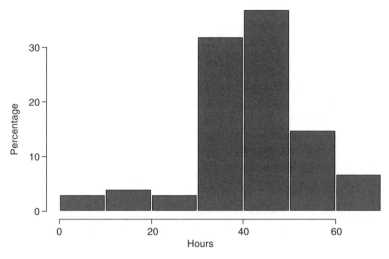

Figure 11.3 Distribution of the number of hours worked per week by men in the United Kingdom, 1997.

In the investigation we will select 100 samples at random, each with 150 observations. For each sample the sample mean is calculated and the 95% confidence interval calculated using the limits, where $\bar{x} \pm 1.96\sigma_{\bar{x}}$, where $\sigma_{\bar{x}} = \sigma/\sqrt{n}$. The different samples are plotted along the x axis in Figure 11.4 and the number of hours worked per week along the y axis. The horizontal dotted line represents the population mean of 41.8 hours. For each sample the vertical line gives the range of the 95% confidence interval while the horizontal tick in the middle of this vertical line gives the sample mean.

One thing to note about this graph is that the lines are all the same length. This occurs because we are always using the known value of the population standard deviation to give

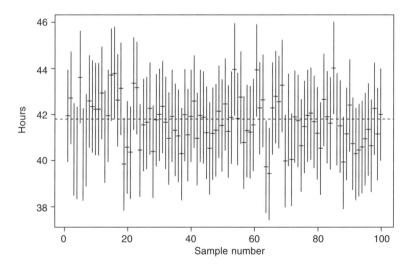

Figure 11.4 Confidence intervals from random samples.

us the width of the confidence interval. The upper 95% confidence limit is $\bar{x} + 1.96\sigma_{\bar{x}}$ and the lower limit is $\bar{x} - 1.96\sigma_{\bar{x}}$. This means that the width of the 95% confidence interval is $2 \times 1.96\sigma_{\bar{x}} = 2 \times 1.96 \times \sigma/\sqrt{n}$, which is the same for all samples of size n.

A second thing to notice is that the confidence intervals vary from one sample to another, as the location of the confidence interval is determined by the sample mean. The sample means from the 100 samples vary over the number of hours worked. As a consequence the confidence intervals vary over the different samples. By contrast, the population mean is a fixed, but in most cases an unknown quantity, which cannot vary from one sample to another as it is a population quantity. Thus a confidence interval is a random quantity affected by the sampling distribution.

If you scan along the confidence intervals for the different samples you will see that in most cases the confidence interval covers the population mean but in a few instances it misses it completely. This is the case in samples 54, 61, 64, 65 and 85. Sample 16 has a confidence interval which just covers the true value. In this simulation investigation there were 100 random samples and the 95% confidence interval contained the true population mean on 95 occasions and did not on the remaining 5 occasions. If we were to repeat the whole investigation again we would not obtain exactly the same results as different samples would be selected and the influences of sampling variability would be observed. However, similar results would be observed and out of every 100 samples we would expect to find that the 95% confidence interval contains the true population mean 95% of the time and does not contain it 5% of the time.

The investigation of the 95% confidence interval for the sample mean when the population standard deviation is known has shown the following:

- The confidence interval is random, and so you will get a different confidence interval should you select a different sample. It also shows you that the interval is a random quantity affected by sampling variability in the same way as the sample mean is affected by sampling variability. The population mean, on the other hand, is not a random quantity as its value is fixed.
- The 'confidence' in the 95% confidence interval is related to our confidence that the interval contains the true population value somewhere between the lower and upper limits of the interval.
- On average, you would expect 95% of the confidence intervals to contain the true value of the population mean. In a given situation when you have one sample and one confidence interval, either the interval contains the true mean or it does not. You do not know which of these two alternatives is true. All you know is that there is a 95% chance that the confidence interval contains the true mean and a 5% chance that it does not.
- If the population standard deviation is known, then for a fixed sample size the 95% confidence intervals are all of the same width. If the sample size is increased then the width of the confidence interval will decrease. This occurs because the width of the 95% confidence interval is $2 \times 1.96 \times \sigma/\sqrt{n}$ which decreases as the sample size, n, increases.

11.2.3 Confidence intervals other than the 95% confidence interval

So far only the 95% confidence interval has been used. This is certainly the most common interval used in practice, but occasionally intervals of different confidences are required. They are constructed in exactly the same way as the 95% confidence interval.

The 95% confidence interval for the sample mean when the population standard deviation is known is

$$(\bar{x} - 1.96\sigma_{\bar{x}}, \quad \bar{x} + 1.96\sigma_{\bar{x}}).$$

The constant value used in the calculation is 1.96, which was taken from the tables of the normal distribution. The normal distribution was used because the distribution of the sample mean is normal in large samples. The values ± 1.96 are the percentage points of the normal distribution which have a probability of 0.95 between these two limits, a probability of 0.025 above 1.96 and a probability of 0.025 below -1.96.

The 2.5% point of the normal distribution is denoted $z_{0.025}$ and has a value equal to 1.96. In the terminology the subscript 0.025 denotes the probability above the value. In general, the $100\alpha\%$ point is denoted z_α and this has a probability of α above it. Using this terminology, the 95% confidence limits

$$\bar{x} \pm 1.96\sigma_{\bar{x}}$$

can be written as

$$\bar{x} \pm z_{0.025}\sigma_{\bar{x}}.$$

This means that a general $100(1 - \alpha)\%$ confidence interval can be written as

$$\bar{x} \pm z_{\alpha/2}\sigma_{\bar{x}}.$$

The other common confidences are 90%, where $\alpha = 0.1$ and the percentage point is $z_{\alpha/2} = z_{0.05} = 1.645$, and 99%, where $\alpha = 0.01$ and the percentage point is $z_{\alpha/2} = z_{0.005} = 2.58$. Occasionally you may see 50% confidence intervals, corresponding to $z_{\alpha/2} = z_{0.25} = 0.67$, and 68% confidence intervals, where $z_{\alpha/2} = z_{0.16} = 1$.

As the confidence increases, say from 90% to 95% to 99%, the percentage point also increases, from 1.645 to 1.96 to 2.58. This in turn means that the width of the confidence interval increases. For a fixed confidence the width of the confidence interval tells you something about the precision of the estimate. For a fixed confidence, the larger the sample size the narrower the confidence interval. However, changing the width of the confidence interval by changing the confidence is not related to the precision of the estimate. For a given sample, a 90% confidence interval will always be narrower than a 95% one which in turn will always be narrower than a 99% interval.

11.2.4 Population standard deviation unknown

The above sections have dealt with the case where the population standard deviation is known and yet the mean of the population is unknown. This is often an unrealistic situation as you need to know the population mean in order to evaluate the population standard deviation. The more realistic setting is that both the population mean and standard deviation are unknown and this requires a slight modification to the equations used to calculate the confidence intervals. This modification involves the replacement of the population standard deviation with the sample standard deviation and the replacement of the percentage point of the normal distribution with the percentage point of the t distribution.

Replacing the population standard deviation with the sample standard deviation gives the estimated standard error of the sample mean as $\hat{\sigma}_{\bar{x}} = s/\sqrt{n}$. The formula for the $100(1 - \alpha)\%$ confidence interval for the population mean when the population standard deviation is unknown is

$$\bar{x} \pm t_{\alpha/2}\, \frac{s}{\sqrt{n}},$$

where $t_{\alpha/2}$ is the percentage point of the t distribution on $n - 1$ degrees of freedom.

The replacement of the population standard deviation with the sample standard deviation is the standard estimation procedure in statistics. If a population value is unknown replace it with the best estimate you can make from the sample. The use of the t distribution is required because we have to take into account not only the sampling variability in the sampling distribution of the sample mean but also the sampling variability in the sampling distribution of the sample standard deviation.

11.2.5 The *t* distribution

This is a probability model for a continuous variable, and the shape of the probability density is similar to the shape of the standard normal probability density function. Like the standard normal distribution, the t distribution is symmetric about zero and has a mean at zero. The t distribution has one parameter, known as the **degrees of freedom**.

The probability densities of the t distributions on 1, 2, 5 and 20 degrees of freedom are shown in Figure 11.5. These distributions are all symmetric about zero like the standard normal distribution. You can see that as the degrees of freedom get larger the t distribution

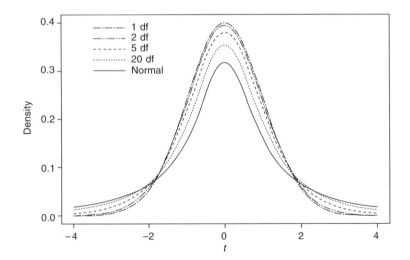

Figure 11.5 Probability densities of the *t* distribution.

becomes more like the normal distribution. Indeed there is very little visual difference between the standard normal density and the density of the t distribution on 20 degrees of freedom.

The major differences between the t distribution and the normal distribution take place at small degrees of freedom, which correspond to small sample sizes. In this case the t distribution has more probability in the tails than the normal distribution, and this is reflected in the percentage points which are shown in Appendix B. The upper percentage points for the t distribution are always greater than the corresponding upper percentage points for the standard normal distribution.

In the table in Appendix B, the rows correspond to the degrees of freedom and the columns correspond to the probability. To find the value for $t_{0.025}$ on 10 degrees of freedom you locate the row corresponding to 10 degrees of freedom:

Degrees of freedom	Probability in upper tail					
	0.2	0.1	0.05	0.025	0.01	0.005
10	0.8791	1.3722	1.8125	2.2281	2.7638	3.1693

The value corresponding to a probability of 0.025 is 2.2281. Thus $t_{0.025} = 2.2881$ on 10 degrees of freedom. Percentage points for the t distribution are only given as far as 30 degrees of freedom. Above this value you can use the percentage points for the normal distribution as an approximation.

11.2.6 Population standard deviation unknown (t intervals)

As the t distribution has more probability in the tails than the standard normal distribution, a 95% confidence interval for the mean based upon the t distribution will be larger than the corresponding interval based upon the normal distribution even if the sample standard deviation and population standard deviation took exactly the same value. The t distribution takes into account the sampling variability in the sample standard deviation as well as the sampling variability in the sample mean. The percentage points for the t distribution are further away from zero than the corresponding percentage points in the standard normal distribution and so the confidence intervals are wider.

Example 11.1: Telephone waiting times

Many large organizations now use telephone answering services to help manage their business. Examples of these include British Airways telephone booking service, the Automobile Association insurance service and the Alliance and Leicester Building Society Mortgage Assistance Line. A common problem is the delay in connection to an appropriate assistant. If the delay is long, then there is a greater chance that the customer will give up and resort to another service. This problem will only exist when all the connection lines are busy and a new caller arrives. The data in the table are based upon a study of answering delays at a telephone help line. The times are all measured in seconds and give the time from connection to answering, but only for calls which were connected at a time all assistants were busy. The data are available for two time periods, 10–11 a.m. and 2–3 p.m.

10–11	43 73 57 19 60 34 33 55 22 30 25 33 40 51 23 30 15 43 47 31 10 32 33 29 66
2–3	46 55 42 48 26 16 39 4 25 24 69 61 44 1 36 68 11 56 62 46 63 23 27 30 52

The summary statistics for the two distributions are as follows:

	Number	Mean	Standard deviation
10–11	25	37.36	16.0023
2–3	25	39.96	19.7051

As the population standard deviation is unknown the t distribution needs to be used in the calculation of the confidence intervals. Part of the process involves making sure that the assumptions on which the process is based are correct. Normal probability plots are shown in Figure 11.6, and as the points are plotted in a fairly straight line the assumption of a normal distribution appears to be a reasonable one. There are 25 observations in each group and the degrees of freedom of the t distribution are 24. For a 95% confidence interval the 2.5% point is needed. For the t distribution on 24 degrees of freedom the percentage point is $t_{0.025} = 2.0639$.

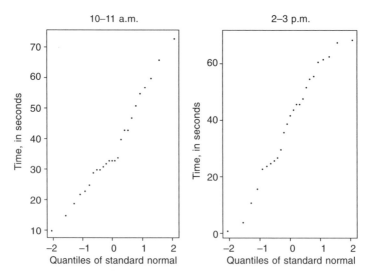

Figure 11.6 Normal QQ plots for the telephone waiting times.

For the 10–11 a.m. sample the 95% confidence interval is

$$37.36 \pm 2.0639 \frac{16.0023}{\sqrt{25}} = 37.36 \pm 2.0639 \times 3.2005 = 37.36 \pm 6.6055$$

$$= (30.75, 43.97) \text{ seconds.}$$

The 95% confidence interval for the 2–3 p.m. sample is (30.83, 47.09) seconds.

For the 10–11 a.m. period the best estimate of the mean waiting time for all calls is 37.4 seconds but the 95% confidence interval is large, from 30.7 seconds to 44.0 seconds, a width of 13.3 seconds. For the other period the mean is 40.0 seconds with a confidence interval from 30.8 seconds to 47.1 seconds. The mean is slightly higher than for the 10–11 a.m. period, but this is not really very important when you consider that the confidence intervals are 13.3 seconds and 16.3 seconds wide, respectively.

The width of the confidence interval provides information on the precision of the estimated value. As the sample size is 25 the precision is not very great. Although there is a 2.6 second difference between the means for the two periods we would be unwise to make any claims about a greater delay time in the afternoon than in the morning, bearing in mind the precision of the estimates.

11.2.7 Interpretation of *t* intervals

Example 11.1 should show you that the use of confidence intervals is more informative than the use of a mean, standard deviation and sample size. Confidence intervals give a range of values and so are much more useful than the single measures and their associated variabilities. If the confidence interval is narrow then the estimate of the mean is precise. If the interval is wide then the estimate is imprecise. 'Narrow' and 'wide' are soft descriptions, and there are no rules for deciding if an interval is narrow or wide. It largely depends on the application and the variable under study.

If the mean time to answer a call is 30 seconds and the observations go from 10 to 80 seconds with a standard deviation of 15 seconds then a confidence interval of 28 to 32 seconds would be considered quite narrow, and one of 20 to 40 seconds quite wide. On the other hand, if the mean response time was 90 seconds with observations from 30 to 240, then a confidence interval from 80 to 100 would be considered narrow.

As the percentage points of the *t* distribution tend towards the percentage points of the normal distribution the *t* and normal distributions can be used interchangeably in large samples. Generally samples in excess of 30 are considered to be large enough not to bother too much about the difference between the normal and *t* distribution. This really only refers to situations where you may need to calculate a confidence interval by hand, say from a published report where only summary data are presented. If you are analysing your own data using a statistical computer program, then you may as well always use an interval based upon the *t* distribution if the population standard deviation is unknown. It is as easy for the computer program to calculate the percentage points of the *t* distribution as it is to calculate the percentage points of the normal distribution.

The graph in Figure 11.7 shows the results of an investigation of *t* confidence intervals, or *t* intervals. This study is similar to that in Figure 11.4, except that the experiment is based upon the telephone waiting times with a population mean of 40 seconds. The samples were of size 25 with unknown population standard deviations, so the *t* distribution was used instead of the normal distribution. The investigation shows exactly the same points. In the graph 95 of the 95% confidence intervals contain the true mean, while five intervals miss it completely, samples 15, 19, 22, 45 and 66; the confidence intervals are random and are not the same in all samples. Unlike Figure 11.4, the confidence intervals here all have slightly different widths. This occurs because the sample standard deviations

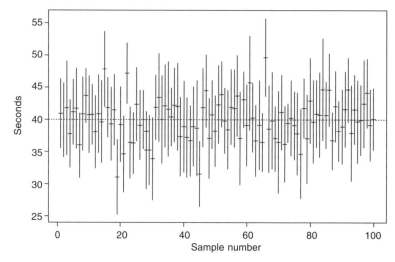

Figure 11.7 Simulation study of *t* intervals.

are not the same and the *t* intervals take this sampling variability into account as well as the sampling variability in the sample mean.

11.3 Confidence interval for a population proportion

The general shape of the formula for the confidence interval for a population mean is

$$\text{Estimate} \pm (\text{Distribution quantile}) \times (\text{Standard error}).$$

The confidence interval for a population proportion is of the same shape. Indeed, all confidence intervals discussed in this book will have the same shape as they will all be based upon distributions which are symmetric. The distribution quantile depends upon the confidence level and the sampling distribution of the estimate. In the simple cases discussed in this book (means and proportions) it will either come from the *t* distribution or the standard normal distribution. The standard error measures the precision of the estimate. Generally, this will be an estimated quantity but there are some rare cases in which the standard error is known from the population.

In terms of sample surveys for market research or other aspects of business, the sample proportion is the most important summary measure. In March 2001, MORI carried out a survey of 1918 adults and reported that 30% of households had access to digital television. In a market research study of 997 individuals for Tripod, a leading provider of web authoring tools, 20% of individuals with internet access stated that they wanted to design their own web pages. Iomega Corporation, a global leader in data management solutions, commissioned a MORI survey of about 1400 working personal computer or laptop users in Great Britain, France and Germany to examine the digital behaviour of Europeans at home and in the workplace. Nearly 75% of Britons, 59% of the French and over one-third of the Germans indicated that they would like to be able to store all their personal data on one portable device. All examples are taken from http://www.mori.com, June 2001.

In each of the surveys mentioned above, there is a random sample of n observations and a sample proportion which will be denoted $\hat{p}$. The 95% confidence interval for the proportion in the population from which the sample is drawn is

$$\hat{p} \pm z_{0.025}\, \hat{\sigma}_{\hat{p}},$$

where

$$\hat{\sigma}_{\hat{p}} = \sqrt{\frac{\hat{p}(1 - \hat{p})}{n}}$$

is the estimated standard error of the sample proportion $\hat{p}$. It is necessary to use the estimated standard error as p is unknown. This formula is only valid in large samples, under the same circumstances as when the sample proportion follows a normal distribution, i.e. $np > 5$ and $n(1 - p) > 5$. Even though the standard error is estimated here we do not use the t distribution. This is so because we are dealing with large samples. If we have small samples and wish to find a confidence interval for the population proportion then we need to calculate exact confidence intervals based upon the binomial distribution. This is technically quite difficult and the methods are not discussed here. The principles and the interpretations are exactly the same. Many computer programs will have the facility for calculating the exact binomial confidence intervals in small samples.

Example 11.2: Politicians' privacy

One of the estimates from the survey cited in the opening paragraph of this chapter was that 49% of the public think that politician's private lives should not enjoy legal protection. In a similar survey one year previously 53% of the public thought that politician's private lives should not enjoy legal protection.

The sample size is $n = 1222$ and the observed proportion in the sample is $\hat{p} = 0.49$. The estimated standard error of $\hat{p}$ is

$$\hat{\sigma}_{\hat{p}} = \sqrt{\frac{\hat{p}(1 - \hat{p})}{n}} = \sqrt{\frac{0.49 \times 0.51}{1222}} = \sqrt{0.000\,204\,5} = 0.0143.$$

This means that the 95% confidence interval is

$$\hat{p} \pm z_{0.025}\hat{\sigma}_{\hat{p}} = 0.49 \pm 1.96 \times 0.0143 = 0.49 \pm 0.0280 = (0.462, 0.518).$$

The 95% confidence interval for the population percentage is 46.2% to 51.8%. This is a relatively narrow interval, indicating a high precision over a range of 5.6%. Roughly speaking, the published percentage is accurate to ±3%, and this will be the case for most sample surveys which are based upon sample sizes in the range 1000–1500, which is the range of sample sizes for the majority of such studies.

A similar survey the previous year published a figure of 53% of the public who thought that politicians should not enjoy special legal protection over their private lives. The information on the sample size is not provided for this study. What we do know is that this percentage is subject to sampling variability and should also have a confidence interval around it. It is likely that the sample size will have been of the same order of magnitude, and we will make a conservative assumption that the sample size is 1000. Based upon this assumption, the 95% confidence interval is from 49.9% to 56.1%. Although there has been a change of 4% in the percentage over the period of one year the 95% confidence intervals overlap to a large extent, and this suggests that there is not likely to have been a major shift in public opinion.

11.4 Difference of means based on independent samples

11.4.1 Market research example

Many surveys and investigations aim to compare two groups of individuals or to compare the same groups at different times. For example, many opinion polls publish information on the opinion about the government separately for men and women, and separately for different age groups. Market research surveys into marketing strategies will look for subgroups of consumers. Different subgroups of consumers can be identified on the basis of their responses to certain attitude questions.

In the mid-1990s, the pharmaceutical company SmithKline Beecham developed a nicotine patch to assist smokers to stop smoking. This product was then tested in clinical trials and it was shown to be of use to smokers who wished to stop smoking. The company now wanted information which would help them to sell the product, information which would enable the company to try to have the product classed as a medical aid, so that it might possibly be available on prescription, and information which would enable them to target the advertising for the product. In 1998, SmithKline Beecham commissioned a market research survey of smokers in the European Union, Poland and Russia. This survey provides the market research information for this product. It also contains information on the characteristics of smokers in Europe, and so provides good epidemiological data on the attitudes of smokers.

Much of the data collected referred to the attitudes of smokers to government policies which aimed to encourage them to give up. One of the questions asked the smokers to rate on a scale from 1 to 10 their agreement with the statement 'It's easy to stop smoking on your own, without help from a doctor or a stop smoking product'. Smokers who agreed with this statement would give a score close to 10, those who disagreed with it completely would give a score of 1, while those who were not influenced by this would give a score of 5 or 6. The means and standard deviations of the responses are tabulated according to the responses to a question asking the individual smoker if he or she is likely to buy a product that has been scientifically proven to help smokers to stop smoking.

	Number	Mean	Standard deviation
Yes definitely	2157	3.19	2.91
Yes possibly	2345	3.77	2.94
No probably not	1475	4.51	2.94
No definitely not	2838	5.27	3.37
Maybe/Don't know	903	4.26	3.02
Maybe/Don't know, depends on cost	360	4.16	3.17
Maybe/Don't know, depends on product	212	3.90	2.64

Not surprisingly, there is a trend in the means among the first four groups, with those who would definitely buy the product having a lower agreement than those who would definitely not buy it. Although there appear to be some differences in mean agreement here, we do not know if these differences are important until we have assessed the effects of sampling variability and taken into account the precision of the estimates. All of the means are estimates based upon large samples. We have to decide whether there are real differences in the estimated effects or whether the precision of the estimates is so small that any differences could have arisen by chance.

In this example we have a quantitative variable which is compared over two or more groups. In the comparison we have to take into account the sampling variability in the estimated mean responses in order to interpret the difference in mean response and assess the significance of the difference with regard to the situation under study.

11.4.2 Statistical method and model

We consider the simplest case where there are just two groups to compare (Figure 11.8). We have two independent random samples from two populations, where the unknown true means are denoted μ_1 and μ_2. The standard deviations in the population are denoted σ_1 and σ_2, and these will generally also be unknown. The random samples are of size n_1 and n_2, and the sample means are denoted $\bar{x}_1$ and $\bar{x}_2$, with the sample standard deviations s_1 and s_2.

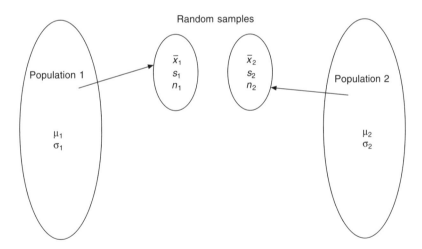

Figure 11.8 Independent samples from two populations.

The difference in the means is calculated when you wish to estimate how much bigger, or how much smaller, the average value in one group is than that in the other group. The 95% confidence interval for the difference in the means ($\mu_1 - \mu_2$) has exactly the same form as the other intervals discussed above:

$$\text{Estimate} \pm (\text{Distribution quantile}) \times (\text{Standard error})$$

Here the estimate is the difference between two sample means and the standard error then has to be the standard error of the difference between two independent sample means.

The difference between the two population means is estimated as $\bar{x}_1 - \bar{x}_2$, and the estimated standard error of this difference is (Sections 9.1.5, 9.4.4)

$$\hat{\sigma}_{\bar{x}_1 - \bar{x}_2} = \sqrt{s^2 \left(\frac{1}{n_1} + \frac{1}{n_2} \right)},$$

where the pooled estimate of the population variance is given by

$$s^2 = \frac{(n_1 - 1) s_1^2 + (n_2 - 1)s_2^2}{n_1 + n_2 - 2}.$$

This is applicable only if the variances in the two populations are the same, $\sigma_1 = \sigma_2$. The 95% confidence interval for the difference between two population means is

$$(\bar{x}_1 - \bar{x}_2) \pm t_{0.025}\hat{\sigma}_{\bar{x}_1 - \bar{x}_2},$$

where the degrees of freedom of the t distribution are given by $n_1 + n_2 - 2$.

Example 11.3 Telephone call waiting times

In the data of Example 11.1 there are two samples selected independently from two separate populations – an hour in the morning and an hour in the afternoon. Using the summary statistics there, we calculate the 95% confidence interval for the difference in mean waiting time between the afternoon and the morning. First of all, we calculate the pooled variance:

$$s^2 = \frac{(n_1 - 1) s_1^2 + (n_2 - 1)s_2^2}{n_1 + n_2 - 2} = \frac{24 \times 16.0023^2 + 24 \times 19.7051^2}{25 + 25 - 2} = 322.1823.$$

This means that the pooled standard deviation is $s = 17.94$ seconds. The estimated standard error is then given by

$$\hat{\sigma}_{\bar{x}_1 - \bar{x}_2} = \sqrt{s^2 \left(\frac{1}{n_1} + \frac{1}{n_2} \right)} = \sqrt{322.1823 \left(\frac{1}{25} + \frac{1}{25} \right)} = 5.0769.$$

The degrees of freedom are 48, and the 2.5% point of the t distribution is 2.0106. The 95% confidence interval is

$$(\bar{x}_1 - \bar{x}_2) \pm t_{0.025}\hat{\sigma}_{\bar{x}_1 - \bar{x}_2} = 39.96 - 37.36 \pm 2.0106 \times 5.0769 = 2.6 \pm 10.1.$$

The interval goes from –7.5 seconds to +12.7 seconds. This is a wide interval spanning 20.2 seconds when the difference in the means is only 2.6 seconds. This illustrates that the difference is imprecisely estimated in this example using samples of 25. In order to increase the precision of the estimate of the difference of the means, much larger samples would be required. This illustrates a general point about the investigation of subgroups in a study, namely that in order to estimate differences between subgroups you need to have large samples in the subgroups.

The calculations above were based upon the assumption that the standard deviations in the two populations were the same. This is reasonable in this example as the two sample standard deviations are similar to each other. A formal test of the equality of variances (known as an F test) is available, but we will not go into it in this book. A rough rule of thumb is to look at the ratio of the variances and, if this is more than 2 and the two sample sizes are both greater than 20, to conclude that there is reason to believe that the assumption of equal population standard deviations may not be true. In such a case the t interval based upon the pooled standard deviation is not completely valid and you should adopt another t interval, where the estimated standard error of the difference of the means is

$$\hat{\sigma}_{\bar{x}_1 - \bar{x}_2} = \sqrt{\left(\frac{s_1^2}{n_1} + \frac{s_2^2}{n_2}\right)}.$$

Note that the degrees of freedom are not now given by $n_1 + n_2 - 2$. Most computer programs will have an option to use this interval, sometimes known as the **unpooled** t interval. Generally you will use a computer program to do the calculations for your own survey or experimental data. However, if you wish to make comparisons with the work of others who only present summary information then you need to use the formula directly.

11.5 Difference of two population proportions

11.5.1 Two independent samples

We have two independent random samples from two populations, where the unknown true proportions are denoted p_1 and p_2. The random samples are of size n_1 and n_2, and the sample proportions are denoted $\hat{p}_1$ and $\hat{p}_2$. The 95% confidence interval for the difference in the proportions $(p_1 - p_2)$ is

$$(\hat{p}_1 - \hat{p}_2) \pm z_{0.025}\hat{\sigma}_{\hat{p}_1 - \hat{p}_2},$$

where

$$\hat{\sigma}_{\hat{p}_1 - \hat{p}_2} = \sqrt{\frac{\hat{p}_1(1 - \hat{p}_1)}{n_1} + \frac{\hat{p}_2(1 - \hat{p}_2)}{n_2}}.$$

The principle of the formula is exactly the same as in the previous cases with the confidence interval for the difference of two means. In virtually all statistical programs there are no simple commands for producing confidence intervals for proportions, and these have to be calculated by hand.

Example 11.4: Smoking prevalence in Australia
In a study of smoking prevalence in Australia in 1992 a random sample of 3063 women and 2983 men were interviewed. Of the men 28.2% currently smoked, and of the women 23.8% did so. The 95% confidence interval for the proportion of men smoking in the population is obtained as follows. Here we are only using one sample, namely the men, and so we use a one-sample confidence interval for a proportion. The sample size is very large and so we can use the normal distribution to give the interval. Since $\hat{p} = 0.282$, we have

$$\hat{\sigma}_{\hat{p}} = \sqrt{\frac{0.282(1 - 0.282)}{2983}} = 0.008\ 238\ 7,$$

and therefore

$$\hat{p} \pm z_{0.025}\hat{\sigma}_{\hat{p}} = 0.282 \pm 1.96 \times 0.008\,238\,7 = (0.266, 0.298).$$

In terms of percentages, the confidence interval is $(26.6, 29.8)\%$. This has a range of 3.2%, which gives an indication of the precision of the estimate. The strict interpretation of this interval is that we are 95% confident that the interval contains the true percentage of Australian men who smoked in 1992. Roughly speaking, it gives a range of plausible values. Although the best estimate we have is 28%, the sample percentage, we would not be surprised if the true value was 29% or 26%. We would be mildly surprised if it turned out to be 25% or 31%, and it is very unlikely that it would be as low as 20% or as high as 35%.

The confidence interval for the difference in smoking prevalence between men and women will provide information on how much more men smoke relative to women. This refers to two samples, which were independently selected, and the confidence interval is calculated as

$$(\hat{p}_1 - \hat{p}_2) \pm z_{0.025}\hat{\sigma}_{\hat{p}_1 - \hat{p}_2},$$

where $\hat{p}_1 = 0.282$ and $\hat{p}_2 = 0.238$. Thus

$$\hat{\sigma}_{\hat{p}_1 - \hat{p}_2} = \sqrt{\frac{0.282(1 - 0.282)}{2983} + \frac{0.238(1 - 0.238)}{3063}} = 0.011\,273$$

and so

$$(0.282 - 0.238) \pm 1.96 \times 0.011\,273 = (0.0219, 0.0661).$$

The difference in the prevalence of smoking among men and women in Australia in 1992 is 4.4%, with a 95% confidence interval from 2.2% to 6.6%. This suggests that smoking prevalence is higher among men (we would be surprised if the true difference was as low as 0% or 1%).

One of the features of this example is that we assume that the two samples are independent of each other. Specifically, this means that in the process of collecting the data the interviewers did not ask husbands and wives. In such a case we would have dependent samples – and different, and much more complex, techniques would have to be used to estimate the difference in the proportions and the confidence interval. This process of calculating a difference in proportions from dependent samples is precisely the sort of calculation that needs to be done in opinion polls when trying to provide a confidence interval for the lead of one party over another.

11.5.2 Dependent samples, confidence intervals and opinion polls

Generally in an opinion poll the percentage of the sample agreeing with a statement is quoted. The sample size should always be published, along with a statement saying how the sample was selected. The percentages and sample size can be used to provide a confidence interval and hence take sampling variability into account when discussing the results of the poll.

The theory of confidence intervals rests on the assumption that the sample has been randomly selected. This is seldom the case for opinion polls, which are almost always quota samples. In quota sampling the selection of the sample is based on the interviewer's subjective choice, and the resulting sample may have unmeasured bias. However, for the purposes of this application of confidence intervals it will be assumed that the opinion poll is a random sample.

The results of a number of surveys of voting intentions carried out over several months by NOP were published in the *Independent* on 27 September 1991. The poll carried out in late September was based on a sample size of 1596, whereas the one in early September had a sample size of 1468. In late September 42% intended to vote Labour and 39% Conservative. These percentages are calculated excluding the don't knows, refusals and so on. The number of excluded responses is not given, so it is not possible to use the correct number of respondents on which the percentages of 42% and 39% are based. Consequently the full sample sizes will be used.

Table 11.1 NOP voting intentions surveys (%, excluding don't knows etc.)

	Jun	July	Aug	Sept 6–7	Sept 21–23
Labour	45	43	44	39	42
Conservative	37	39	38	41	39
Liberal Democrat	14	13	12	15	14
Others	4	4	4	5	5
Lead	Lab 8	Lab 4	Lab 6	Con 2	Lab 3

Information on methods used to conduct the poll is included on the book's website.

The 95% confidence intervals are as follows. For Labour,

$$\hat{p} = 0.42,$$

$$\hat{\sigma}_{\hat{p}} = \sqrt{\frac{0.42 \times 0.58}{1596}} = 0.0124,$$

$$\hat{p} \pm z_{0.025}\hat{\sigma}_{\hat{p}} = 0.42 \pm 1.96 \times 0.0124 = (39.6, 44.4)\%.$$

For the conservatives,

$$\hat{p} = 0.39,$$

$$\hat{\sigma}_{\hat{p}} = \sqrt{\frac{0.39 \times 0.61}{1596}} = 0.0122,$$

$$\hat{p} \pm z_{0.025}\hat{\sigma}_{\hat{p}} = 0.39 \pm 1.96 \times 0.0122 = (36.6, 41.4)\%.$$

These intervals overlap to a moderate extent, which means that it is possible that there is no real difference in the percentages who intend to vote for the two main parties. This could be tested using significance tests (see Chapter 12).

Much of the important interpretation of opinion polls involves a comparison of two proportions – for example, the percentages who intend to vote Labour in two separate opinion polls, or the percentages voting Labour and Conservative in a single poll. If there are two separate polls or if two independent subgroups within one poll are to be compared then the procedure is exactly the same as in Section 11.5.1. If there is only one poll and two related percentages are to be compared, then the standard deviation of the difference in the percentages has to be calculated taking the dependence between the samples into account.

In this case the Labour lead over the Conservatives in the late September poll is to be estimated. Let $\hat{p}_L$ and $\hat{p}_C$ be the two sample proportions respectively; with $\hat{p}_L = 0.42$ and $\hat{p}_C = 0.39$ the difference (lead) is 0.03.

The same procedure as for independent samples cannot be used here, as intending to vote Labour and intending to vote Conservative are mutually exclusive. Consequently, the two proportions cannot be independent and are related. To calculate the variance of the difference in proportions it is necessary to add the variances of the two individual proportions and also to take into account the covariance between them. The variance of the difference of two dependent proportions is (see Section 9.1.5)

$$\text{Var}[\hat{p}_L - \hat{p}_C] = \text{Var}[\hat{p}_L] + \text{Var}[\hat{p}_C] - 2\text{Cov}[\hat{p}_L, \hat{p}_C],$$

where there is a negative covariance between the two proportions of

$$\text{Cov}[\hat{p}_L, \hat{p}_C] = -\frac{\hat{p}_L \hat{p}_C}{n}$$

n being the sample size. This standard error is estimated by

$$\sqrt{\left(\frac{0.42 \times 0.58}{1596}\right) + \left(\frac{0.39 \times 0.61}{1596}\right) + 2\left(\frac{0.42 \times 0.39}{1596}\right)} = 0.0225.$$

The 95% confidence interval for the difference is

$$0.03 \pm 1.96 \times 0.0225 = (-1.4, 7.4)\%.$$

This interval is wide, indicating that this difference is not estimated with much precision. As the interval contains 0%, this suggests that there is no real difference between the two percentages. To all intents and purposes we cannot conclude from the sample that one party is significantly ahead of the other party.

The use and interpretation of confidence intervals has not suggested that the percentage Labour vote is different from the Conservative vote. This contrasts with the conclusion of the journalist.

11.6 Large and small samples

The methods discussed here for proportions are applicable in large samples only. If you have a qualitative random variable which gives rise to a binomial distribution and you have only a small sample, then you cannot use the large-sample procedures to give you confidence intervals for the proportion of positive responses. You need to use exact methods based upon the binomial distribution.

It is not quite such a big issue with quantitative variables. If the variable follows an approximately normal distribution, then you use the methods based upon the t distribution outlined in this chapter. If you do not have a normal distribution but do have a large sample, then you can use the same procedures, but with the standard normal distribution, replacing the t distribution. If the sample size is small and you do not have a normal distribution, then the procedures discussed here are not applicable. You can try a transformation of the random variable, e.g. by taking logarithms, to try and get a normal distribution (see Chapter 6). If that fails then you need to use other methods such as those based on non-parametric techniques.

The non-parametric techniques are the one-sample Wilcoxon confidence interval, which is based upon the ranks of the observations rather than their numerical values, and the one-sample sign test. For two samples there is the Mann–Whitney confidence interval based upon the ranks of the data, like the Wilcoxon procedure. All of these facilities are available in many statistical packages and all provide a confidence interval for the median, or difference in medians, rather than means. The interpretation of the intervals is exactly the same as for the intervals for means or proportions.

11.7 Sample size calculations for confidence interval width

If the confidence interval is wide, then the estimation is relatively imprecise. This leads on to another way of determining the sample size based on the width of the 95% confidence interval. If you are able to specify an acceptable range for the confidence interval, then you can calculate the sample size required to achieve this.

11.7.1 Estimating a mean to a specified precision

The consumption of fruit and vegetables is thought to be related to the risk of a number of cancers, particularly breast cancer and colorectal cancer. A small pilot study revealed that the mean number of times vegetables were eaten per week by a group of 52 women aged 50 and over in New York State, USA, was 5.3 with a standard deviation of 3.5. The investigator wishes to see if there are differences in vegetable consumption between a group of cancer patients and a group of control subjects without cancer, and is able to specify that an acceptable width for the 95% confidence interval is 0.5 in each group. How large a sample is needed?

We work with the large-sample approximation for the 95% confidence interval, $\bar{x} \pm z_{0.025} \, s/\sqrt{n}$. The width of the interval is

$$\left(\bar{x} + z_{0.025} \, \frac{s}{\sqrt{n}} \right) - \left(\bar{x} - z_{0.025} \, \frac{s}{\sqrt{n}} \right) = 2 z_{0.025} \, \frac{s}{\sqrt{n}}.$$

If we let w denote the specified width, then the sample size can be calculated as

$$2 z_{0.025} \, \frac{s}{\sqrt{n}} = w$$

$$\therefore \qquad \sqrt{n} = 2 z_{0.025} \, \frac{s}{w}$$

$$\therefore \qquad n = \left(2z_{0.025}\,\frac{s}{w}\right)^2.$$

In the example we have a 95% confidence interval, so $z_{0.025} = 1.96$ and the acceptable width is 0.5. The best estimate of the standard deviation comes from the pilot study, and we have $s = 3.5$. Thus

$$n = \left(\frac{2 \times 1.96 \times 3.5}{0.5}\right)^2 = \left(\frac{13.72}{0.5}\right)^2 = 753.$$

To estimate the mean number of times vegetables are eaten per week by women aged over 50 in New York State so that the width of the 95% confidence interval is 0.5 times, a sample of 753 subjects is required.

The calculated sample size is large, but this is not uncommon. In essence, it is large because the specified precision is small relative to the variability in the population. In many production experiments in business you may find that the standard deviation of your measurements is much smaller than the mean. This will lead to smaller samples sizes required for a specified precision.

It is possible to use other confidence interval widths, though the 95% is the most commonly used in practice. Other values used are 90%, 80% and sometimes 50%. The smaller you make the confidence the smaller the sample size you will require, but this is not a valid procedure to adopt. If you have a small confidence then you have a lower chance of the interval containing the mean. If you go down to a 50% interval then the calculated interval is as likely to contain the mean as you are to get a head when you toss a coin.

The distribution of the number of times you eat vegetables per week must be discrete, taking values such as 0, 1, 2, 3, ... It is still valid to use the normal distribution to calculate the sample size as we get a large sample and the distribution of the sample mean is normal in large samples.

11.7.2 Sample sizes for a proportion

This procedure can be used for proportions as well as for means. The width of a 95% confidence interval for a proportion is

$$2z_{0.025}\,\sqrt{\frac{\hat{p}(1-\hat{p})}{n}} = w,$$

from which we obtain

$$\sqrt{n} = 2z_{0.025}\,\frac{\sqrt{\hat{p}(1-\hat{p})}}{w}$$

$$\therefore \qquad n = \left(2z_{0.025}\,\frac{\sqrt{\hat{p}(1-\hat{p})}}{w}\right)^2.$$

With proportions it is necessary to have a rough idea of the expected sample proportion before you are able to work out how large a sample you need to be able to estimate the proportion with the specified precision. This is a little circular, and it does not matter a great deal what the exact value of the sample proportion is provided it is in the range 0.2 to 0.8. This is so because the term $\sqrt{\hat{p}(1-\hat{p})}$ does not vary a great deal within these limits. In fact, it varies from 0.4 to 0.5. When you are dealing with rare or very common events then you will need smaller sample sizes than for events which occur with probabilities in the central range.

Example 11.5: Percentage of smokers

On of the aims of the study by SmithKline Beecham, previously described in Section 11.4.1, was to estimate the percentage of smokers who would buy medication to help them. If we imagine ourselves at the design stage of the survey, then one of the important questions is how large a sample we need. This is dictated by the cost of collecting the information, the questions being asked and the precision with which the answer is required. Precision is a statistical idea, and we look at the design of the study from that point of view.

If we say that we wish to estimate the proportion of smokers who will buy medication to help them stop smoking with a 95% confidence interval of width 0.10, then we can use the above equations to give a sample size. In order to use the equation we need a guess or previous estimate of the value of the proportion expected. Suppose we believe that this is around 0.30. Then the required sample size is

$$
n = \left(2z_{0.025} \, \frac{\sqrt{\hat{p}(1-\hat{p})}}{w} \right)^2 = \left(2 \times 1.96 \times \frac{\sqrt{0.3(1-0.3)}}{0.1} \right)^2 = 323.
$$

If the precision is increased then the sample size will quickly increase. Specifying a precision of $\pm 1\%$ (a confidence interval of width 0.02) leads to a sample size in excess of 8000 smokers. This is going to lead to a study with sampling costs way above the budget for many studies.

Width of confidence interval of proportion	0.1	0.05	0.02	0.01
Sample size	323	1291	8067	32 269

The sample size will also change depending upon the anticipated value of the proportion. For a precision of $\pm 5\%$ there is not a great deal of effect of an increase in the sample size as the anticipated proportion goes from 0.3 to 0.5. If the anticipated proportion was much smaller at 0.05, then a smaller sample size is required:

Proportion	0.05	0.10	0.20	0.30	0.40	0.50
Sample size	73	138	246	323	369	384

When designing surveys to measure a percentage you need to be a little aware of the anticipated values. Often this information can be obtained from small pilot studies or from reports of similar studies carried out previously.

In the study is was decided to select a sample size of 600 smokers in each of 17 countries. Assuming a 30% response, the anticipated precision of the estimate in each country was ±3.6%. In the survey the percentage of smokers who said that they would be willing to pay for medication to help them stop smoking was 43.8%.

12

Significance tests

- Understanding the concepts involved in a significance test
- Knowing about the two types of errors that can be made in a test
- Knowing how to specify the null and alternative hypothesis in a variety of situations
- Knowing and using the correct test statistic
- Understanding the p-value and its calculation and interpretation
- Being aware of the difference between one and two-sided tests
- Knowing how to carry out a significance test for a mean, the difference of two means, a proportion and difference of two proportions
- Knowing about the critical value approach to significance tests and its relationship to the p-value approach
- Understanding that there are many difficulties in the application of significance tests in practice: some of these are associated with choosing the null hypothesis on the basis of a prior inspection of the data, others are concerned with multiple testing
- Being aware of the concept of the power of a significance test and the use of the power in the calculation of the study size.

12.1 Introducing tests and errors

In Chapter 11 we looked at how sampling distribution information (Chapter 10) can be used to obtain a confidence interval estimate of a population parameter. In this chapter we consider a related problem: how to decide if the population parameter has a particular value, or lies within a specific range of values, on the basis of the available sample data. Which of the two approaches is used depends on what questions are to be asked. If we need to have an actual value of (say) sales to use as the basis for a plan, then an interval estimate would be appropriate. On the other hand, if what we want to know is whether sales have changed, then we would use a significance test.

Significance tests, also known as **hypothesis tests**, are a valuable and powerful statistical tool which helps the decision-making process. There are several fairly subtle issues involved in their use, plus some rather opaque terminology, and it is little wonder that the significance test is often one of the more confusing areas for those learning statistics. In this chapter we will start by introducing the key concepts by way of an example, and then lay out the terms and common features of significance tests, before looking at the most common and useful ones in turn.

In this section we will investigate the key concepts underlying significance tests.

Rather than plunge directly into the details of such tests themselves, we will make use of a digression into a simpler problem (relating to classification) which we hope will make it easier for the central ideas to be understood.

12.1.1 Customer classification example

Suppose that you wish to identify new customers who have a good chance of providing return business. In a pilot study you randomly select a group of new customers, and have your sales team contact them to try to promote further custom. In due course you find out which of the customers in the study came back for more; you can now look at all of the records of the customers in the study, to see which characteristics best identify the two groups of interest following the promotion: 'returns' who come back and 'non-returns' who do not. Ideally you would like some perfect separation of these groups according to a simple characteristic, but in reality you are more likely to find a situation such as that portrayed in Figure 12.1, where there is reasonable but incomplete separation between the two groups; here it was found that the amount spent by a customer at first purchase was a reasonable indicator of whether or not they could be persuaded to bring return custom – those who spent less being less likely to buy again. There is overlap between the groups, so no single value of the amount spent allows perfect discrimination.

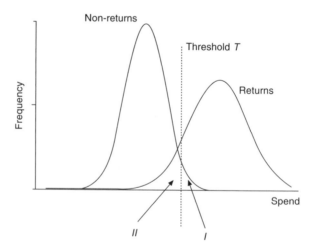

Figure 12.1 Distribution of amounts spent by two groups of customers. T is the value below which we classify a customer of unknown type as a 'non-return', and above which as a 'return'. The region marked I corresponds to non-returns who are misclassified, and has area (and hence probability) a. The region marked II corresponds to returns who are misclassified, and has area (and hence probability) b.

Note that we have glossed over some important marketing details for the sake of simplicity. For example, we would in practice have to compensate for the number who would return without the sales effort. We would also have to allow the method to take account of changing prices. The amount spent, the variable 'Spend' in the Figure 12.1, could be given as a ratio to a standard value expected to vary in the same way as your prices (say the average cost of similar services in the sector, which can be obtained from trade journals and magazines).

In Figure 12.1 a proposed threshold T has been marked. This provides a criterion for distinguishing between the two groups: all of those with Spend less than T would be considered to be non-returns, while all of those with Spend at least T would be considered returns. To the right of the value T is an area under the non-returns frequency curve labelled I: this represents people who would be wrongly classified as returns; we will denote by α the fraction of actual non-returns this applies to. To the left of T is an area under the returns frequency curve labelled II: this represents people who would be wrongly classified as non-returns; we will denote by β the fraction of actual returns this applies to. Apart from these two fractions, everybody else gets classified correctly. We will use these fractions as error probabilities, when a customer of unknown type is classified: α is the probability of classifying a non-return as a return, and β is the probability of classifying a return as a non-return.

By increasing or decreasing the value of T in Figure 12.1, we see that the two errors α and β can be changed, but that trying to make one small only makes the other larger. This trade-off between the two types of error can be visualized as in Figure 12.2. When α is small, the threshold T is towards the right and this means that β will be large. When the threshold T is towards the left, α will be large and β will be small.

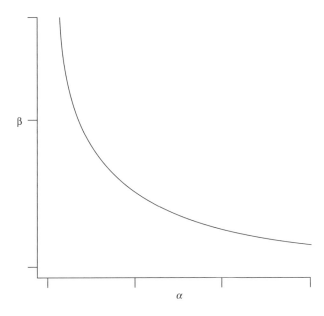

Figure 12.2 Trade-off between the two types of error.

Which value of the threshold T gives the best choice of α and β? There is no unique answer: the criteria for what is 'best' depend not on statistical issues but upon the objectives of the user. For example, suppose that the sales promotion effort was exceptionally expensive; then you would not want to waste money by including too many non-returns. This means that you would select T so that α was as small as possible, and β (returns you would miss targeting) acceptably small. By contrast, if the promotion was inexpensive, but the income per return was gratifyingly large, then you could accept targeting some non-returns (that is, α need not be very small) but would not like to miss too many returns by setting T too high (so β needs to be as small as possible). A choice of T is implied by your particular decision about the trade-off between the two types of risk.

In this example of classification we can see that each decision about classifying an individual is accompanied by two errors, which are portrayed in Table 12.1. There is a true state of the customer, known only to him or her, and a decision made by the company about that customer. Each customer is in one of the two true states, and for each state there is an associated error. Significance tests are concerned with making decisions and you should be aware that each test is associated with the same two types of errors.

Table 12.1 Classification errors and correct decisions

True state of the customer	Decision made by classification rule	
	Single purchaser	Repeat purchaser
Non-return purchaser (single purchaser)	Correct	Error, with probability α
Return purchaser (repeat purchaser)	Error, with probability β	Correct

12.1.2 Simple binomial test: tasting experiment

When I was lecturing in business statistics at Strathclyde University I tried, once in a while, to have interactive lectures. In order to try to liven up the introduction to significance tests I used to run a taste-testing experiment to see if a sample of students could tell the difference between Diet Pepsi and Diet Coke. I don't like either, so my prior belief was that you could not really tell the difference. This was usually quite good fun, for a statistics lecture, and in the end-of-study feedback questionnaires was frequently reported as one of the most memorable lectures. However, the subject of the lecture, significance tests, was usually reported as the most difficult concept so brightening up the lecture had no effect on understanding of the concepts.

The experiment involved giving one glass containing Coke and another containing Pepsi to each of 10 students. The students did not know which glass contained the Coke or the Pepsi and were asked to select the glass containing the Coke. The responses by the 10 students were as follows:

Student	1	2	3	4	5	6	7	8	9	10
Response	C	C	W	C	C	W	C	C	C	C

In this experiment all the students were being asked to do was to differentiate between the tastes of Coke and Pepsi, and out of the 10 students 8 did this correctly (denoted by C) and 2 did not (denoted by W). The purpose of the experiment is to find out if students can tell the difference and my prior belief is that they cannot. This represents my prior hypothesis and we will use the results of the experiment to carry out a significance test to see if my prior hypothesis is reasonable or not.

We have already come across data of this kind, where there are a fixed number of independent trials and each trial has only two outcomes, in Chapter 9. These data are commonly modelled by using the binomial distribution which has a common probability, p, of a correct response on each trial, and the random variable is the number of correct responses. For a binomial distribution all we need is the value of p in order to carry out some probability calculations.

My prior hypothesis is that students cannot tell the difference between Coke and Pepsi. This means that when they are forced to choose one glass or the other glass as the glass containing Coke, all they are really doing is guessing. If you guess one of two possible outcomes then the probability of each outcome is going to be $p = 1/2$. This represents my prior hypothesis. This is usually known as the **null hypothesis** of the test.

As in Section 12.1.1, where there was a choice between two types of customers, here we have a choice between two hypotheses. The other hypothesis is usually known as the **alternative hypothesis** – that the null hypothesis is false. If students are able to correctly pick out the Diet Coke then we would expect that they are not guessing, and consequently the probability of a correct response is going to be larger than 0.5, i.e. $p > 1/2$. If everyone could tell the difference then clearly we would expect $p = 1$, but this is not the case, and the reality is that we do not really know the exact value of p. All we know is that it is larger than 0.5 but not the exact value.

If the null hypothesis is true then we will be able to work out probabilities based upon the binomial distribution (because we know what the value of p is). If the alternative hypothesis is true, then we cannot work out probabilities because we do not know the value for p. This means that all our calculations are based upon the assumption that the null hypothesis is true. It also means that we can only make a decision about whether or not we believe the null hypothesis to be true.

In the example we observed that 8 out of 10 students were correctly able to pick out the glass containing Coke. The outcome of this experiment, like all experiments, is a random variable. As 80% of the students correctly chose the Coke, it looks as though the null hypothesis that they were guessing is going to be false. However, we also know that even if the true probability of choosing the Coke is equal to 0.5 it is still quite possible to get 8 out of 10 students with the correct choice. The crux of the decision is how likely it is, and this is measured by a probability.

The probability that 8 out of 10 students choose correctly, assuming that the probability that any one of them chooses correctly is 0.5, is

$$P(X = 8) = \binom{10}{8}\left(\frac{1}{2}\right)^8\left(\frac{1}{2}\right)^2 = \frac{45}{1024} = 0.0439.$$

This is quite a small probability, less than 1 in 20, but it is not the whole story. If we had observed that 9 out of the 10 students chose correctly or, indeed, that all 10 did, then we would be even more likely to believe that the null hypothesis was false. The probability of 9 correct is

$$P(X = 9) = \binom{10}{9}\left(\frac{1}{2}\right)^9\left(\frac{1}{2}\right)^1 = \frac{10}{1024} = 0.0098,$$

and of 10 correct

$$P(X = 10) = \binom{10}{10}\left(\frac{1}{2}\right)^{10}\left(\frac{1}{2}\right)^0 = \frac{1}{1024} = 0.000\,976,$$

and so the probability of observing 8 or more correct is

$$P(X \geq 8) = \frac{56}{1024} = 0.0547.$$

The quantity that we have just calculated is known as the p-value of the test, and this quantity pervades many statistical reports. We have assumed that the null hypothesis is true and have calculated the probability of that 8 or more students out of 10 can choose correctly. Recall that we have observed 8 out of 10 in the experiment so, in this case, the p-value is the probability of finding observations which are as extreme as the data in the direction of the alternative hypothesis.

Figure 12.3 shows the probabilities for all possible outcomes of the tasting experiment, assuming that the null hypothesis is true. If the null hypothesis is true then we would expect to get 5 students who were able to guess correctly. Judging by the heights of the bars, the probabilities, we would not be surprised if we observed between 3 and 7 students with the correct choice. If the alternative hypothesis is true then we would expect to have larger numbers of students with the correct choice, around 8–10. The bars shaded black represent the p-value.

The p-value is a probability calculated assuming that the null hypothesis is true. Thus it conveys information about the null hypothesis only. It does not convey any information

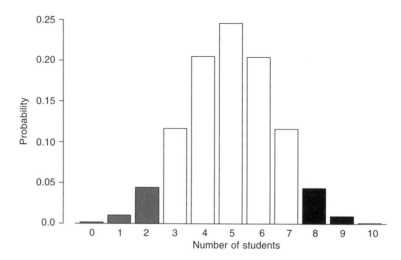

Figure 12.3 The distribution of the number of students with the correct choice, assuming that the null hypothesis is true.

about the alternative hypothesis. Thus our interpretation of the p-value will enable some decision to be made about the null hypothesis.

The p-value for this significance test is 0.0547, which is quite a small probability. We have to decide to say that we believe the null hypothesis to be true or that we believe it to be false. A decision has to be made taking into account the evidence. If the null hypothesis is indeed true then $p = 0.5$, and by observing 8 out of 10 students with a correct choice we have by chance observed an event (8 or more students with a correct choice) with a probability of 0.0547. If the null hypothesis is false then $p \geq 0.5$, and by observing 8 out of 10 students with a correct choice we have observed an event (8 or more students with a correct choice) which has a probability of occurring greater than 0.0547.

The actual decision made then rests upon how we interpret the probabilities. If the p-value is very small, say 0.01 (one in a hundred) or 0.001 (one in a thousand), then most people would be quite happy to conclude that this probability is so small that it is unlikely that the null hypothesis is true, so we conclude that the null hypothesis is false. On the other hand if the p-value is large, say 0.2 (one in five), then again most people would accept that this represents an event which is quite likely and so we believe that the null hypothesis is true.

Some possible observed values and p-values for the tasting experiment are listed in Table 12.2. We can see that if all 10 students chose correctly then the p-value would be 0.0010 and we would have no hesitation in rejecting the null hypothesis and concluding that the students could tell the difference between Coke and Pepsi. Similarly, if 9 students chose correctly then the p-value would be 0.0107 and we would come to the same conclusion. If 7 students chose correctly then the p-value would be 0.1719 which is quite high, and we would say that there is no evidence to reject the null hypothesis that the students were just guessing.

Table 12.2 Observed values and p-values

Observed, x	$P(X = x)$	$P(X \geq x)$ p-value
7	0.117 19	0.1719
8	0.043 95	0.0547
9	0.009 77	0.0107
10	0.000 98	0.0010

Traditionally statisticians have worked with a cut-off value for the p-value of 0.05, saying that if the p-value is below this value then there is evidence to reject the null hypothesis; if the p-value is greater than this value then there is no evidence to reject the null hypothesis. The cut-off value of 0.05 is known as the **significance level** of the test, and 0.05 is an arbitrary value which works in practice to distinguish between events which are unlikely and those which are likely. It corresponds to an event which is likely to happen on about one occasion in 20 – an event which is slightly less likely than getting four heads in a row when tossing a coin. If the p-value is smaller than the significance level, we say that the result of the test is statistically significant.

In this example, the difficulty comes with the observed value of 8 which has a p-value of 0.0547. As this is greater than the accepted cut-off value (0.05) we would say that there is no evidence to reject the null hypothesis at the significance level of 0.05. With the p-value we do know that the evidence against the null hypothesis is quite strong but it is not strong enough to conclude that there is a statistically significant difference, at a significance level of 0.05.

If I had selected 11 students and 9 of them had chosen correctly then the p-value would have been 0.033, which would have been reported as a significant result. Thus you can see that the sample size has some influence on the results as well. This is concerned with another very important concept known as the power of the test, which will be investigated later.

12.1.3 Key points

In summary, we have seen in this section that a significance test results in a choice between two hypotheses, known as the null hypothesis and the alternative hypothesis. We observe some data and calculate the probability of observing data as extreme as these relative to the null hypothesis. This probability is calculated assuming that the null hypothesis is true and is known as the p-value of the test. The p-value is compared to the significance level of the test, which is specified in advance, and a decision is taken about rejecting the null hypothesis.

The examples have introduced most of the key concepts needed to understand significance tests. There are two points to stress, however. First, a significance test does not generally involve a single measurement such as the 'Spend' quantity for each customer used above. Instead, it deals with data from one or more samples, and attempts to make inferences about population means, proportions and variances, using the data and some statistical theory. Thus, in the second example of the taste-testing experiment, the parameter was the probability of choosing correctly – or, equivalently, the population proportion of students who are able to choose correctly. Statistical theory suggests that the binomial distribution would be an appropriate model for the responses.

Second, we generally are not in the fortunate position outlined in the customer classification example, where we had detailed knowledge of the frequency distributions for both alternatives (the two classifications of customer in that case). Much more commonly, we are quite ignorant about one of the distributions, for reasons that will become clear in the next section, but have good theoretical knowledge of the other one. One source of this knowledge is discussed in Chapter 10: because we are dealing with sample proportions, sample means and the like, we have a statistical theory which tells us what their sampling distributions ought to be. Another is our prior beliefs about the population or experiment, which tell us about the values of the parameters of these distributions. In the tasting example my prior beliefs led me to believe that students could not tell the difference between Coke and Pepsi and so would be guessing, which means that the probability of choosing correctly is 0.5.

In terms of the customer classification example, this second point of course means that only one of α and β can be available to us when we set about making our classification! In the tasting example it was clear that we could work things out under the null hypothesis where we had a value for p, but not under the alternative hypothesis. In other words, we can quantify and control one of the risks involved (say, α), but have no initial idea of how to quantify the other (β). It is important to note, however, that even so we can be assured that a relationship such as that shown in Figure 12.2 must hold true: keeping α very low automatically means letting β grow. The word 'significance' in this chapter arises directly from such considerations: the significance level of a test will be precisely the value of α produced by applying a threshold value to the known distribution.

12.2 Components of a significance test

12.2.1 Hypotheses

When a new customer is classified in the customer classification example, there are two possibilities: the customer is either a 'return', or a 'non-return'. It is important to realize that when we do so, we do not know for sure whether our classification is correct, due to the existence of the two types of error. Hypotheses are the key here: under one hypothesis the customer is a 'return', while under the other the customer is a 'non-return'. We are using statistical methods to help decide which of these two hypotheses is best supported by the data (in this case the Spend value for the customer). Note that the two competing hypotheses are mutually exclusive and cover all of the options for the customer; this is a feature of all of the hypotheses used in significance tests.

As in the tasting example, the two hypotheses are labelled the null hypothesis, often denoted H_0, and the alternative hypothesis, denoted H_1 or H_A. The null hypothesis represents our prior beliefs – and in the simplest significance tests, based upon a single sample of data, generally involves specifying an explicit value for the parameter of a distribution. This parameter will generally be a mean or proportion, but could also be a variance. Also, you will normally be able to specify the null hypothesis before any data are collected, though you may need to use the data to check some of the distributional assumptions.

The alternative hypothesis represents what we will believe if the evidence from the data is that null hypothesis should be rejected. Generally this hypothesis is not specified exactly, as in the tasting example, and includes a range of values for the parameter. The sets of null and alternative hypotheses that we will consider in this chapter are listed below for one- and two-sample tests of proportions and means.

		Two-sided	One-sided	
One sample	Proportion	$H_0: p = p_0$ $H_1: p \neq p_0$	$H_0: p = p_0$ $H_1: p > p_0$	$H_0: p = p_0$ $H_1: p < p_0$
	Mean	$H_0: \mu = \mu_0$ $H_1: \mu \neq \mu_0$	$H_0: \mu = \mu_0$ $H_1: \mu > \mu_0$	$H_0: \mu = \mu_0$ $H_1: \mu < \mu_0$
Two samples	Proportions	$H_0: p_1 = p_2$ $H_1: p_1 \neq p_2$	$H_0: p_1 = p_2$ $H_1: p_1 > p_2$	$H_0: p_1 = p_2$ $H_1: p_1 < p_2$
	Means	$H_0: \mu_1 = \mu_2$ $H_1: \mu_1 \neq \mu_2$	$H_0: \mu_1 = \mu_2$ $H_1: \mu_1 > \mu_2$	$H_0: \mu_1 = \mu_2$ $H_1: \mu_1 < \mu_2$

For the one-sample tests our interest is usually in testing if the mean or proportion is equal to a specified value, denoted p_0 or μ_0. The alternative is either that the parameter is not equal to the specified value, for a two-sided test, or greater than or less than the specified value for a one-sided test. When we have two samples then the common null hypothesis is that the two parameters are equal to each other, with the alternatives being not equal for a two-sided test or one parameter greater than or less than the other for a one-sided test. These are not the only types of null and alternative hypotheses, but are the most common of the simpler ones.

12.2.2 Test statistics

The decision from a significance test is either to reject the null hypothesis in favour of the alternative or not to reject the null hypothesis. This decision is based upon the calculation of a probability, the p-value, and if this probability is small then we believe that we have observed an event which is unlikely under the null hypothesis. This leads us to doubt the validity of the null hypothesis and so we reject it in favour of the alternative hypothesis. If the p-value is large then there is no evidence to reject the null hypothesis. This was the process we went through in the testing example.

In order to calculate the p-value we need to have a fully specified probability distribution. This means that we need to know the type of distribution, and explicit values of any parameters of the distribution. The values will come from the null hypothesis, so this probability distribution is only valid under the null hypothesis.

In the tasting example we had only one observation, but this is unusual and we normally have several of them in a sample. The hypotheses which are involved in the significance test involve properties of a population mean, proportion, etc. (for example, 'sales are up 20%' versus 'no they're not'), and we try to find which is best supported by the sample data. The trick is to find some function of the sample data which has a known distribution *if* the null hypothesis is true; this allows us to find out about α, the probability of discarding that hypothesis when it is in fact correct. This function of the sample data and the null hypothesis value of the parameters is known as the **test statistic**.

In the tasting experiment the test statistic was just the number of students with the correct choice. Under the null hypothesis of guessing we assumed, based upon our knowledge of the experiment, that this had a binomial probability distribution with a parameter $p = 0.5$. Consequently, we could work out the p-value and so make a decision.

In other cases the test statistic will be different and so will have a different distribution. In this chapter it will either be the normal or t distribution. When we have large-sample tests on proportions or tests on means, if the population variance is known then the normal distribution will be appropriate. When we have tests on means with unknown population variance, a t distribution is required.

To demonstrate that we can know the exact distribution of the test statistic under the null hypothesis in other situations, in Figure 12.4 I have used simulation methods to generate a large number of samples (each of 10 observations) and calculate the distribution of the test statistic, which involves both the sample mean and sample standard deviation for each sample. The histogram shows the frequency distribution of the test statistic, while the smooth curve is the theoretical frequency curve for this quantity. It can be seen that theory and simulation are in excellent agreement.

Although we may not know the distribution of the individual measurements in a sample, the theory of sampling distributions means that we can with some confidence

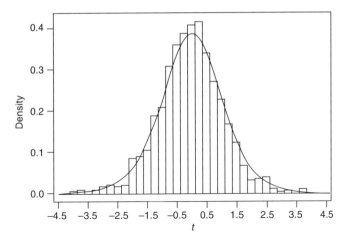

Figure 12.4 Simulation to show that test statistics have a known distribution. The histogram was produced by simulating 1000 samples of size $n = 10$ from a normal distribution with mean 10 and standard deviation 2, and then calculating the test statistic $t = (\bar{x} - 10.0)/(s/\sqrt{n})$ for each sample. The smooth curve is the corresponding theoretical distribution, a t distribution with 9 degrees of freedom.

know the distribution of summary quantities such as the sample mean or sample proportion based on an entire sample's data. In large samples, the sample mean and the sample proportion both have an approximate normal distribution and so the significance tests can be based upon this distribution.

12.2.3 The *p*-value

In any significance test you need to make a decision either to reject the null hypothesis or not to reject it. This decision is based upon the *p*-value which is the probability of obtaining a more extreme value of the test statistic than that observed. We illustrate this graphically, in Figure 12.5, with a one-sided test of $H_0: \mu = \mu_0$ against $H_1: \mu > \mu_0$ based upon a random sample of size n from a normal distribution, where the population variance σ^2 is known. We know that the quantity

$$Z = \frac{\bar{X} - \mu_0}{\sigma/\sqrt{n}}$$

follows a standard normal distribution if the null hypothesis H_0 is true. This means that the distribution of Z is known exactly, and so Z is a suitable test statistic. We also know that if H_0 is true then the observed value of Z will be around zero and if H_1 is true then the observed value of Z will be greater than zero.

In Figure 12.5 the *p*-value is the area to the right of the observed value of the test statistic under the curve of the distribution of the test statistic. This area is calculated without any reference to the alternative hypothesis. Thus the *p*-value only measures evidence for or against the null hypothesis. It does not have anything to do with the alternative hypothesis, and so cannot give any information on the probability that the alternative is true.

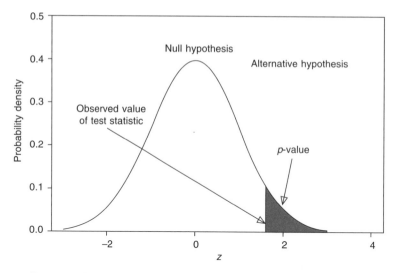

Figure 12.5 Illustration of p-value.

If the null hypothesis is true then the distribution of the test statistic is given in Figure 12.5. Thus the p-value is simply the probability that the test statistic could have given a more extreme value than that observed. Another way of looking at it is to imagine taking lots of samples from a population where you know that the null hypothesis is true. The p-value represents the proportion of samples in which you would get a test statistic greater than the one you observed. This implies that for some samples the p-value will be small and less than the significance level, while for other samples the p-value will be larger than the significance level. So even if the null hypothesis is true it is possible to find a small p-value. This is not going to happen very often, but it can do so.

If we have a two-sided test, rather than a one-sided test just illustrated, then the alternative hypothesis is going to be $H_1: \mu \neq \mu_0$. This means that both large and small values of the test statistic will imply that the null hypothesis is to be rejected. For such tests the p-value will have to measure both the upper and lower tails of the distribution of the test statistic. The p-value for a two-sided test is illustrated in Figure 12.6. In this example we imagine that the observed value of the test statistic is $z = -2$, and the values which are more extreme are all those values of z less than -2 and also all values of z greater than 2. Thus the p-value will be given by the probability that Z is less than -2 or greater than 2, which is

$$P(Z < -2 \text{ or } Z > 2) = 2P(Z < -2) = 2P(Z > 2).$$

These equalities arise because the distribution of the test statistic is symmetric about zero and so $P(Z < -2) = P(Z > 2)$, as can be seen from Figure 12.6.

All that you really need to know is that if you have a two-sided test then you need to calculate the p-value taking probabilities from both tails of the distribution of the test statistic. If you have a one-sided test then the p-value will come from just one of the tails. In this chapter all the test statistics will have a symmetric distribution, assuming that the null hypothesis is true, and this means that to calculate the p-value for a two-tailed test all you need to do is to take the p-value for one side and multiply it by 2.

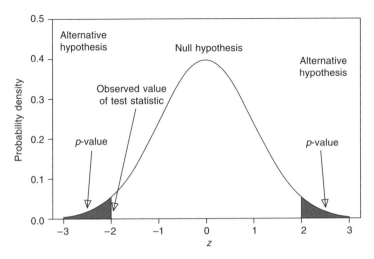

Figure 12.6 p-value for a two-sided test.

In a significance test the decision that you make is based upon a comparison of the p-value, calculated from information in the null hypothesis and the observed data, with the significance level. The significance level is specified in advance, at the same time as you specify the null and alternative hypotheses, and represents the probability of making one type of error associated with significance tests. This error represents concluding that the null hypothesis should be rejected when, in fact, it is true.

12.2.4 Overview of the components of a significance test

This section has been devoted to introducing the key concepts behind significance tests. The customer classification example was used to acquaint you with the idea of having to decide between a pair of competing hypotheses, on the basis of available data. The taste-testing example introduced you to the concepts of the null and alternative hypothesis and the use of the p-value to help you decide between them. Provided that the distribution of a test statistic is known under the assumption that one of the hypotheses is true, then one of the error probabilities can be calculated. This probability can be controlled to specify the level of acceptable risk of error in the test, and is known as the significance level of the test.

The key steps in any significance test are as follows:

1. The specification of the null hypothesis. In this chapter this will take the form of specifying an exact value for a population parameter or that the values of the same parameter in two populations are equal. The null hypothesis will also include specifications about the distribution of the variable and the assumption of random sampling.
2. The specification of the alternative hypothesis. Here you need to decide if the test is to be a one-sided or two-sided test. Normally it will be a two-sided test, unless you have very strong reasons for expecting only one direction of deviation from the null hypothesis.

3. The specification of the significance level. This is the probability of making the error of deciding that the null hypothesis is to be rejected when it is true. This is normally set at a probability of 0.05, usually called the 5% significance level. It is not possible to eliminate the possibility of this type of error occurring completely, and by specifying a small significance level we are saying that the chance of this error occurring is small.

4. The calculation of the test statistic. This is a quantity whose probability distribution is known exactly, assuming the null hypothesis is known. It is a function of the information in the observed data and the null hypothesis.

5. The calculation of the p-value. This is based upon the distribution of the test statistic, assuming the null hypothesis is true, and its observed value. The alternative hypothesis only comes into consideration for the distinction between a one-sided or two-sided test.

6. The decision is made by comparing the p-value to the significance level. If the p-value is less than the significance level then the decision is to reject the null hypothesis in favour of the alternative hypothesis. If the p-value is larger than the significance hypothesis then the null hypothesis is not rejected. There are some extremely subtle points here, which we will come back to at the end of the chapter, regarding the acceptance of the null hypothesis. A large p-value does not mean that the null hypothesis is true, it just means that there is no evidence to say that it is false.

12.3 One-sample t test for a mean

12.3.1 Method

The data in Table 12.3 give the price changes for a random sample of 39 stocks taken from the *New York Herald Tribune* at the last weekend of June 1998. The main reason for this analysis is to see if there was evidence of a general increase in price over the weekend, as opposed to no increase. The data for the 39 changes in prices are displayed in Figure 12.7, and it is clear that a normal distribution is an appropriate assumption for

Table 12.3 The change in price (US dollars) of 39 stocks quoted on the New York Stock Exchange between close on Friday 26 June 1998 and close on Monday 29 June 1998

−0.187	−0.500	0.625
0.313	−0.062	−0.312
−0.312	0.187	0.937
1.188	0.875	1.375
0.250	2.312	0.562
−0.500	0.375	−0.625
0.186	0.562	−1.750
−1.188	0.000	0.250
−0.937	0.188	1.063
0.125	−0.375	−0.937
−0.625	0.500	1.125
2.125	−0.188	0.000
−1.125	−0.375	−1.000

Source: New York Herald Tribune, 30 June 1998.

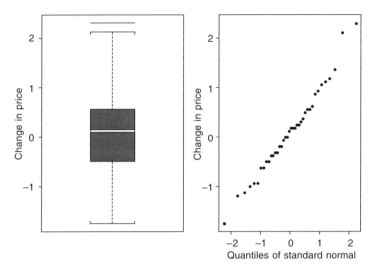

Figure 12.7 Boxplot and normal probability plot of stock price change data.

the distribution of the changes in price as the boxplot is symmetric and the normal probability plot has all the points more or less along a straight line.

In terms of statistical models we would write down:

- The change in stock price, represented by the random variable, X.
- $X \sim N(\mu, \sigma^2)$, where μ represents the population mean price change over all stocks in the population from which the 39 were randomly selected and σ represents the population standard deviation in the price changes. It is important to note that both parameters refer to the population from which the sample is selected and not to the sample itself.
- From the data, we have

$$n = 39, \; \sum x = 4.125, \; \sum x^2 = 29.285, \; \bar{x} = 0.1058,$$

$$s^2 = \frac{1}{39 - 1}\left(29.285 - \frac{1}{39} \times 4.125^2\right) = 0.7592$$

$$\therefore \qquad s = 0.8713.$$

As we have seen in Chapter 11, we can use both $\bar{x}$ and s to estimate the population quantities.

The purpose of the analysis is to see if there is evidence of an increase in price, as opposed to no increase. This fairly general statement has to be turned into a precise statistical hypothesis, involving specifying values for the parameters. If there is an increase in the price then this is going to manifest itself through the value of the mean of the distribution μ and so the hypotheses are going to be phrased in terms of the population mean. If there is no increase in price then clearly $\mu = 0$, and if there is an increase in price then we would expect $\mu > 0$. This gives us the null hypothesis $H_0: \mu = 0$, where the value of the parameter is specified exactly, and the alternative hypothesis, $H_1: \mu > 0$, where μ is unspecified but positive. This will be a one-sided test, as we are looking to see if there has been an increase in price.

In terms of the significance test, we now have the two hypotheses, we know the probability distribution, we have the data, and all we need to get is the test statistic. The hypotheses are concerned with the population mean and the sample mean is an estimate of the population mean, so it is clear that the test statistic is going to link the sample mean and the population mean.

Now the test statistic has to have a distribution which is known exactly assuming the null hypothesis is true, i.e. there are no unknown parameters in the distribution. We know from Chapter 10 that if we have a random sample of n observations from a normal distribution, $X \sim N(\mu, \sigma^2)$, then the sample mean also has a normal distribution, $\overline{X} \sim N(\mu, \sigma^2/n)$ and that the quantity, $Z = (\overline{X} - \mu)/(\sigma/\sqrt{n})$ has a standard normal distribution, $N(0, 1)$. This could be used as a test statistic if we knew what the population value of σ is. We only know the sample estimate, s, and this is not the same thing. In Chapter 11 we discovered that if the population standard deviation was not known but was estimated using the sample standard deviation, then the quantity $t = (\overline{X} - \mu)/(s/\sqrt{n})$ followed a t distribution with $n - 1$ degrees of freedom. This is the required test statistic in this instance.

Let us recap on the derivation. As the null hypothesis concerns the population mean we use the sample mean, which has a normal distribution. The difference between the sample mean and the population mean $(\overline{X} - \mu)$ is standardized by dividing by the standard error of the sample mean, $\sigma/\sqrt{n}$. As σ is unknown it is estimated by the sample standard deviation s, and this leads us to the t distribution.

If the null hypothesis is true, then the population mean is $\mu = 0$ and so the value of the test statistic becomes $t = (\overline{x} - 0)/(s/\sqrt{n})$, and on substituting the values for $\overline{x}$ and s we have

$$t = \frac{0.1058 - 0}{0.8713/\sqrt{39}} = 0.758.$$

The sampling distribution of this test statistic is shown in Figure 12.8, and the p-value is $P(t > 0.758)$, where t has 38 degrees of freedom. Using a computer this is calculated as 0.227. There is no evidence to reject the null hypothesis at the 5% significance level, as the p-value is clearly much larger than $\alpha = 0.05$.

In the example the specified value of μ under the null hypothesis was 0. Normally, μ will take other values and the general terminology is to use a subscript of 0 to indicate the value of a parameter under the null hypothesis. Thus we write $H_0: \mu = \mu_0$ with the alternative hypotheses as $H_1: \mu \neq \mu_0$, $H_1: \mu < \mu_0$, or $H_1: \mu > \mu_0$. The test statistic is written as $t = (\overline{X} - \mu_0)/(s/\sqrt{n})$, and provided the sample of observations is a random sample from a normal distribution then t will follow a t distribution on $n - 1$ degrees of freedom, assuming that the null hypothesis is true. We can then calculate the p-value and compare it to the significance level.

12.3.2 *Investors Chronicle* example

For a second example we use data from 26 June 1998 issue of the *Investors Chronicle*, in which details of various then current takeover bids were given. In 16 cases, the terms of the takeover involve shareholders being paid a bounty per share held (in four other cases,

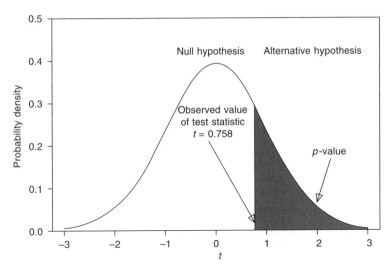

Figure 12.8 Distribution of *t* statistic, on 38 df, for stock price change data.

the bounty was either a share equivalence or in an overseas currency). We will evaluate a rule of thumb relating value and bounty: specifically, we will test the hypothesis that the value of the company to be taken over (V in £ millions) and the bonus per share (p, in £), are on average related by $V = 10^p$. Another way to write this is $p = \log_{10}(V)$, so if we define the random variable $X = p/\log_{10}(V)$ we have $n = 16$ observations to test the hypothesis that the average X is equal to 1.0 against the alternative that it is not.

The values of X for the 16 companies are listed in Table 12.4. We see at once that none of the X-values equals 1.0 exactly, and that they range from 0.262 to 2.081. But it is the average value of X that is our concern; if the average value of X is denoted by μ, then we are formally stating the pair of hypotheses

$$H_0: \mu = 1.0,$$

$$H_1: \mu \neq 1.0.$$

We make the assumption that in the population of all companies about to be taken over the values of X are normally distributed with a standard deviation, σ, which is unknown

Table 12.4 *Investors Chronicle* data

Firm	X	Firm	X	Firm	X	Firm	X
Cliveden	0.582	Bellwinch	0.262	EIS	2.081	Hambros Insurance	0.682
Oriel	0.714	La Senza	0.807	Christies Itnl	1.386	Hambros	0.703
Trust Motor	1.542	Metsec	1.602	Capital Group	1.304	Trafford Park	0.878
SDX Business Sys	1.552	Hunters Arnley	0.823	Brunner Mond	0.879	Courtaulds	1.379

to us. We can check that the distribution in the sample is reasonably normal using the same types of graph as in Figure 12.7. These are shown in Figure 12.9. These may indicate that the normal distribution assumption is not completely valid, but as the sample size is only 16 it is difficult to be absolutely sure and we will proceed with the test. The distribution is a little skew and there is a gap in the normal plot, with no observations between 0.9 and 1.3.

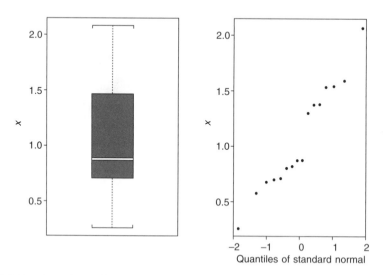

Figure 12.9 Boxplot and normal probability plot of *Investors Chronicle* data.

We will test the null hypothesis at the 5% significance level. Now from the data $n = 16$, $\sum x = 17.176$ and $\sum x^2 = 21.955$. Hence $\bar{x} = 1.074$ and $s = \sqrt{0.2344} = 0.484$. Thus the value of the t statistic is

$$t = \frac{1.074 - 1.0}{0.484/\sqrt{16}} = 0.607.$$

As this test is a two sided test the p-value has to reflect both tails of the t distribution. Thus the p-value is $p = 2P(t > 0.607) = 2 \times 0.2764 = 0.5528$. Thus, we find no evidence in the sample to reject the null hypothesis in favour of the alternative. We conclude that the hypothesis of $\mu = 1.0$ is consistent with the data. In other words, there is ostensibly nothing in this set of takeover data to contradict the idea that on average $V = 10^P$, providing a measure of support for the rule of thumb.

The 95% confidence interval for the population mean is (0.82, 1.33). This is only based upon a sample of 16 observations, and so is relatively wide. With small samples you cannot really expect to estimate populations quantities very precisely.

This is not the whole story, of course. Close inspection of the data (possibly using a stem-and-leaf plot, but also visible in the normal plot of Figure 12.9) reveals that there is a substantial range of X-values, from about 0.9 to 1.3, where no values lie. If the sample size were much larger we might begin to suspect that the data are bimodal. If that were true, then the assumptions underlying the t test have been severely violated, and our conclusion about the rule of thumb would vanish into thin air. One possibility is that the rule of thumb is missing a key factor, such as the type of company to be taken over. In Table 12.4 we can observe a tendency for finance-related institutions to have low X-values, and for technology-related ones to have high X-values, for example. This takes us into an area of statistical modelling to be covered in Chapter 15.

12.3.3 Calculating *p*-values for *t* tests using tables of percentage points of *t* distributions

Generally you will use a computer for calculating the test statistic and the *p*-value in any test. Many of these calculations can be carried out in spreadsheets, though you often need specialized statistical software for more advance tests. As we can see from the graphs of the distributions of the test statistics (Figures 12.3, 12.5, 12.6 and 12.8), the *p*-value is calculated as the area under a curve for continuous variables, which requires integration, or as a sum of probabilities for discrete test statistics. Computer programs can do these integrations and sums very easily and you do not really need to bother much about them. If you do not have access to a computer to calculate the *p*-value for a *t* test you can still do the test using the tables of percentage points of the *t* distribution.

For the stock price change data, $t = 0.76$ on 38 degrees of freedom. Going through the table of percentage points of the *t* distribution on 38 degrees of freedom, we find the following information:

Degrees of freedom	Probability in upper tail					
	0.2	0.1	0.05	0.025	0.01	0.005
30	0.8538	1.3104	1.6973	2.0423	2.4573	2.7500
Normal	0.8416	1.2816	1.6449	1.9600	2.3263	2.5758

We know that as the degrees of freedom get larger the *t* distribution becomes like the normal distribution, and we can see that the percentage points of the *t* distribution on 30 degrees of freedom are not too different from the corresponding values for the normal distribution. In this case we can estimate the percentage points of the *t* distribution on 38 degrees of freedom using either of the above. For illustration we use the row corresponding to 30 degrees of freedom.

The 20% point of the distribution is $t_{0.2} = 0.8538$, which means that $P(t > 0.8538) = 0.2$. For the *p*-value we need to calculate $P(t > 0.758)$, and from the tables we know that this is going to be greater than 0.2. Thus we can report $p > 0.2$ for the *p*-value of the test, and often this is sufficient. You will often find *p*-values reported in scientific papers in this fashion.

When the degrees of freedom of the *t* test are large we can use the tables of probabilities for the normal distribution to calculate the *p*-values. From these tables we can see that $P(t > 0.758) \approx P(Z > 0.76) = 1 - 0.7764 = 0.224$, which is not very far away from the exact value of 0.227.

In the data from the *Investors Chronicle* example there were 15 degrees of freedom and the *t* statistic was 0.607. From the percentage points of the *t* distribution on 15 degrees of freedom we see that this must correspond to $P(t > 0.607) > 0.2$. This was a two-sided test, so this probability has to be doubled to get the p-value. Thus we would report $p > 0.4$.

Degrees of freedom	Probability in upper tail					
	0.2	0.1	0.05	0.025	0.01	0.005
15	0.8662	1.3406	1.7531	2.1314	2.6025	2.9467

If the test statistic had turned out to be 3.0 rather than 0.607 then we would have had $p = 2P(t > 3.0) < 2 \times 0.005 = 0.001$, as 3.0 lies to the right of 2.9467 in the table and so has a smaller probability. Multiplying by 2 arises because it is a two-sided test. If we had found $t = 2.5$ then we would be reporting $p = 2P(t > 2.5) < 2 \times 0.025 = 0.05$ as 2.5 lies to the right of 2.1314 and so has a smaller probability. Sometimes you will find ranges reported for the p-value, and for $t = 2.5$ we could report $0.05 > p > 0.02$ as $p = 2P(t > 2.5) > 2 \times 0.01 = 0.02$ and 2.5 lies to the left of 2.6025 and so has a larger probability. If we had $t = 2.0$ then we would report $0.1 > p > 0.05$ using the same type of argument.

12.3.4 Critical value approach

As the calculation of a p-value from the t distribution was not particularly easy to do until the ready accessibility of computer programs, an alternative mechanism for carrying out significance tests grew up. This mechanism is known as the critical value approach and is linked to the use of tables of the percentage points of probability distributions. We will illustrate this approach with the *Investors Chronicle* data.

The critical value approach is the same as the p-value approach in most details except that the p-value is not calculated. Instead the significance level is used to work out a region of values of the test statistic which is consistent with the null hypothesis and a region which is not consistent with it. This is illustrated in Figure 12.10. For the *Investors Chronicle* data the test statistic follows a t distribution on 15 degrees of freedom. and the observed value was 0.607. The test is a two-sided test at the 5% significance level with hypotheses

$$H_0: \mu = 1.0, \qquad H_1: \mu \neq 1.0.$$

If the null hypothesis is true then we would expect values of the test statistic around zero and if the null hypothesis is not true then we would expect either large or small values of the test statistic. On this basis we can partition the possible values of the test statistic into two distinct regions. One region is known as the **critical region** and is the region associated with the alternative hypothesis, while the other region is associated with the null hypothesis.

The critical region is evaluated by using the significance level, which in this case is $\alpha = 0.05$. We then find the area under the curve representing the distribution of the test statistic corresponding to this probability. As we have a two-sided test we have to split the significance level between the two ends of the distribution, and so we use $\alpha/2 = 0.025$ in either tail of the distribution.

From the tables of the percentage points of the *t* distribution of 15 degrees of freedom we find that the 2.5% point is 2.131, i.e. $P(t > 2.131) = 0.025$. As this distribution is symmetric about zero we know that $P(t < -2.131) = 0.025$. These are known as the **critical values** as they split the possible values of *t* into the region of rejection of the null hypothesis, the critical region, and the region where the null hypothesis is not rejected (Figure 12.10).

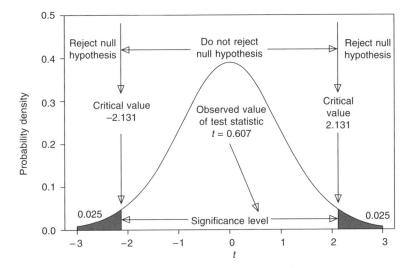

Figure 12.10 Critical values for a significance test.

Using the critical values the possible values of the test statistic *t* are split into two regions. The critical region in this example is in two-parts because we have a two-sided test and is given by all values of *t* such that $t > 2.131$ and $t < -2.131$, while the region in which the null hypothesis is not rejected is given by the remaining values $- 2.131 < t < 2.131$. The decision of the test is then based upon which region the observed value of *t* falls into. In this case $t = 0.607$ which falls in the region in which the null hypothesis is not rejected, and we conclude that there is no evidence to suggest that the mean is not equal to 1.

In the illustration a two-sided test was used, and this means that the significance level is split equally between the two tails of the distribution. When you are carrying out a one-sided test then you assign all of the significance level to the one tail of the distribution. This means that you look up the 5% point of the distribution of the test statistic for a one-sided test, whereas you look up the 2.5% point for a two-sided test.

The *p*-value and critical value approaches will always lead to the same decision as they are mathematically equivalent. As you will normally do the tests using a computer the *p*-value approach is likely to be the way that you will do the tests. The critical value approach is not as important nowadays as it does not provide quite as much information as the *p*-value method. When you calculate the *p*-value you have a measure of the amount of evidence against the null hypothesis, which you do not have with the critical value approach.

For example, if you have $p = 0.003$ then you will reject the null hypothesis at the 5% level of significance, but you have extra information that the p-value is very small, so the evidence against the null hypothesis is very strong. If $p = 0.049$ then you still reject the null hypothesis, but you can see that the evidence is weak.

In both these situations you had a test statistic which gave rise to the p-value. Now suppose that you don't calculate the p-value but carry out a straight comparison with the critical value instead. The latter is the same in each case as it depends on the significance level of 5%. Your test statistic will lie in the critical region in both situations, but you now have no quantitative measure of the strength of the evidence against the null hypothesis.

12.4 *t* test for the difference of two independent means

12.4.1 Small samples

The data in Table 12.5 give the earnings per share and the dividend per share for 15 companies listed in the Extel financial database. Extel is part of Financial Times Information and gives a range of financial information on 500 companies. This information is available for several years and the data in the table refer to the latest accounts, either 1995 or 1996. There is information on 14 companies in the breweries, pubs and restaurants industry and 17 in the hotel and leisure industry. Random samples of 7 and 8 companies, respectively, were selected.

Table 12.5 Earnings and dividends per share

Breweries, pubs and restaurants			Leisure and hotels		
Company	Earnings per share, pence	Dividend per share, pence	Company	Earnings per share, pence	Dividend per share, pence
Bass	0.504	0.25	Rank	0.603	0.1575
Greenhalls	0.3205	0.154	Stanley	0.186	0.0665
Mansfield	0.2062	0.059	Thistle	0	0
Wolverhampton Dudley	0.457	0.17	Millennium Copthorne	0	0
Compass	0.316	0.086	Airtours	0.4913	0.16
Vaux	0.2093	0.106	Ladbroke	0.0514	0.06
Whitbread	0.4608	0.2185	First Leisure	0.1764	0.0772
			Stakis	0.0579	0.0215

Source: Extel data from Biz/Ed, http://bizednet.bris.ac.uk:8080/dataserv/datahome.htm.

In the analysis of these data we will see if there is any evidence to suggest that the earnings per share and the dividend per share are different in these two groups of industries. In a number of senses the statistical analysis of these data will differ from a financial analysis. From the statistical point of view we will look to see if there are broad differences between companies operating in slightly different areas. From this point of view we do not care very much about the specific companies and are only really interested in them as members of the two groups. From a financial and accounting point of view the main interest is likely to be in the individual companies themselves.

For both of these financial ratios we want to see if there is any difference between the companies in the two groups of industries. In order to do this we proceed as usual, using statistical models to describe the variables in terms of parameters, using the data to estimate these parameters and setting up the hypothesis to test.

The variables are both continuous, and if we had looked at boxplots for each industry we would have seen that the distributions are reasonably symmetric, although we have very small samples. Consequently, we make the assumption that the earnings per share and the dividend per share are both normally distributed.

When looking for a difference in the financial ratios between the companies in the different industries we are principally interested in seeing if the average is higher in one industry than in the other. We set up a statistical model in which X_i represents the earnings per share, with the subscript i taking the value 1 for the brewing industry and 2 for the hotel and leisure industry. We assume that $X_i \sim N(\mu_i, \sigma^2)$ and that the two groups of companies are independent of each other. This means that we believe that there may be differences between the population means of the two distributions, but that the standard deviations are the same. We could postulate a model in which the standard deviations over the two group are different, but this introduces another level of complexity which we will avoid for the moment.

If we specify a null hypothesis that the two means are equal and an alternative hypothesis that they are different, then we have a means of deciding if the average earnings per share is or is not the same in the two industries. These are written as

$$H_0: \mu_1 = \mu_2, \qquad H_1: \mu_1 \neq \mu_2,$$

though an equally valid way of writing them is

$$H_0: \mu_1 - \mu_2 = 0, \qquad H_1: \mu_1 - \mu_2 \neq 0.$$

In the latter framework we are back in a familiar situation similar to that for the one-sample test where we have a population parameter equal to a specified value.

As we are testing if two means are equal we will use the sample means to estimate the unknown population means. As the population standard deviation is unknown we will use the sample standard deviations to estimate this. There are two samples and so there are two sample standard deviations, but only one population standard deviation. Thus we will have to combine the two standard deviations to get the estimate of the population standard deviation.

From our knowledge of sampling distributions (Chapter 10), we know that $\overline{X}_i \sim N(\mu_i, \sigma^2/n_i)$, where there are n_i observations in sample i. In Chapters 10 and 11 we also saw that $\mathrm{Var}(\overline{X}_1 - \overline{X}_2) = \sigma^2(1/n_1 + 1/n_2)$ if the two samples are independent of each other. Furthermore, the distribution of $\overline{X}_1 - \overline{X}_2$ is also normal. In order to test the hypothesis we will use a test statistic which is based upon the difference between the two means.

We have obtained a test statistic in previous examples by using a function of the data to estimate the function of the parameters in the null hypothesis, together with a measure of the sampling variability in the estimate. For many test statistics this is achieved by standardizing them:

$$\frac{\text{Sample estimate} - \text{Population value}}{\text{Standard error of the sample estimate}}.$$

The null hypothesis here is that the difference of two population means takes a specified value, H_0: $\mu_1 - \mu_2 = 0$. The function of the data to estimate this difference will be the difference of the sample means, $\bar{X}_1 - \bar{X}_2$. We know what the variance of the difference of the sample mean is, $\sigma^2(1/n_1 + 1/n_2)$, and so we can standardize. This gives as a possible test statistic the quantity

$$Z = \frac{(\bar{X}_1 - \bar{X}_2) - (\mu_1 - \mu_2)}{\sigma\sqrt{\left(\dfrac{1}{n_1} + \dfrac{1}{n_2}\right)}}.$$

If the population standard deviation was known then Z would have a standard normal distribution and could be used as a test statistic. As σ is unknown we will need to estimate it from the sample and use

$$t = \frac{(\bar{X}_1 - \bar{X}_2) - (\mu_1 - \mu_2)}{s\sqrt{\left(\dfrac{1}{n_1} + \dfrac{1}{n_2}\right)}},$$

where s is the pooled sample standard deviation calculated from both the samples. This is the test statistic for the pooled two-sample t test, and if the null hypothesis is true then this statistic will follow a t distribution on $n_1 + n_2 - 2$ degrees of freedom. This procedure is very similar to that for the one-sample t test.

We have already come across the pooled sample standard deviation used in the estimation of the common standard deviation in both populations, in Chapter 11. The pooled variance is given by

$$s^2 = \frac{(n_1 - 1)s_1^2 + (n_2 - 1)s_2^2}{n_1 + n_2 - 2},$$

where s_i^2 is the sample variance in the ith sample.

Earnings per share

The summary statistics for the earnings per share are given in Table 12.6. We are now in a position to calculate the test statistic. The pooled standard deviation is

Table 12.6 Summary statistics for the earnings and dividend per share

	Industry	Number	Mean	Standard deviation
Earnings per share	Brewers, etc.	7	0.3534	0.1224
	Hotels and leisure	8	0.1958	0.2300
Dividend per share	Brewers, etc.	7	0.1491	0.0700
	Hotels and leisure	8	0.0678	0.0633

$$s = \sqrt{\frac{(n_1 - 1)s_1^2 + (n_2 - 1)s_2^2}{n_1 + n_2 - 2}} = \sqrt{\frac{6 \times 0.1224^2 + 7 \times 0.2300^2}{13}} = 0.1881,$$

and the estimated standard error of the difference between the two means is

$$s\sqrt{\left(\frac{1}{n_1} + \frac{1}{n_2}\right)} = 0.1881\sqrt{\left(\frac{1}{7} + \frac{1}{8}\right)} = 0.0974.$$

This means that the value of the *t* statistic assuming that the null hypothesis, H_0: $\mu_1 - \mu_2 = 0$, is true is

$$t = \frac{(\overline{X}_1 - \overline{X}_2) - (\mu_1 - \mu_2)}{s\sqrt{\left(\frac{1}{n_1} + \frac{1}{n_2}\right)}} = \frac{(0.3534 - 0.1958) - 0}{0.0974} = 1.618.$$

The degrees of freedom are $n_1 + n_2 - 2 = 7 + 8 - 2 = 13$ and the *p*-value is obtained by calculating $P(t > 1.618) = 0.065$ and multiplying by 2 because the alternative hypothesis is two-sided. This gives the *p*-value as $2 \times 0.065 = 0.130$.

As the *p*-value is greater than 0.05 we conclude that at the 5% level of significance there is no evidence in these data to reject the null hypothesis that the means of the two populations are equal. This means that we conclude there is no difference between the two groups of companies as regards the earnings per share. This conclusion is based upon a number of assumptions, among them that we have a random sample of companies from the populations, that the variance of the earnings per share is the same in the two populations, and that the normal distribution is valid. If these assumptions are not valid, then the test is not valid.

Dividend per share

Using the data on the dividend per share we have the same null and alternative hypotheses:

$$H_0: \mu_1 - \mu_2 = 0, \qquad H_1: \mu_1 - \mu_2 \neq 0.$$

Using the summary data in Table 12.6, we calculate the pooled standard deviation of $s = 0.0665$ and a *t* statistic of 2.363. This corresponds to a two sided *p*-value of 0.034. As $p < 0.05$ we reject the null hypothesis at the 5% significance level and conclude that there is evidence in the sample to suggest that the dividends per share are different in the two industry groups. From the means in Table 12.6 we can see that the dividends per share are higher among companies in the breweries, pubs and restaurants sector than in the hotel and leisure sector.

As the difference in the means is significantly different from zero we are interested in estimating this difference. Our estimate is $0.149 - 0.068 = 0.081$p, with a 95% confidence interval of $(0.007, 0.156)$p. For one share there is not a great deal of difference but if you have thousands of them, as you might have if you were managing an investment fund, then the financial implications of the difference are more considerable.

12.4.2 Large samples

The summaries in Table 12.7 are based upon data taken from a quota sample of 533 men aged 16–65 (see Romaniuk *et al.* 1999). The survey was carried out in five European countries by a market research company on behalf of some manufacturers of personal care products. All the panellists were asked to complete a diary for seven days, beginning on the first Monday after they were recruited into the study and ending on the following Sunday. For each product under study the panellists were asked to record the day and time of use as well as other associated information about the product. Demographic information about the panellists was also collected.

Table 12.7 Shampoo use per week: means and standard deviations

	Number	Mean	Standard deviation
Age group			
16–44	322	4.11	2.95
45–65	211	3.31	2.87
Social class			
A, B, C1	251	3.85	2.80
C2, D, E, Unclassified	282	3.75	3.07

Note: These figures are not presented in Romaniuk *et al.* (1999). They were obtained by simulation based upon parameters and data reported in the paper. Thus these summaries are only representative of the data obtained in this study and in other similar studies of consumers' use

For the whole sample the mean and standard deviation of the number of times shampoo was used in the week were 3.78 and 2.94, respectively. This corresponds to using shampoo every two days, on average. In this analysis we are going to see if there is any evidence of a different frequency of shampoo use among younger men than among older men, and in the different social classes.

We follow more or less the same procedure as above, the only difference now being that there are large samples. In this case we do not have to be overly concerned about the distribution of the variable as we know that in large samples the sample mean will follow a normal distribution. For the application of the *t* test in small samples we need to have a normal distribution, but in large samples this is not absolutely necessary.

In fact, the mean and standard deviation above are both of similar magnitude and, furthermore, you know that negative values for the number of times are impossible. This means that the distribution is likely to be very skew, with a long tail towards large numbers. This is the case in Figure 12.11, where you can also see that there is a peak at 0 corresponding to men who did not use the products and another one at 7 corresponding to men who used the products daily. This type of distribution is common in studies of consumer use where there is a subgroup who never use the product and another subgroup with extremely regular use. For more sophisticated analyses more complicated models will be needed taking into account the distribution, but in this example we can make do with the simple assumption of a normal distribution for the sample mean in the large samples.

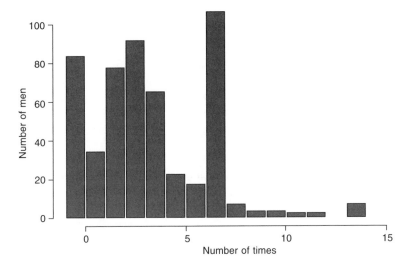

Figure 12.11 Shampoo use.

Age group

There are two populations of men, one aged 16–44 and the other 45–65. We assume that we have random samples from each of these two populations. If X represents the number of times shampoo is used per week then we further assume that the mean number of uses is μ_i and that the standard deviation is the same in both populations, σ. As we have large samples we do not need to worry about the distribution of X and assume $\overline{X}_i \sim N(\mu_i, (\sigma^2/n_i))$, where there are n_i observations in each sample.

The null and alternative hypotheses are

$$H_0: \mu_1 - \mu_2 = 0, \qquad H_1: \mu_1 - \mu_2 \neq 0.$$

corresponding to a two-sided test as we have no information on the direction of departure from the null hypothesis. Most tests will be two-sided as we cannot rule out differences in either direction.

The pooled sample standard deviation is

$$s = \sqrt{\frac{(n_1 - 1)s_1^2 + (n_2 - 1)s_2^2}{n_1 + n_2 - 2}} = \sqrt{\frac{321 \times 2.95^2 + 210 \times 2.87^2}{531}} = 2.919$$

and the estimated standard error

$$s\sqrt{\left(\frac{1}{n_1} + \frac{1}{n_2}\right)} = 2.919\sqrt{\left(\frac{1}{322} + \frac{1}{211}\right)} = 0.259.$$

The *t* statistic is thus

$$t = \frac{(\overline{X}_1 - \overline{X}_2) - (\mu_1 - \mu_2)}{s\sqrt{\left(\frac{1}{n_1} + \frac{1}{n_2}\right)}} = \frac{(4.11 - 3.31) - 0}{0.259} = 3.089.$$

Up until this point all the calculations have been exactly the same as before. At this stage we use the normal distribution to calculate the p-value as the t distribution on 531 degrees of freedom is to all intents and purposes indistinguishable from the standard normal distribution. The p-value is $2P(t > 3.089) = 2 \times 0.001 = 0.002$. As $p < 0.05$ we reject the null hypothesis at the 5% significance level and conclude that there is very strong evidence to suggest that the mean shampoo use is not the same in the two age groups. Shampoo use is more common among younger men by an estimated $4.11 - 3.31 = 0.8$ times per week with a 95% confidence interval of $(0.3, 1.3)$ times.

Social class

If we repeat the procedure for the two social class groups then we find that the pooled standard error is 2.95 with an estimated standard error of the difference of the two means of 0.26 and a t-value of 0.38 corresponding to a p-value of 0.71. As this is greater than 0.05 there is no evidence to suggest that the mean frequency of shampoo use is different in the two social class groups.

In terms of the marketing interpretation of this analysis, we know that men use shampoo about once every two days, on average. There is no social class effect so men from social classes A, B, C1 use shampoo as frequently as men from classes C2, D, E. There is an effect of age and younger men use shampoo just under once per week more than older men. This is entirely reasonable – they have more hair and tend, for example, to take part in more sports. This might imply that advertising campaigns for shampoos be targeted at younger men rather than older men as the sales are greater in this segment of the market.

12.4.3 Unequal standard deviations

In the analysis of the consumer use data on shampoos we assumed that the standard deviations in the two populations were the same. This is quite reasonable based upon the data in Table 12.7. If we have large samples, equal variances do not constitute a very important assumption and can be omitted, leading to a slight modification to the test statistic.

If we believe that the standard deviations in the two populations (σ_i) are not the same then we use s_i to estimate σ_i and calculate the estimated standard error of the difference in the two sample means as

$$\sqrt{\left(\frac{s_1^2}{n_1} + \frac{s_2^2}{n_2}\right)},$$

and the test statistic becomes

$$t = \frac{(\bar{X}_1 - \bar{X}_2) - (\mu_1 - \mu_2)}{\sqrt{\left(\frac{s_1^2}{n_1} + \frac{s_2^2}{n_2}\right)}}.$$

If we have a large sample then we can just refer t to the standard normal distribution to calculate the p-value. Using the example of shampoo use by age group gives a t statistic of 3.113 and a *p*-value of 0.0018 with the above formula compared to 3.089, $p = 0.0020$, using the pooled sample standard deviation. In this case you come to exactly the same conclusions with either procedure, and this is the general situation.

If you have small samples from normal distributions with different population standard deviations then you can use the above test statistic. In this case the normal distribution cannot be used to give you the *p*-value and you must use a *t* distribution. However, the degrees of freedom are not given by $n_1 + n_2 - 2$ but are calculated from a formula based on terms for the number of observations in each group and the standard deviation in each group. This will usually result in degrees of freedom which are not whole numbers and means that you will have to use a computer program to carry out the test.

You should find that most computer programs for carrying out the calculations for the two-sample *t* test have an option to choose between the pooled (equal variances) version and the unpooled (unequal variances) version. You should normally choose the pooled version unless one of the sample variances is more than twice as large as the other one. This is a rough rule and it is possible to carry out a formal test of the equality of two variances, using what is known as an *F* test. This is not covered in this book – all we will say is that the style of the test is the same as for the tests for a difference of two means, except that it is based upon a ratio of two variances rather than their differences. This means that you need to use another probability distribution, known as the *F* distribution.

12.5 *t* test for paired differences

12.5.1 Example of dependent samples

The financial liquidity of a company is an important measure for accountants. Roughly speaking, it is the ability of a company to meet any short-term debts with short-term assets. There are two common accounting ratios used to measure liquidity. One of these is the current ratio, given by current assets divided by current liabilities. If this ratio is above 1 then the company has sufficient assets to pay its debts. The other measure is the acid test ratio, which is similar to the current ratio except that stock is substracted from current assets and the answer divided by current liabilities. The stock of a company is its raw materials and components, work-in-progress and finished goods; as it may take a long time before these can be converted into cash their value may not be used to pay current liabilities. Thus the current ratio may overestimate the liquidity of a company, and the acid test ratio is a more rigorous measure of liquidity.

The data in Table 12.8 give the acid test ratios for a group of companies for 1996 and 1997. The dividend per share is also given. This measures the amount of money paid out by the company per share to its shareholders. The purpose of the analysis is to see if the acid test ratio or the dividend per share have changed over the two years studied.

In Table 12.8 there are data from 11 companies. In reality these are not a random sample of all companies registered at the London Stock Exchange, but for the purposes of this example we will assume that they are. We wish to know if there is evidence to suggest that the acid test ratio has changed between 1996 and 1997, and similarly for the dividend per share.

Although there are two samples of values, those for 1996 and those for 1997, this is

Table 12.8 Financial ratios for selected companies

	Dividend per share (p)		Acid test ratio		Difference
	1996	1997	1996	1997	(d)
Manchester Utd	5.20	6.20	1.18	1.72	0.54
Sheffield Utd	0.00	0.00	0.56	0.64	0.08
Hi-Tec	NA	1.20	0.57	0.77	0.20
Yates Brothers	4.32	3.60	0.16	0.29	0.13
Glaxo	34.00	35.00	0.65	0.76	0.11
Tate and Lyle	17.00	17.00	0.99	0.92	−0.07
Zeneca	35.00	38.50	1.02	0.91	−0.11
Cadbury-Schweppes	17.00	18.00	0.52	0.79	0.27
Harry Ramsden's	5.00	5.00	0.42	0.39	−0.03
Unilever	8.00	8.40	0.67	1.29	0.62
Kwik-Save	20.00	20.00	0.16	0.18	0.02

$$\Sigma\, d = 1.76$$
$$\bar{d} = 0.16$$
$$s_d = 0.2369$$

Source: Biz/Ed company reports (http://bizednet.bris.ac.uk:8080), where detailed interpretations of the financial ratios can also be found.

not a two-sample t test as in Section 12.4 as the two samples are related to each other. Each company has two observations, one for 1996 and the other for 1997. This means that the two samples cannot be independent of each other and so one of the key assumptions for the use of the two-sample t test is violated. In order to use the two-sample t test you need to have two samples which are completely unrelated to each other.

12.5.2 Paired t test

To carry out the test we have to take into account the dependence between the two observations for each company, and this is achieved by considering the differences in the ratios from 1996 to 1997. If we let X_1 be the acid test ratio in 1996 and X_2 the ratio in 1997, then the difference is, $d = X_2 - X_1$ (Table 12.8). These differences have a mean in the population of companies of μ_d and a standard deviation of σ_d. The are estimated by the sample mean of the differences, $\bar{d}$, and the sample standard deviation of the differences, s_d.

In the null hypothesis we assume that there is no difference between the 1996 and the 1997 ratios and, in terms of the separate measurements for the two years, this can be expressed as H_0: $\mu_1 = \mu_2$ or H_0: $\mu_1 - \mu_2 = 0$. The analysis is in terms of the differences, so we must write the null hypothesis in terms of the differences. Since $d = X_2 - X_1$, then $\mu_d = \mu_2 - \mu_1$ and we have

$$H_0: \mu_d = 0, \qquad H_1: H_d \neq 0,$$

the alternative being two-sided since we are interested in any change in the ratios over the two years.

By calculating the differences in Table 12.8, the two-sample problem has been reduced to a one-sample problem, exactly the same as in Section 12.3. The test statistic is thus a t statistic, based upon the differences, and is given by

$$t = \frac{\bar{d} - \mu_d}{s_d/\sqrt{n}} = \frac{0.16 - 0}{0.2369/\sqrt{11}} = 2.24.$$

If we assume that the differences are normally distributed, and the companies are a random sample from all companies, then this *t* statistic should follow a *t* distribution with $n - 1$ degrees of freedom, where n is the number of differences.

In this example there are 11 companies and therefore 10 degrees of freedom. The *p*-value for the test is thus $2P(t > 2.24) = 0.049$. As this is less than 0.05 we conclude that there is evidence at the 5% significance level to conclude that there has been a change in the acid test ratio in companies from 1996 to 1997. The mean difference is 0.16 and the 95% confidence interval is (0.001, 0.319). Within each company we estimate that there has been an increase in the ratio of 0.16 units.

If we repeat the calculations for the dividend per share we see that there are only 10 companies which can be used as no dividend per share is reported for Hi-Tec in 1996. The test statistic is $t = 1.68$ on 9 degrees of freedom. The two sided *p*-value is 0.13, and as this is in excess of 0.05 we have no evidence to reject the null hypothesis of a mean difference of zero in the population at the 5% level. There is no evidence that the mean dividend per share has increased.

12.5.3 Relationship with *t* test of the means of two independent samples

The difference between the paired *t* test and the two-sample *t* test in Section 12.4 is important. You can only use the (pooled) two-sample *t* test if the samples are independent, and this test is not valid for the company data in Table 12.8 because there are two observations on each company. Consequently the two samples are related to each other.

If we think in terms of some statistical theory about expected values and variances we can see why the use of dependent samples, specifically two observations on the one individual, may be useful. In Section 9.1 we looked at the variance of the difference of two dependent random variables and at the variance of the difference of two independent random variables (see also Sections 11.4 and 11.5). We will use these and other similar results here to look at situations in which you might want to use dependent samples and situations in which you might use independent ones.

If we assume that $X_i \sim N(\mu_i, \sigma^2)$ and that we have two independent random samples each of size n, then for the sample means we have $\bar{X}_i \sim N(\mu_i, \sigma^2/n)$. The expected value and variance of the difference in the two means are

$$E[\bar{X}_1 - \bar{X}_2] = \mu_i - \mu_2,$$

$$\text{Var}[\bar{X}_1 - \bar{X}_2] = \text{Var}[\bar{X}_1] + \text{Var}[\bar{X}_2] = \frac{\sigma^2}{n} + \frac{\sigma^2}{n} = \frac{2\sigma^2}{n}.$$

If the two random variables are not independent, then this means that there is some association between them; this is measured by the correlation, ρ, between them (see Chapters 6 and 15). If X_1 is the first value for an individual and X_2 is the second then the difference is $d = X_1 - X_2$. The expected value and variance of the difference are

$$E[d] = E[X_1 - X_2] = \mu_1 - \mu_2,$$

$$\text{Var}[d] = \text{Var}[X_1 - X_2] = \text{Var}[X_1] + \text{Var}[X_2] - 2\,\text{Cov}[X_1, X_2]$$

$$= \sigma^2 + \sigma^2 - 2\rho\sigma\sigma = 2(1 - \rho)\sigma^2.$$

This means that the expected value and variance of the mean of the differences is

$$E[\bar{d}] = \mu_1 - \mu_2,$$

$$\text{Var}[\bar{d}] = \frac{2(1 - \rho)\sigma^2}{n}.$$

The expected value of the mean difference is the same as the expected value of the difference in the means. However, the variance of the mean difference is not the same as the variance of the difference in the means, due to the correlation between the two variables.

If there is no correlation, $\rho = 0$, then $\text{Var}[\bar{d}] = \text{Var}[\bar{X}_1 - \bar{X}_2]$ and the two tests are equivalent. Usually, if two observations on the same quantity are taken on the same individual at different times there will be a positive correlation between them. This means that $\rho > 0$ and so $\text{Var}[\bar{d}] < \text{Var}[\bar{X}_1 - \bar{X}_2]$, i.e. the variance of the mean difference is smaller than the variance of the difference in means. The opposite would occur if the correlation were negative, $\rho > 0$, but this is not likely to occur.

From the point of view of doing a study there is a benefit from taking two observations on the same individual. If a cosmetics company is testing a sun protection cream then they might consider an experiment in which two groups of individuals, one with the cream and the other without the cream, are exposed to the sun. The amount of sunburn would be measured after exposure to the sun, possibly by some measure of skin damage. This has the set-up for a two-sample test as there are two groups of individuals and one measurement will be taken on each.

Individuals will exhibit varying reactions and, without any sun cream, it is quite possible to find two individuals exposed to exactly the same amount of sun but with differing amounts of sunburn. People with red or blond hair tend to have skin which is more susceptible to sunburn than people with black hair, for example. This means that some of the differences in mean sunburn between the two groups of individuals will be due to differences in skin type as well as differences in the use of the sun cream. The experiment could be more carefully controlled by, for example, making sure that the two groups of individuals were as similar as possible.

The most extreme form of control would be to use each individual as his or her own control by using the cream on one half of the back but not on the other. With this study design there will be two measurement for each individual and the analysis would be along the lines of a paired test. This should be a better study design than comparing two independent groups of individuals because the amount of exposure to the sun will be exactly the same on each half of the back. Also the skin type will be the same. Thus the differences between the sunburn measurements from the half of the back with cream and the half of the back without cream will eliminate any effects of the individual person such as the skin type and length of time in the sun. These effects will not be eliminated from the study with two independent groups.

This inplies that paired experiments should have a greater precision than ones with two independent samples. Using two observations on one individual eliminates differences between individuals as a source of variability. When you have two independent groups then some of the variability in the responses is due to variation among the individuals.

12.5.4 Relationship with binomial (sign) test

In Table 12.8 there were three negative and eight positive values among the differences. Under the assumption that there was no change in the acid test ratio from 1996 to 1997 among these 11 companies, we would expect as many positive differences as negative ones. Another way of putting this is to say that we would expect the probability of a positive difference to be 0.5. If there was a change then there would either be more positive values or more negative values. Thus we can phrase a null and alternative hypothesis in terms of the probability of obtaining a positive or negative difference, p. These can be written as

$$H_0: p = 0.5, \qquad H_1: p \neq 0.5.$$

As there are 11 companies and each can give rise to a positive difference or a negative difference (ignoring for the minute the possibility of a difference of zero), the binomial distribution might be a good probability model for the number of positive differences. We have to further assume that the companies are independent. This is exactly the same set-up as in the tasting example in Section 12.1.2. Because we are counting the number of positive (or alternatively the number of negative) signs, this test is known as the **sign test**.

If the null hypothesis is true then the number of positive differences out of 11 has a binomial distribution with $p = 0.5$. We observe 8 and $P(X \geq 8) = 0.1133$. As this is a two-sided test we have to double this value for the p-value to get $p = 0.2266$. As this is greater than 0.05 we conclude from the sign test that there is no evidence to suggest that the acid test ratio has increased from 1996 to 1997.

We have a dilemma between the conclusion from the sign test and that from the paired t test. The paired t test has a p-value of 0.049 and we concluded there was some evidence to suggest a change in the ratio at the 5% level. The sign test has a p-value of 0.227 and we conclude no difference. The two tests purport to be testing essentially the same hypothesis, yet we get quite different results.

The solution to the dilemma lies in the different assumptions underlying the test. In the sign test we assumed that the companies were independent and measured only an increase or a decrease in the acid test ratio. In the paired t test we assumed that the differences in the acid test ratios were a random sample from a normal distribution. Implicit in this assumption is another, that the companies were a random sample from the population of companies.

In the sign test we only use the sign of the difference, not its value as in the paired t test. Thus the sign test uses less information. In both we are assuming that the companies are independent of each other but in the t test we are assuming that the companies are a random sample from a larger population. Thus the t test is based upon extra assumptions. You will generally find that the p-value from the t test is smaller than that from the sign test.

On one hand the paired t test is a more powerful test as it uses more information, but on the other it is less robust as it is based upon more assumptions. If the assumptions are valid then the t test is the better test to use. If the assumptions of normality and a random sample from a population are not valid then the sign test is more appropriate.

In the example it is clear that the companies are not a random selection of all companies. This suggests that the assumptions of the paired t test are not completely valid. This means that the sign test is probably the more reliable one to use here.

The sign test is one of a number of tests known as **randomization** or **exact** test which are based upon working out the exact distribution of the test statistic by considering all possible values of it given the sample data. It is the only one we will discuss, but others you might come across are the Wilcoxon rank sum test and the Wilcoxon–Mann–Whitney U test, which are both based upon the ranks (ordinal information of the data and not the observed values. These tests do not rely on having a random sample from a population and are based only on the sample data.

The t tests in this chapter are all parametric tests as they are based upon making some assumption about a probability distribution, such as a normal distribution for the observations or the sample mean, and having a random sample of observations from a population. The most common statistical tests are all parametric and to use them you need to be sure that the assumptions on which they are based are valid. Generally slight violations of the distributional assumption are not too serious, so you can use the t test even if there is a slight skew in the distribution or if there are a few outliers.

12.6 Testing proportions

12.6.1 One sample test

The sign test in Section 12.5.4 is an example of a one-sample test on a proportion. It is used in the special case when the proportion is 0.5. If the null hypothesis specifies another value for the proportion then it is possible to carry out a test using the binomial distribution. In practice, most of the time you are likely to be testing proportions based upon large samples and so can use the normal approximation to the binomial distribution to evaluate the distribution of the test statistic. As a result the tests for proportions are based upon the normal distribution, and so are similar to the tests on means above.

In an article in the *Guardian* on 9 June 1998 entitled 'Strong backing for on-the-spot drug fines' the results of a opinion poll survey by ICM were discussed. The survey was designed to elucidate public opinion on the question of on-the-spot fines for the possession of cannabis. The survey was carried out by telephone between 5 and 6 June 1998, and 1201 adults aged 18 and over were interviewed. The sample results were adjusted for non-response bias by using sample weights to match the sample demographic profile with that of the propulation. Some of the points made were as follows:

> More than 75 per cent of the public believe that drug awareness lessons should be given in primary schools, demolishing fears that parents would be shocked by it.
> A significant minority (47 per cent) also believe the illegality of such [soft] drugs actually encourages teenagers to experiment with them. Only 13 per cent believe that criminality actually deters teenagers from trying them. Among 18 to 24 year olds, the

proportion who believe that illegality is part of the attraction rises to 64 per cent, against 8 per cent who think that it is a deterrent.

I will use this example to see if, in fact, there is evidence that a minority of the public believe that the illegality of such drugs encourages teenagers to experiment with them.

Let us assume that in the population of adults a proportion, p, believe that the illegality of drugs encourages teenagers to experiment with them. If this proportion is less than 0.5 then we have a minority, while if it is greater than 0.5 we have a majority of adults; if it is 0.5 then there is neither a majority nor a minority. This means that we can specify the null hypothesis as $H_1: p = 0.5$ and the alternative hypothesis will be one-sided as we are looking for a minority, $H_1: p < 0.5$.

If we have a random sample of size n, then the number of people who believe that the illegality of drugs encourages teenagers to experiment with them will follow a binomial distribution with $p = 0.5$ assuming the null hypothesis is true. In this case the sample size is large, and so we can use the normal approximation to the binomial. This tells us that in large samples the sample proportion $\hat{p} \sim N(p, p(1 - p)/n)$. If we standardize the sample proportion to get

$$Z = \frac{\hat{p} - p}{\sqrt{p(1 - p)/n}},$$

then Z will follow a standard normal distribution, assuming that the null hypothesis is true. Consequently, we can use Z as the test statistic as its distribution is known exactly under the null hypothesis. In this instance we expect to find values of Z around zero if the null hypothesis is true, and large negative values of Z will provide evidence that the null hypothesis is not true.

In the sample we find that 47% of the public believe the illegality of drugs actually encourages teenagers to experiment with them. This means that the sample proportion is $\hat{p} = 0.47$. The sample size is $n = 1201$. If the null hypothesis is true then $p = 0.5$ and so the standard error of the sample proportion is

$$\sqrt{\frac{p(1 - p)}{n}} = \sqrt{\frac{0.5 \times 0.5}{1201}} = 0.0144.$$

The calculated value of the test statistic from the sample is

$$Z = \frac{\hat{p} - p}{\sqrt{\frac{p(1 - p)}{n}}} = \frac{0.47 - 0.5}{0.0144} = -2.083.$$

The p-value is therefore $P(Z < -2.083) = 0.019$. As this is less than 0.05 we conclude that there is evidence at the 5% significance level to suggest that the proportion of the public who believe that the illegality of drugs actually encourages teenagers to experiment with them is less than 0.5. This means that it is still a minority of the public.

The article also mentions that only 13% of the public believe that criminality actually deters teenagers from trying drugs. If we had some prior hypothesis about this value, say that it was bigger than 10%, then we could perform another significance test. However, there is no compelling reason for doing so unless, for example, the 10% came from a very large previous study, or was derived from a census or from another population. None of this is the case here.

Generally with opinion polls you will find that you are less interested in testing a single value against a prespecified value than in estimating a value. As such the methods described in Chapter 11 using confidence intervals are more useful. On the other hand, you will often be faced with the problem of comparing percentages from two sub-populations. In the above example it would be natural to compare the percentage of 18–24-year-olds who believe that criminality actually deters teenagers from trying drugs with the percentage of people over 24 who do so. This is a two-sample test of the equality of two proportions.

12.6.2 Two sample test

In the article in the previous subsection it is stated that only 8% of 18–24-year-olds believe that criminality actually deters teenagers from trying drugs, compared to 13% of the population. If there are 139 men and women aged 18–24 in the sample and 1062 people aged 25 and over, then we can work out the percentage for the older age group from

$$\frac{139 \times 0.08 + 1062 \times p_{25+}}{1201} = 0.13,$$

which gives $p_{25+} = 0.14$. Out of the 139 people aged 18–24, 8% believe criminality deters teenagers from trying drugs, compared to 14% of the 1062 people aged 25 or over. The difference of 6% looks large, but we need to carry out a statistical significance test to assess the evidence against the hypothesis that the two population percentages are equal to each other.

In our statistical model of the sample data we assume that there are two distinct populations, one with people aged 18–24 and the other with people aged 25 or over. Random samples of size n_i are selected from each population. The population proportion is denoted p_i for population i, and we wish to test the hypothesis that these two proportions are equal to each other, against the alternative that they are not equal to each other. This gives the null hypothesis $H_0: p_1 = p_2$ or $H_0: p_1 - p_2 = 0$, and the two-sided alternative $H_1: p_1 \neq p_2$ or $H_1: p_1 - p_2 \neq 0$.

As with the case of the test for the equality of two means, the test statistic is derived from the difference of the two sample proportions. As the samples are large we know that $\hat{p}_i \sim N(p_i, p_i(1 - p_i)/n_i)$, and as the samples are selected independently of each other the two sample proportions are independent. This means that the standard error of the difference between the two sample proportions will be given by

$$\sqrt{\frac{p_1(1 - p_1)}{n_1} + \frac{p_2(1 - p_2)}{n_2}} = \sqrt{p(1 - p)\left(\frac{1}{n_1} + \frac{1}{n_2}\right)},$$

where p is the common value of the population proportion assuming that the null hypothesis is true, $H_0: p_1 = p_2 = p$. The test statistic is then

$$Z = \frac{(\hat{p}_1 - \hat{p}_2) - (p_1 - p_2)}{\sqrt{p(1 - p)\left(\frac{1}{n_1} + \frac{1}{n_2}\right)}},$$

which has a standard normal distribution under the null hypothesis.

We do not know the common value for the population proportion, p, so we must estimate it from the sample data. We know the two sample proportions, $\hat{p}_1$ and $\hat{p}_2$, and the two sample sizes, n_1 and n_2. The common value for the population proportion is not simply the average of $\hat{p}_1$ and $\hat{p}_2$ as we must take into account the sample sizes. In fact we estimate p from

$$\hat{p} = \frac{n_1 \hat{p}_1 + n_2 \hat{p}_2}{n_1 + n_2}.$$

In this example we already know the common value from the sample as it was included in the article (0.13), but for illustration we calculate

$$\hat{p} = \frac{139 \times 0.08 + 1062 \times 0.14}{1201} = 0.13.$$

This means that the estimated standard error of the differences of the two sample proportions is

$$\sqrt{p(1-p)\left(\frac{1}{n_1} + \frac{1}{n_2}\right)} = \sqrt{0.13 \times 0.87 \left(\frac{1}{139} + \frac{1}{1062}\right)} = 0.0303,$$

and the test statistic is

$$Z = \frac{0.08 - 0.14}{0.0303} = -1.98.$$

The p-value is given by $2P(Z < -1.98) = 0.048$. As this is less than 0.05 we reject the null hypothesis at the 5% level. There is sufficient evidence to conclude that the percentage of people aged 18–24 who believe that criminality actually deters teenagers from trying drugs is different from the percentage of people over 25 who believe this.

The p-value for the test is only just below the significance level, so the evidence from this sample is not very strong. Although the difference between the sample percentages is 6% and the overall sample is large there is only weak evidence for a difference in the percentages. The evidence is weak principally because one of the samples is small for the estimation of a percentage.

12.6.3 Sample size and two-sample test for proportions

A sample size of 1201, as in the previous example, is large enough to estimate any percentage to within ±3% for the 95% confidence interval, but to test for a difference between two percentages the important sample size is the number of observations in the smallest group. In this case this is only 139. The 95% confidence intervals for the percentage of people who believe that criminality actually deters teenagers from trying drugs for the two populations are as follows:

Age group	Number	Percentage	Lower limit	Upper limit
18–24	139	8	4	14
25 and over	1062	14	12	16

You can see that the confidence interval for the older age group is narrow but that for the smaller sample size it is wide.

Although this sample is large for the estimation of percentages based upon the whole population it is not large for some subgroup analyses particularly when some of the subgroups are small as in the case of the 18–24-year-old subgroup. If, for the sake of argument, both subgroups had been the same size, then the 95% confidence intervals would have been

Group	Number	Percentage	Lower limit	Upper limit
1	600	8	6	10
2	601	14	10	16

The estimated standard error of the difference of the two proportions would have been 0.0181 and the test statistic $z = 3.32$, giving a two sided p-value of 0.0009. For two groups of the same size there would have been a highly significant difference between the groups, with exactly the same observed percentages as before. This illustrates that the size of the sample plays a crucial role in the conclusions drawn from a significance test, and in Section 12.8 this issue will be discussed further.

In the article referred to in Section 12.6.1 on which the example used in this section was based there was no mention anywhere about the sample sizes in the 18–24 and 25 + age groups. In fact, the figures of 139 and 1061 were made up for the purpose of using these data to illustrate the test for a difference of two proportions. The absence of sufficient information is a common difficulty when reading the results of statistical studies in the press.

In this case I estimated the sample sizes based upon the data from the 1996 estimates of the population supplied by the Office for National Statistics (1998a) assuming that the ratio of 18–24-year-olds to 25 + was the same in the sample as in the population. As the sample was designed to be a random sample of the adult population aged 18 and over this is a reasonable procedure to adopt for the purposes of illustration. Obviously for more serious work you would want to get the correct data by contacting the company which carried out the study. Often this may not be possible and you will find that you have to make some assumptions in order to carry out your statistical analysis of studies reported in the press.

12.7 General points on the interpretation of significance tests

12.7.1 Type I and Type II errors

In any significance test, there are two types of errors, known as **Type I** and **Type II** errors, and we can recast Table 12.1 as Table 12.9 in terms of these errors.

If the null hypothesis is true and the decision of the test is to reject the null hypothesis then a Type I error has been made. The probability of this is the significance level of the test, α:

$$P(\text{Type I error}) = P(\text{Decide } H_1 | H_0 \text{ true}) = \alpha.$$

Table 12.9 Type I and Type II errors

True state	Decision	
	H_0	H_1
H_0	Correct	Type I error
H_1	Type II error	Correct

The probability of the other error, usually denoted β, is

$$P(\text{Type II error}) = P(\text{Decide } H_0 | H_1 \text{ true}) = \beta.$$

Obviously, when constructing a test we would like to make the probability of both of these errors as small as possible. However, we have already seen in Figure 12.2 that this is not possible.

If you try to construct a test with α as small as possible then β gets larger, while if you try to make β as small as possible then α gets bigger. Trying to reduce the probability of one error leads to an increase in the probability of the other error. This is one of the reasons for fixing the significance level at 5% for reporting a significant result in most significance tests. By fixing the significance level we are specifying the value of α and hence are controlling the probability of a Type I error. Given a fixed value for α we can then go on to find the sample size needed to achieve a small enough value for β. This is discussed further in Section 12.8.

12.7.2 Specifying hypotheses on the basis of looking at the data

When formulating a significance test there is always a null hypothesis to test which is specified precisely in terms of equality of parameters, and an alternative hypothesis which is specified less precisely and often involves a range of values of the parameters. You should normally be in a position to formulate these hypotheses before any sample data are available.

A particular problem is that you should not allow your inspection of the sample data to determine which hypotheses you are going to test. The data in Table 12.10 give the summary statistics for a set of ten financial ratios. This table would reasonably constitute a summary of results such as you might include in a report. If you specify before looking at the results that you want to test for a difference in acid test ratio, then you have the basis for a test and you can carry on as described above. It is not permissible to scan through the mean difference column and the standard error column, and pick out those ratios associated with a low standard error and a big mean difference such as net asset turnover, and then do a test to see if there is a difference between the two periods. This is virtually bound to give you a significant difference, but the basis of the test is completely violated and so the p-value is meaningless. The null hypothesis for the test is that the mean difference is zero but if you look at the data and select the variable to test on the basis of the large mean difference then you know that there is a good chance that the mean difference is not zero. Consequently, your null hypothesis is invalid.

Table 12.10 Means and standard errors of financial ratios for a sample of companies

Financial ratio	Mean, period 1	Mean, period 2	Mean difference	Standard error of mean difference
Earnings per share	22.82	30.97	−8.15	4.06
Dividend per share	14.55	13.93	−0.62	0.25
Return on net assets	49.08	34.93	14.15	6.01
Profit margin	14.90	14.34	0.56	0.57
Net asset turnover	4.25	2.99	1.26	0.39
Current ratio	0.95	1.09	−0.14	0.05
Acid test ratio	0.63	0.79	−0.16	0.05
Debtor days	52.46	47.89	4.57	3.26
Stock days	73.55	72.11	1.44	3.73
Gearing	32.06	29.18	2.88	3.47

If you have noticed a significant difference, say between the mean net asset turnovers in the two periods, then the correct approach is to take another sample from the same population and test if you get a significant difference in that second sample. It is perfectly valid to let one sample influence the hypotheses associated with a second sample. It is not valid to let the data of one sample influence the hypotheses associated with the same sample.

If you have a number of subgroups and have decided to test if there is a difference between two population means on the basis of having seen that the difference between the two sample means is the maximum difference, over all pairs of subgroups, then there are appropriate tests to use. One is known as Tukey's T test. The two-sample t test is not appropriate in this case.

12.7.3 Multiple testing

Another procedure which is often carried out with significance tests refers to the problem of multiple testing. It can also be illustrated with reference to Table 12.10. It is quite possible to add to this table a column of p-values and perform ten significance tests, all of which were specified before looking at the data. Specifying these tests before collecting the sample data gets around the problem of using your data to generate the hypothesis, but it introduces the problem of multiple testing.

The p-value of a test is the probability of obtaining a more extreme value of the test statistic than the observed value, and so is a measure of the strength of the evidence against the null hypothesis. In any test it is quite possible to make an error associated with random variation. In coming to a decision we use the rule that if the p-value is smaller than the significance level then we reject the null hypothesis. We do not know for certain if the null hypothesis is true or false. The significance level sets a limit for this error, and we usually work with a 5% significance level.

Assume that the null hypothesis is true and we work with the 5% significance level. If we do one test then the probability of not rejecting the null hypothesis is 0.95, and of rejecting it 0.05. Now suppose we do two independent tests, say on two unrelated variables. The probability that we will not reject the null hypothesis on both tests is going to be

$0.95^2 = 0.9025$. The probability of rejecting the null hypothesis both times is very small at $0.05^2 = 0.0025$ but the probability of rejecting the null hypothesis in one of the tests but not the other is $2 \times 0.95 \times 0.05 = 0.095$. Thus the probability of rejecting the null hypothesis on at least one of the tests is $1 - 0.95^2 = 0.0975$, which is very nearly at the 10% significance level.

If we were to do three independent tests, then the probability of finding at least one significant result, even if the null hypothesis is true, is going to be $1 - 0.95^3 = 0.14$. For five tests it is 0.23, for 10 it is 0.40 and for 20 it is 0.64. With these probability calculations you can see that the more tests you do the more likely you are to find at least one significant result even if the null hypotheses are all true. This means that by doing multiple tests on different variables you can find significant results even when there are no significant differences.

12.7.4 Adjustment for multiple testing

The fact that the probability of a Type I error will increase the more tests you do does not deter investigators from using statistical methods inappropriately. There are appropriate methods for taking into account this problem of multiple testing, one of which we will now describe. This is based on the observation that if you do two independent tests at the $5\%/2 = 2.5\%$ significance level then the overall probability of a Type I error is $1 - 0.975^2 = 0.049$; if you do three independent tests at the $5\%/3 = 1.67\%$ level then the overall Type I error probability is $1 - 0.983^3 = 0.049$. It is possible to show that if you are going to do k tests then you should do each test at the significance level of α/k in order to preserve an overall Type I error probability of α. This rule is known as the **Bonferroni adjustment**.

While this adjustment is necessary to avoid reporting too many results which may be significant just by chance, it is not used very often in practice. If an investigator carries out ten unplanned significance tests then he or she should really be doing each of them at the 0.5% significance level rather than the 5% level, though it is often very difficult to persuade investigators about this, even if they are aware of the problem of multiple testing, and they will often come to the wrong conclusion about the significance of the tests they have carried out. If the actual p-values are reported then, at least, you will be able to make a judgement on the interpretation of the investigator through the use of the Bonferroni adjustment. The reporting of the p-values is a very rare occurrence and you are generally left wondering if the reports of a significant difference represent the real purpose of the study or just the reporting of the significant results from a number of tests.

The Bonferroni adjustment was only derived for a series of independent tests with the aim of making sure that the investigator does not make too many Type I errors – reporting significant differences when these may have just come about by chance. The problem is most acute in a study in which there are no significant differences and the investigators are searching around for some significant results. It is less acute when the reported significant differences are in accord with results from other similar studies, as the differences are more likely, then, to reflect real differences.

Note that, if the variables on which the tests are carried out are all correlated with each other then the tests will not be independent of each other, and if you get a significant difference on one variable then you are likely to find one also in any related variables. The

Bonferroni adjustment is not really appropriate if the tests are not independent of each other. The effect is to over-correct for the number of tests and use a significance level which is too small. This is one of the reasons why the Bonferroni adjustment is not used all that often.

12.8 Power and sample size calculations

12.8.1 Influence of sample size and alternative hypothesis

In the data presented in Table 12.6 we had 15 companies, seven in one group and eight in the other; summary figures are presented in the first row of Table 12.11. These are the figures we calculated in Section 12.4. In rows 2 and 3 we see the effect of pretending that the sample sizes are larger, with exactly the same values for the summary statistics. You can see that as the sample size increases the t statistic increases and the p-value decreases. With samples of 7 and 8 we have $p = 0.13$, whereas with samples of 14 and 16 the difference between the means is significant at the 5% level with $p = 0.03$. Thus the sample size has an important role in the interpretation of significance tests, and for a fixed set of estimates we are more likely to get a significant result from the test the larger the sample size.

Rows 4–6 in Table 12.11 follow on from the first row, and here we keep the sample sizes fixed at their observed values but this time increase the difference between the two sample means. In the observed data the difference was 0.1576, and as this difference increases to 0.17 (row 4) the p-value decreases from 0.130 to 0.105. As the difference between the means increases, the t statistic increases and the p-value decreases.

Table 12.11 The influence of sample size and mean difference on the pooled two-sample t test

	Brewers, etc			Hotels and leisure				
	n_1	$\bar{x}_1$	s_1	n_2	$\bar{x}_2$	s_2	t	p
1	7	0.3534	0.1224	8	0.1958	0.2300	1.618	0.130
2	14	0.3534	0.1224	16	0.1958	0.2300	2.289	0.030
3	28	0.3534	0.1224	32	0.1958	0.2300	3.238	0.002
4	7	0.36	0.1224	8	0.19	0.2300	1.745	0.105
5	7	0.37	0.1224	8	0.18	0.2300	1.951	0.073
6	7	0.38	0.1224	8	0.17	0.2300	2.156	0.050

Both of these observations are related to a quantity known as the **power** of the test. This is an important aspect of significance tests and an even more important aspect of the design of studies. The **power** of the test is the probability of deciding H_1 when H_1 is the correct hypothesis and so is related to the probability of a Type II error:

$$\text{Power} = P(\text{Decide } H_1 | H_1 \text{ true}) = 1 - P(\text{Decide } H_0 | H_1 \text{ true}) = 1 - \beta,$$

where β is the probability of making a Type II error (see Section 12.7.1). If we fix the significance level of a test, α (see Section 12.7.1), then we can design a study to ensure that the power is high. This is known as a power calculation and is illustrated in the next section.

12.8.2 One-sample test for a mean with the standard deviation known

In this study of the power of a test it is easier to consider the critical value approach than the p-value approach to carrying out the significance test. We will also use a simple one-sample test on a mean of a normal distribution, where the population variance is known, by way of illustration. We are to test H_0: $\mu = 0$ against H_1: $\mu > 0$ using data from a sample of $n = 16$ observations, where the population standard deviation is $\sigma = 4$. In this situation the test statistic is $Z = (\bar{x} - \mu)/(\sigma/\sqrt{n})$, and if H_0 is true then Z is from a standard normal distribution. This test statistic was mentioned in Section 12.3 in connection with the one-sample t test, but it was not used as it is not common to know the population standard deviation. It is used in this section as an illustration as the calculations can easily be carried out, but the same general points will apply to all significance tests.

As this is a one-sided test, if we use a 5% significance level then the critical value for the test will be $z = 1.645$, since $P(Z > 1.645) = 0.05$. By using this critical value we have ensured that the probability of making a Type I error will be 0.05. In order to calculate the probability of making a Type II error we need to make some probability calculations assuming that the alternative hypothesis is true. This poses a few problems as we do not know the value of the population mean under H_1 – all that we know is that it is greater than 0.

If we use μ_1 to denote the value of the population mean assuming that H_1 is true then we can work out the distribution of Z under the alternative hypothesis. The expected value of Z will be

$$E[Z] = E\left[\frac{\bar{x} - \mu}{\sigma/\sqrt{n}}\right] = E\left[\frac{\bar{x}}{\sigma/\sqrt{n}}\right] = \frac{\mu_1}{\sigma/\sqrt{n}},$$

as $\mu = 0$ and $E[\bar{x}] = \mu_1$; furthermore, $Var[Z] = 1$ and Z is normally distributed. The power of the test is the probability of deciding H_1 when H_1 is true and, in this example, will be $P(Z_1 > 1.645)$, where Z_1 follows a normal distribution with a mean of $\mu_1 (\sigma/\sqrt{n})$ and a standard deviation of 1. This is the same as $P(Z > 1.645 - \mu_1/(\sigma/\sqrt{n}))$, where Z has a standard normal distribution. At this stage you should note that the power depends upon the mean under the alternative hypothesis, the standard deviation and the sample size.

These probability calculations are a little complex and we will illustrate the concepts graphically. There are three graphs in Figure 12.12 which are based upon Figure 12.5, with the addition of the distribution of the test statistic assuming a particular value for the mean under the alternative hypothesis. The fourth graph is a power curve, with the power of the test plotted against the possible values of the mean under the alternative hypothesis.

In Figure 12.12a we assume that the value of the mean under the alternative hypothesis is 3. The critical value is 1.645 and the area (in black) to the right of this value, under the distribution for the test statistic assuming that the null hypothesis is true, represents the probability of making a Type I error, i.e. the significance level, which is 0.05. The area (in grey) to the left of the critical value and under the distribution of the test statistic assuming that the alternative hypothesis is true is the probability of a Type II error. In this case it is 0.088. You can see from this graph that if you try to make the probability of a Type I error smaller, by moving the critical value to the right, say to 2 or 2.5, then you will automatically make the probability of a Type II error larger, as illustrated:

Critical value	1.28	1.645	1.96	2.33
Probability of a Type I error	0.10	0.05	0.025	0.01
Probability of a Type II error	0.043	0.088	0.149	0.251

This illustrates quite convincingly that it is not possible to minimize both types of error simultaneously. As the significance level decreases, the probability of making a Type II error increases and so the power of the test will decrease.

Figures 12.12b and 12.12c are similar to Figure 12.12a except for a change in the value of the mean under the alternative hypothesis. If the significance level is kept fixed then, as the mean under the alternative hypothesis moves further away from its value under the null hypothesis, the probability of a Type II error decreases, and so on the power of the test will increase.

Figure 12.12d shows how the power of the test increases as the mean of the population increases under the alternative hypothesis. The horizontal dotted line represents the significance level of 0.05 and the power curve meets this line at the value of the mean

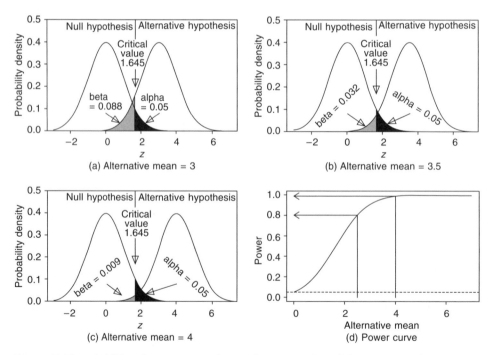

Figure 12.12 Probabilities of Type II error and power for various values of the mean under the alternative hypothesis.

under the null hypothesis (i.e. $\mu = 0$). As μ increases through all its possible values under the alternative hypothesis the power increases; however, the power is over 0.90 only for values of μ greater than 2.93. The power will be large provided there is a large gap between the two hypotheses – that is, if μ, under the alternative hypothesis, is large. If the true value of μ under the alternative hypothesis is not much bigger than 0 then the power will be lower and not much bigger than the significance level.

The influence of sample size was not considered in Figure 12.12, and all the calculations were based upon a sample of 16 observations. We now look at the power curves for the same null hypothesis but with different sample sizes. These are displayed in Figure 12.13 for three samples sizes, 8, 16 and 32. For a fixed value of μ under the alternative hypothesis, the power increases as the sample sizes increases. If, in fact, the true value of μ is 2.5 then the power is 0.55 for a sample size of 8, 0.80 for a sample size of 16 and 0.97 for a sample size of 32. This gives a clear message for designing studies to carry out significance tests – the larger the sample size, the greater the power.

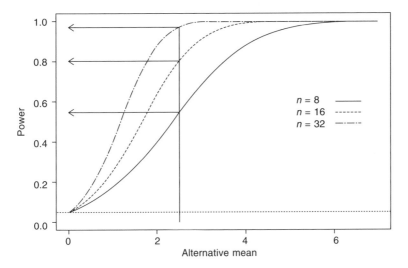

Figure 12.13 Power and sample size.

This small investigation has shown us that the power of a significance test is an important concept. For a fixed sample size and significance level the power of a study will decrease as the mean under the alternative hypothesis becomes closer to the mean under the null hypothesis. For a fixed significance level and alternative mean, the power will increase as the sample size increases.

12.8.3 Power calculations in practice

The power of a test is the probability of rejecting the null hypothesis when it is known to be false. Ideally we would like this to be very high, and values of 0.8 and above are commonly accepted as being large enough in most practical situations. If the power of a test is 0.5 and the significance level is 0.05, then this means that the probability of rejecting the null hypothesis when it is true (Type I error) is low at 0.05 but that the probability of rejecting the null hypothesis when it is false (Type II error) is 0.5. So even if the alternative hypothesis were true, the probability of deciding so on the basis of the test is just the same as tossing a coin. Clearly this is an unacceptable situation, and in order to have faith in the significance test it is necessary to ensure that the power of the test is high (the probability of a Type II error is low). The only way that this can be achieved is by increasing the sample size and using power calculations in the same way as illustrated above.

The probability calculations involved with this can be quite formidable, and the simplest way of finding out the sample size required is by using sample size software. There are a large number of different types, and a list of them can be obtained at the Learning and Teaching Support Network website (http://www.ltsn.gla.ac.uk). One free program for MS DOS is DSTPLAN (http://odin.mdacc.tmc.edu/anonftp/page_2.html) and a commercial Windows-based program is nQuery Advisor (http://www.statsol.ie/nquery/nquery.htm).

As the power of a test on a sample mean depends upon the value of the mean under the null hypothesis, the value under the alternative hypothesis, the standard deviation and the sample size, it is necessary to specify values for these parameters and use them to calculate the sample size. This is an approximate procedure as the exact values for these parameters are seldom known in advance, and so rough estimates have to be used. This is an important aspect of study design, and by doing this type of calculation beforehand you can have some idea of what you can realistically expect from a study.

12.8.4 Case study in power calculations

An article in the journal *Nature* on genetic inheritance (Skuse *et al.* 1997) was widely reported in the newspapers shortly after publication as it sparked off much debate on whether behavioural characteristics were transmitted genetically from fathers to daughters:

> recent scientific findings . . . claim that a 'social gene', or group of genes on the X chromosome, makes women more sensitive and better behaved than men.
>
> Women normally inherit two X chromosomes, one from their mother and one from their father. Men, however, get just one. The result, according to the research, is that girls are blessed with 'feminine intuition' and a code of morality from birth, while boys have to learn good behaviour to stop them going off the rails.
>
> The genetic advantage also makes women sensitive and better communicators who are equipped to cope with social situations than men. Professor David Skuse, a child psychiatrist who led the research, doubted whether doctors would ever be able to alter the gene to give men the same intuitive ability.
>
> (*Sunday Telegraph*, 24 May 1998)

The research was based upon a study of 80 girls and young women who have Turner's syndrome, which is characterized by having only one X chromosome as opposed to the usual XX in women and XY in men. In 55 girls the single X chromosome was inherited from the mother and in 25 girls it was inherited from the father. Impaired social competence and adjustment are common in Turner's syndrome, but a minority have good social skills. Intelligence is usually normal. The study was set up to test if there was a difference in social behaviour between girls with the single X chromosome inherited from the mother and those with the single X chromosome inherited from the father.

Using a social cognition score with a possible range from 0 to 24, the researchers found that the girls with the single X chromosome inherited from their mother had an average of 9.3 (standard deviation 3.7), while those with the single X chromosome inherited from their father had a mean score of 4.2 (2.5). Using a pooled two-sample t test the p-value for the difference between the mean scores is $p < 0.0001$, providing strong evidence of a difference in social cognition scores.

Using these figures, this study has a power of over 0.99 to detect a difference in mean score of 5.1 at the 5% significance level using a two-sided test. If the sample sizes had been smaller, say 25 and 10, then the observed p-value would have been 0.0003, but the power would have been 0.97. The power is not affected very much by this change in the sample size. The reason why this study has a high power is that there is a large effect (the mean difference is 5.1) and the sample size is adequate to detect this difference.

If we were to design a second experiment to replicate these results, how many Turner's syndrome girls would we need to enrol in the study? The previous experiment gives us a value for the pooled standard deviation of 3.4. If we keep the same ratio of maternal to paternal inheritance for the single X chromosome of 5 : 2, and we decide that we wish to detect differences in the score of at least 3 points, then we find that for a power of 0.9 we would need 70 girls in total; for a power of 0.95, 84 girls would be needed.

We use a postulated effect of 3 points as being the smallest difference which it is worthwhile to detect. If we stated that the repeat study had to be capable of detecting a difference of at least 2 points, then 150 girls would be needed for a power of 0.90 and 185 for a power of 0.95. Thus you can see the influence of the anticipated difference on the sample size required. To be confident of detecting small differences you need large sample sizes.

The final decision about the sample size will also be based upon other information such as the cost of collecting the data and the time needed to collect them. These statistical calculations are designed to make sure that the study has sufficient statistical power to detect the effects that are important for the investigator. There is a usually an abundance of speculation and informed judgements about the estimates used in the sample size calculation. It is important not to devote large resources to small underpowered studies. It is always informative in the business context to have as much information as possible available before making decisions, and sample size and power calculations are some of the many pieces of information required.

13

Qualitative variables: goodness of fit and association

- Knowing about the χ^2 probability distribution and using its tables
- Knowing how to carry out the X^2 goodness-of-fit test
- Knowing how to carry out the X^2 test of association
- Understanding about the relationship between the X^2 test of association and the test for the difference of two proportions
- Understanding that no association in a table is the same as equality of row proportions and equality of column proportions
- Knowing how to try and interpret association by looking at large residuals
- Understanding the influence of sample size on the X^2 test of association.

13.1 Goodness-of-fit tests

All of the significance tests in the previous chapter have been carried out within the framework of testing a parameter of a probability distribution. This has been either the mean of a normal distribution or the proportion in a binomial distribution. It would be possible to go on to cover tests for the variance in a normal distribution or the equality of two variances. The concepts that we have already met – hypotheses, test statistics, significance levels and p-values – are exactly the same, though the technicalities are different and a little harder.

When setting up a significance test it is a good idea to specify the assumptions that you are making. These include such matters as random samples, independent samples, equal variances, as well as the distributional assumptions. For a t test on a mean we are assuming a normal distribution, and for a test on a proportion a binomial distribution. The validity of the normal assumption is checked by using normal probability plots, and we now look at a simple and intuitive statistical test which can be used to decide if the postulated probability model is valid. This test can be used to check the validity of the binomial or any discrete distribution. It is known as the X^2 goodness-of-fit test. Although this test can be used with continuous random variables, probability plots are easier, and we will not illustrate the test with a continuous distribution.

The essential features of the X^2 goodness-of-fit test are the initial hypothesis about the probability model; the calculation of the expected behaviour, assuming that the probability model is true; the comparison of the observed responses with the expected responses through a test statistic; and the final conclusion about whether or not the model is valid by calculating the p-value of the test. The description and use of the test will be illustrated with a simple probability model.

13.2 Goodness of fit of a simple discrete probability model

Under the assumption of equally likely outcomes, the proportion of families with two boys in families of size 2 (an event which we will call B) should be 1/4. In a random sample of 200 families the number of families with two boys is 57. We test the hypothesis that the probability distribution is $P(B) = 1/4$ and $P(\overline{B}) = 3/4$ against the alternative that it is not.

Intuitively, the X^2 test is one of the most appealing statistical tests as it is based on the concept of comparing the observed values with the expected values, over all possible outcomes in the experiment. The test will be developed along the same lines as all other ones. First of all, we will set up the hypothesis. Then we will work out the expected values and then derive a test statistic which measures how close the observed values are to the expected values. Values of the test statistic are tabulated, and by referring the calculated values of the test statistic to the tabulated values we can make our decision.

13.2.1 Null and alternative hypotheses

In this experiment there are two outcomes of interest, B and ($\overline{B}$). The null hypothesis is that $P(B) = 1/4$ and hence $P(\overline{B}) = 3/4$, and the alternative hypothesis is that the probabilities are not as stated in the null hypothesis, i.e. $P(B) \neq 1/4$ and $P(\overline{B}) \neq 3/4$.

If the null hypothesis is true, we can calculate what the expected values are (Table 13.1). There are 200 independent families and each family either has two boys or does not. In this kind of situation the binomial distribution is often appropriate as there are independent trials (families) and each has only two outcomes. The expected value is

$$\text{Expected value} = \text{Number} \times \text{Probability}.$$

Table 13.1 General terminology for the X^2 goodness-of-fit test, and observed and expected values for two boys in a family

| | Example | | General terminology | |
	B	($\overline{B}$)	Class 1	Class 2
Observed number (O)	57	143	O_1	O_2
Expected number (E)	50	150	E_1	E_2
$(O - E)^2/E$	49/50	49/150	$(O_1 - E_1)^2/E_1$	$(O_2 - E_2)^2/E_2$

13.2.2 Test statistic

The X^2 test is appealing and intuitive, as it is accomplished by comparing the expected values with the observed ones. If the null hypothesis is true then the observed values should be close to the expected ones, and if the null hypothesis is to be rejected then the observed values should be quite different from the expected ones. All we need is a test statistic which measures the agreement between these two sets of values.

When there are just the two outcomes the X^2 goodness-of-fit test statistic is

$$X^2 = \frac{(O_1 - E_1)^2}{E_1} + \frac{(O_2 - E_2)^2}{E_2}$$

This takes the value zero if $O_1 = E_1$ and $O_2 = E_2$, and if there is a great discrepancy between these two values then X^2 is large and positive. This means that the X^2 goodness-of-fit test is always going to be a one-sided test. The expected values are on the denominator of each comparison because it is the relative differences between the observed and expected values which are more important than the actual differences. For example, a difference between O and E of 10 is more important if E is 50 than it is if E is 500.

13.2.3 *Chi-squared* distribution

If the null hypothesis is true then the distribution of X^2 is known exactly. In Figure 13.1 a histogram of 1000 values of X^2 is plotted. These values were calculated by simulating data from 1000 surveys of 200 families based upon the discrete probability model assuming that the probability of having two boys was 0.25. For each of the 1000 surveys the X^2 test statistic was calculated. The curve in Figure 13.1 is the probability density function for a Chi-squared (χ^2) distribution on 1 degree of freedom, and you can see that there is a very good agreement between the histogram and the curve.

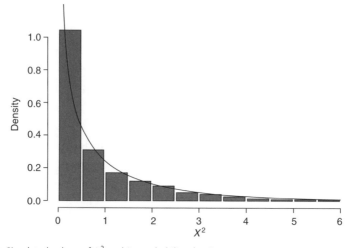

Figure 13.1 Simulated values of X^2 and its probability density.

If the null hypothesis is true then the X^2 test statistic will follow a χ^2 distribution. The degrees of freedom are calculated as

Degrees of freedom = Number of cells − 1 − Number of parameters estimated,

and this tells you which row of the table of the percentage points of the χ^2 distribution to use. In the example there are 2 cells and no parameters were estimated from the data, so the degrees of freedom are 1. (The value of the probability was specified based upon equally likely outcomes and did not require any use of the observed data.)

The χ^2 distribution is a continuous probability distribution. It has one parameter, known as its degrees of freedom. This is often denoted by the letter v. The probability density functions of several χ^2 distributions are illustrated in Figure 13.2. This distribution is skew, especially when the degrees of freedom are small, and always takes positive values. The expected value of a χ^2 distribution with v degrees of freedom is v and the variance is $2v$.

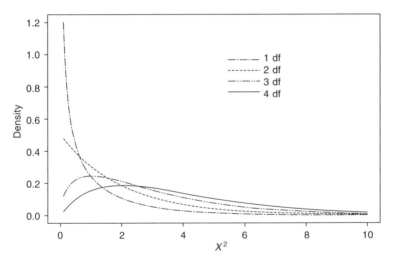

Figure 13.2 Probability density functions of χ^2 distributions.

Tables of the χ^2 distribution are given in Appendix C. Unlike the normal distribution, where the probabilities were listed corresponding to specified points, the percentage points of the distribution are given in the table corresponding to certain specified probabilities. Thus the table is similar to the tables of the t distribution. The first row of the table gives these probabilities. The column that will be used most often in connection with the X^2 test is the column headed 0.95. This gives the percentage points of the χ^2 distributions which have a probability of 0.95 below and 0.05 above.

The second row of the table gives the percentage points corresponding to the χ^2 distribution on 1 degree of freedom. The 5% point is 3.8415 and is illustrated in Figure 13.3. For a random variable following a χ^2 distribution on 1 degree of freedom there is a probability of 0.05 of observing a value greater than 3.8415 and a probability of 0.95 of observing a value less than 3.8415.

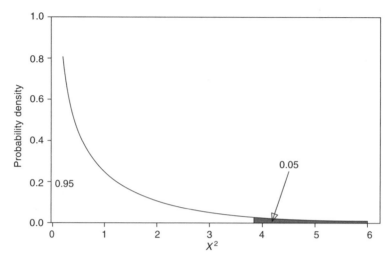

Figure 13.3 Five per cent point of the χ^2 distribution on 1 degree of freedom.

13.2.4 Critical region

The calculated value of the test statistic in Table 13.1 is $X^2 = 49/50 + 49/150 = 1.307$. The critical value on 1 degree of freedom is 3.84, and as the calculated value is less than this we conclude that the value of X^2 is acceptably small. Thus there is a good agreement between the observed and expected values, and this means that the original hypothesis is valid. So, on the basis of this sample, we would conclude that the probability of two boys in families of size 2 was indeed 1/4, thus upholding the assumption of equally likely outcomes.

13.2.5 Relationship to the one-sample test on a proportion

The concepts underlying the example used in this section are no different from those used in Section 12.6.1. The number of families out of 200 with two boys should follow a binomial distribution with $p = 0.25$. We can test the hypothesis H_0: $p = 0.25$ against the alternative H_1: $p \neq 0.25$. The test statistic is

$$Z = \frac{\hat{p} - p}{\sqrt{\dfrac{p(1 - p)}{n}}}$$

which has a value of 1.143, as $\hat{p} = 57/200 = 0.285$. Although it might seem just a huge piece of mathematical trickery, you can demonstrate algebraically that $Z^2 = X^2$, and numerically we have $Z^2 = 1.143^2 = 1.307 = X^2$. The critical value for the Z test at the 5% level is 1.96, and $1.96^2 = 3.84$, which is the critical value for the X^2 test.

When the discrete distribution has just two values and is binary the X^2 goodness-of-fit test is exactly the same as the Z test for a proportion, and they can be used interchangeably. If the discrete distribution has three or more possible values then you have to use the X^2

goodness-of-fit test and it is wrong to do a sequence of Z tests, one for each of the different values. This leads to a problem of multiple testing Section 12.7.2.

13.3 X^2 goodness-of-fit test

If there are more than two outcomes, then with k outcomes the test statistic is

$$X^2 = \frac{(O_1 - E_1)^2}{E_1} + \frac{(O_2 - E_2)^2}{E_2} + \ldots + \frac{(O_k - E_k)^2}{E_k}$$

and the degrees of freedom are $k - 1$ – number of parameters estimated.

This is a large-sample test, and to ensure its validity you need to have large enough samples. The rule of thumb for deciding if the sample is large enough is that the E_i are each greater than 5. This is exactly the same rule of thumb as for the use of the normal approximation to the binomial distribution (Chapter 9) and the use of the Z test for a proportion (Chapter 12). In fact, the test is still valid even if one of the expected values is lower than 5, but all must be bigger than 1. If there are too many low expected values then you need to combine adjacent cells until you get the expected values large enough. If there are four cells, for example, with expected values 10, 7, 3, 3, then you should combine the last two cells to give a single cell with expected value 6 before calculating the X^2 statistic.

13.3.1 Goodness-of-fit of a binomial distribution

The data in Table 13.2 were previously introduced in web resource (summarising, section binomial) and refer to the number of pill boxes with missing pills. The firm supplied a drug in boxes of 28 pills which are supposed to be taken once per day. These pills are packaged in a 4×7 grid labelled with the days of the week. During a time of machine malfunction a large number of boxes had fewer than 28 pills. The data refer to a sample of 371 boxes in which an average of 0.3127 pills per box were missing. As the expected value of the binomial distribution is $\mu = np$ the probability that a pill was missing from any one of the 28 locations can be estimated from the data (see the book's website) by setting $np = 0.3127$, and so $p = 0.3127/28 = 0.0112$ as there are $n = 28$ trials.

We use the X^2 goodness-of-fit test to test if the binomial model is a reasonable one, using the value of p which has been estimated from the data. Considering the random

Table 13.2 Number of missing pills per box in a quality control example from a pharmaceutical company

Number of missing pills	Frequency	Expected	X^2 contribution
0	272	270.65	0.0067
1	84	85.83	0.0393
2+	15	14.51	0.0165
Total	371	370.99	0.0625

variable, we might expect a binomial distribution to be appropriate. The main reason why the binomial distribution would not be appropriate is likely to be the assumption of independence of events. This might manifest itself as there being more boxes with a very large number of missing pills, or more boxes with no missing pills, than we would expect according to the binomial distribution.

The null hypothesis for this test is

$$H_0 : P(x) = \binom{n}{x} p^x (1-p)^{(n-x)},$$

and the alternative is that the binomial probability distribution is not appropriate. The probabilities are calculated assuming that the null hypothesis is true using the binomial probability function

$$P(x) = \binom{n}{x} p^x (1-p)^{(n-x)}$$

with $n = 28$ and $p = 0.0112$. Although the original data had 2 boxes with three missing pills we calculate the expected values for two or more rather than three or more to avoid having expected values less than 5.

The observed values and expected values match up very well in this case. The X^2 statistic is low at 0.06. Since there are three cells and one parameter is estimated, the degrees of freedom are $3 - 1 - 1 = 1$ and the critical value is 3.84 at the 5% significance level. The calculated value of X^2 is well below this, so we can be quite happy with the use of the binomial distribution in this case.

13.4 Association

In Chapter 3, the relationship between two quantitative variables was investigated by the use of graphs such as scatter plots. This is relevant for investigation of the association between, for example, market share and advertising expenditure. A time series plot would be used to investigate the relationship between market share and time.

To investigate the relationship between market share and the different product groups of a company, separate boxplots of market shares may be drawn for each product group. The product group is a qualitative variable and must be treated as such in the plots. Scatter plots are not appropriate in this case as a qualitative variable does not have any ordinal information and generally only takes a limited number of distinct values. The association between two qualitative variables is now investigated.

13.4.1 Interpretation of tables

When data on qualitative variables are recorded from a survey it is common to present the results in the form of a table. Many government surveys collect a considerable amount of qualitative information – the General Household Survey is a good example. The responses to two variables are tabulated in Table 13.3 The first variable is the sex of the head of household. The classification of head of household depends on the income of the various

Table 13.3 Head of household and bedroom standard

	2+ below standard	1 below standard	Standard	1 above standard	2 above standard	3+ above standard
Male	6	35	252	409	258	44
Female	0	16	62	72	32	4

Source: Teaching data set, General Household Survey 1991, ESRC Data Archive.

adults in the household and ownership or tenancy of the house and is normally, but not necessarily, the 'man of the house'. The bedroom standard is a measure of the extent to which a house might be overcrowded or underoccupied. The number of adults and children living in the house, the ages of the children and the relationships among the adults are used to estimate the number of bedrooms required. A married couple with no children require one bedroom. If they have a boy age 10 and a girl of 14 then they require three bedrooms.

Although there is information from 1190 households in Table 13.3, it is not easy to see anything important other than the fact that most households have a man as head. The entries show you that in 6 households the head was a man and there were two or more bedrooms below standard. Although there were 44 households with a man as head and three or more bedrooms above standard and only 4 households with a woman as head and three or more bedrooms above standard, it is not easy to make any sensible conclusions about the association between bedroom standard and the sex of the head of household. There are between 5 and 10 times as many households with a male head as households with a female head.

The reason for the construction of the table is to compare the distribution of bedroom standard among households with a man or a woman as head. The tabular presentation is to provide useful information on the association between the sex of the head of household and the bedroom standard of the accommodation. The principal aim of this comparison would be to see if households with a woman as head had more or fewer bedrooms in relation to the standard. This question cannot be answered by looking at the raw numbers and it is necessary to convert the entries in Table 13.3 into row or column percentages so that fair comparisons of like with like can be made.

13.4.2 Row and column percentages

The **row percentages** in a table are simply the observations written as a percentage of the row total, while the **column percentages** are the observations written as a percentage of the column totals. Which one to use depends upon the interpretation that needs to be made. Often both provide useful information, but for clarity of presentation it is better to present them in separate tables if both are to be used.

The row percentages relating to Table 13.3 are given in Table 13.4. There are 1004 households with a man as head, of which 44 had three or more bedrooms above the standard. Thus the percentage is given by $(44/1004) \times 100 = 4.383$, which is rounded to 4.4%. The others are calculated in a similar fashion. They are expressed in the table to one

Table 13.4 Head of household and bedroom standard: row percentages

	2+ below standard	1 below standard	Standard	1 above standard	2 above standard	3+ above standard	Toal number
Male	0.6	3.5	25.1	40.7	25.7	4.4	1004
Female	0.0	8.6	33.3	38.7	17.2	2.2	186

decimal place and it is seldom necessary to use any more decimal places – but you need to be aware of the possibility of introducing rounding error into your calculations, and the risk of this is greater the fewer decimal places you use.

The column percentages are presented in Table 13.5. There are 48 households with three or more bedrooms above standard, of which 4 had a woman as head and 44 a man. Thus the column percentages are $(4/48) \times 100 = 8.3\%$ and $(44/48) \times 100 = 91.7\%$.

Table 13.5 Head of household and bedroom standard: column percentages

	2+ below standard	1 below standard	Standard	1 above standard	2 above standard	3+ above standard
Male	100.0	68.6	80.3	85.0	89.0	91.7
Female	0.0	31.4	19.7	15.0	11.0	8.3
	6	51	314	481	290	48

If there were no association between the sex of the head of household and the bedroom standard then the percentages in the male row would be equal to the percentages in the female row. From the comparison of the two rows in Table 13.4 it is relatively easy to see that the households with a woman as a head tend to have a greater percentage of bedrooms at the standard, with lower percentages in each of the three categories of number of bedrooms above standard. Where a woman is head, 8.6% of households have accommodation with bedrooms below the standard number, compared to 4.1% of households where a man is head.

The column percentages in Table 13.5 can also be used to interpret the association between head of household and the number of bedrooms above or below standard. If there were no association then all the percentages along the first row in Table 13.5 would be equal. Similarly, all the percentages in the second row would be equal. Clearly, the percentages are not equal to each other – observe that 69% of households with one bedroom below standard have a man as the head whereas 92% of the households with three or more bedrooms above standard have a man as head. With the exception of the first column, which is based on only 6 observations, the proportion of households with a man as head increases as the number of bedrooms above standard increases.

Of the two presentations above it is the one with the row percentages which is possibly the easiest to interpret and compare here, though there is not much to choose between them. In both cases the interpretation is made by looking to see if the percentages are equal. In Table 13.4 the equality of the percentages in each of the rows is the issue, while in Table 13.5 it is the equality of the percentages across the columns.

13.5 X^2 independence test for 2 × 2 tables

In Table 13.4 the visual impression among the row percentages is that there is evidence that the bedroom standard of the accommodation is related to the sex of the head of the household. If the head is female then there are a greater percentage of households with below standard bedroom accommodation, while if the head of the household is male then a greater proportion of households have bedroom accommodation above standard. In making this conclusion only the row percentages are considered. However, these percentages are derived from a survey which is subject to sampling variability. Moreover, the number of households with a female as head is 186, which is a much smaller sample than that of the number of households with a male as head (1004). Consequently, the effect of sampling variability is going to be greater in the percentages for female heads of household than for male heads of household. The X^2 test of independence is a significance test which is designed to assess if the association between two qualitative variables is statistically significant or not.

The test is illustrated first with a small example involving two qualitative variables each with only two levels. This is referred to as a **2 × 2 table**. The main aim of the test is to see if the two sets of column percentages are equal to each other. This is identical to saying that the two sets of row percentages are equal to each other and to saying that the two variables are independent of each other. This test can also be extended to compare more than two proportions and to cases in which there are more than two groups.

The data in Table 13.6 were derived from Table 13.3 by combining the six levels of Bedroom Standard into just two levels, below standard and standard or above. The hypothesis is that overcrowded accommodation is related to the gender of the head of the household. We wish to test if the proportion of households with below standard accommodation is the same if there is a man as head or if a woman is head. If this is so then there is no evidence of any association between housing standard and gender, whereas if the hypothesis is rejected then there will be evidence of some association. Such evidence does not mean that there is any causal relationship, just that gender is associated with bedroom standard.

Table 13.6 Head of household and two-level bedroom standard

	Below standard	Standard or above
Male	41	963
Female	16	170

13.5.1 Null hypothesis

The null hypothesis represents the hypothesis of no association between the two variables in the table, i.e. there is no association between gender and bedroom standard. If there were no association then the probability of having a house below standard would be the same for men as women. In the whole study 57 households have housing below standard and so the proportion is $57/1190 = 0.0479$. Thus if there was no association then the row percentages would be expected to be as follows:

	Bedroom standard		Total
	Below	Standard or above	
Male	4.79%	95.21%	100%
Female	4.79%	95.21%	100%
Total	4.79%	95.21%	100%

The null hypothesis can be written, in terms of the row conditional probabilities, as $P(\text{Below} | \text{Male}) = P(\text{Below} | \text{Female})$. These probabilities refer to equality of the row proportions in Table 13.6. The alternative hypothesis is that these probabilities are not equal, which means that $P(\text{Below} | \text{Male}) \neq P(\text{Below} | \text{Female})$, which implies also that $P(\text{Standard or above} | \text{Male}) \neq P(\text{Standard or above} | \text{Female})$. Anything other than equality of the proportions is consistent with there being an association between bedroom standard and gender of the head of household.

The hypothesis could equally well be expressed in terms of the column percentages, or equivalently the column conditional probabilities. If there is no association between bedroom standard and gender then $P(\text{Male} | \text{Below}) = P(\text{Male} | \text{Standard or above})$ and $P(\text{Female} | \text{Below}) = P(\text{Female} | \text{Standard or above})$. As there are 1004 males and 186 females, the probability of selecting a man at random from the 1190 individuals in the study is $1004/1190 = 0.8437$. If there was no association between gender and bedroom standard then the column percentages would be the same in the two columns:

	Bedroom standard		Total
	Below	Standard or above	
Male	84.37%	84.37%	84.37%
Female	15.63%	15.63%	15.63%
Total	100.00%	100.00%	100.00%

These null hypotheses written in terms of equality of row or column proportions are exactly equivalent to the definition of independent events in Chapter 8. This means that independence of events is the same as no association which is the same as equality of proportions. The X^2 test can be interpreted in any of these terms, which are largely interchangeable in this context.

13.5.2 Expected values and test statistic

In order to work out the expected values we need to have a value for the probability of below standard accommodation under the null hypothesis. If the null hypothesis is true then gender and bedroom standard are independent and so $P(\text{Below} | \text{Male}) = P(\text{Below} | \text{Female}) = P(\text{Below})$. Thus the best estimate is

$$p = P(\text{Below}) = 57/1190 = 0.0479.$$

We can now work out the expected values in each of the four cells in the table, assuming that the null hypothesis is true, by multiplying together the number in the sample (the number of male or female heads) by p in the case of below bedroom standard accommodation and $1 - p$ for standard or above. Thus:

Gender	Standard	Expected value
Male	Below	$1004 \times 0.0479 = 48.1$
Male	Standard or above	$1004 \times 0.9521 = 955.9$
Female	Below	$186 \times 0.0479 = 8.9$
Female	Standard or above	$186 \times 0.9521 = 177.1$

In general, the expected values are given by

$$E = \frac{R \times C}{N},$$

where R is the row total, C is the column total and N the total number of observations. In the above example the row totals are 1104 and 186, the column totals are 57 and 1133, and the total number of observations is $N = 1190$.

Combining our observed and expected values, we obtain Table 13.7. The test statistic is

$$X^2 = \sum \frac{(O - E)^2}{E} = 7.04,$$

which is compared to appropriate values from the tables of the χ^2 distribution to judge if there is significant evidence of association or inequality of proportions.

Table 13.7 Expected values, observed values and contributions to X^2

		Below standard	Standard or above	Totals
Male	Observed	41	963	1004
	Expected	48.1	955.9	
	$(O - E)^2/E$	1.05	0.05	
Female	Observed	16	170	186
	Expected	8.9	177.1	
	$(O - E)^2/E$	5.66	0.28	

13.5.3 Degrees of freedom

If the observed value in the below standard and male cell of the table is x then all the other values in the table are automatically known, given the information in the marginal totals (Table 13.8). Although there are four entries in the table there is only one piece of independent information in it, represented by the x. The number of pieces of independent information is known as the degrees of freedom. For a table with 2 rows and 2 columns there is thus 1 degree of freedom.

Table 13.8 Degrees of freedom

	Bedroom standard		Total
	Below	Standard or above	
Male	x	$1004 - x$	1004
Female	$57 - x$	$(1133) - (1004 - x)$	186
		$= 186 - (57 - x)$	
		$= 129 + x$	
Totals	57	1133	1190

13.5.4 Significance levels, critical values and *p*-value

As pointed out in Chapter 12, we can assess our significance test in two equivalent ways. The 5% level critical value of χ^2 on 1 degree of freedom is 3.84. As the value of X^2 calculated in Section 13.5.2 is greater than this, we have evidence to reject the hypothesis of no association between gender and bedroom standard, where the latter is classified as either below bedroom standard or not. We conclude that the proportion of households with a man as head who have accommodation below bedroom standard is not the same as the proportion of households with a woman as head who have accommodation below bedroom standard. This is the critical value approach, which you will recall tells us nothing about the strength of our evidence.

The other way is to calculate the *p*-value, the probability of obtaining a more extreme value of the test statistic than the calculated value, assuming that the null hypothesis is true. In order to calculate this exactly you need to have access to the cumulative distribution function of χ^2 distribution. The mathematics required to do this is well beyond the scope of this book! However, the function is available as a standard function in most statistical packages and spreadsheets. For our observed value of $X^2 = 7.04$ you should obtain a value of 0.008.

If you do not have access to a computer then you can use the table in Appendix C. There you will see that the 0.99 point of the χ^2 distribution on 1 degree of freedom is 6.6349; our observed value of X^2 is greater than this so we know that $P(\chi^2 > 7.03) < 0.01$. We reject the null hypothesis, as we did with the critical value approach, but we now have the additional information from our very small *p*-value that there is very strong evidence to back our decision.

13.5.5 Continuity correction and validity

The test we have carried out in this section is an example of a X^2 independence test for a 2×2 table. You will also see it in other textbooks and computer programs in a version which has a continuity correction. The test statistic with the continuity correction is

$$X_{cc}^2 = \sum \frac{\left(\mid O - E \mid - \frac{1}{2} \right)^2}{2}.$$

The only difference relative to the version without the continuity correction is the subtraction of $1/2$ from the absolute value of the difference between the observed and expected values. This continuity correction is needed if there are few observations in one of the cells of the table. When you are using statistical programs to analyse association in 2×2 tables you should normally use the version with the continuity correction as it is applicable over a wider variety of tables than the version without the continuity correction. In the example the value of X^2 with the continuity correction is 6.07, which is slightly smaller than the value without the continuity correction.

The continuity correction is needed for the X^2 test for a 2×2 table because the table is inherently discrete, which means that the X^2 test statistic must take only a finite set of discrete values and the χ^2 distribution is continuous. In fact, as the smallest row or column entry in Table 13.6 is 57 there are only at most 58 possible values of X^2 corresponding to x going from 0 to 57 in Table 13.8. When there are large numbers of possible values of X^2 the continuity correction has little effect, but when there are relatively few values then the continuity correction becomes necessary.

Referring the calculated value of the test statistic to tables of the χ^2 distribution is only valid provided the sample is large. You need to ensure that the sample is large enough, and the rough rule of thumb for determining this is if all of the expected values are bigger than 5. If you have a 2×2 table in which one of the expected values is smaller than 5 then the test will not be reliable. Many statistical tests are based upon large samples and their use is not valid in small samples. Generally business and social statistics surveys are based upon large samples, and you do not usually have to worry about the applicability of the large-sample rule.

In Chapter 12 you will find a Z test for the equality of two proportions based on the equations for the confidence interval for the difference of two independent proportions discussed in Chapter 11. Like the X^2 goodness-of-fit test for a binary random variable, which is identical to the Z test for one proportion, the X^2 test of independence in a two-way table is identical to the Z test for the equality of two proportions. Again you will find that the Z statistic is equal to $\sqrt{X^2}$.

13.6 X^2 independence test for $r \times c$ tables

The test of the previous section can be easily extended to tables which have r rows and c columns. As before, the general formula for the expected values is $E = RC/N$, where R is the row total, C is the column total and N the total number of observations. Now, however, the degrees of freedom are given by $(r-1)(c-1)$. In fact the test for 2×2 tables is just a special case of that for $r \times c$ tables, with degrees of freedom $(2-1)(2-1) = 1$.

Let us return to the example presented in Table 13.3. In order to test the hypothesis that bedroom standard and gender of the head of household are not associated we use the X^2 test of independence, but now with 2 rows and 6 columns. In terms of probabilities and proportions, this hypothesis can be expressed as

$$P(2+ \text{ below standard} \mid \text{Male}) = P(2+ \text{ below standard} \mid \text{Female}),$$

$$P(1 \text{ below standard} \mid \text{Male}) = P(1 \text{ below standard} \mid \text{Female}), \text{ etc,}$$

with the alternative hypothesis that these probabilities, or proportions, are not all equal.

The observed and expected values are given in Table 13.9. Applying this test to the original table which has 2 rows and 6 columns yields $X^2 = 21.47$ on 5 degrees of freedom. However, the expected values in a number of cells are smaller than 5 and so the use of tables of the percentage points of the χ^2 distribution is not appropriate.

Table 13.9 Head of household and bedroom standard: observed and expected values

		2+ below standard	1 below standard	Standard	1 above standard	2 above standard	3+ above standard
Male	O	6	35	252	409	258	44
	E	5.1	43.0	264.9	405.8	244.7	40.5
Female	O	0	16	62	72	32	4
	E	0.9	8.0	49.1	75.2	45.3	7.5

The problem lies with the first column, corresponding to 6 households with 2+ bedrooms less than the standard. The way to overcome the problem caused by this small expected value is to combine two of the columns together. In this case the variable, bedroom standard, is an ordinal variable and so it is only relevant to combine adjacent columns. Thus we combine 2+ bedrooms below standard with 1 below standard (see Table 13.10).

The test statistic is now $X^2 = 17.46$ on 4 degrees of freedom. From the tables we can see that this corresponds to $p < 0.01$ (the exact p-value is 0.0016). This is very small and indicates that the hypothesis that the proportions are equal is not valid. These data provide evidence that there is some association between the severity of over crowding (level of underoccupancy) and the gender of the head of household.

Table 13.10 Housing standard grouped

	1+ below standard	Standard	1 above standard	2 above standard	3+ above standard
O					
Male	41	252	409	258	44
Female	16	62	72	32	4
E					
Male	48.1	264.9	405.8	244.7	40.5
Female	8.9	49.1	75.2	45.3	7.5
$(O - E)^2/E$					
Male	1.05	0.63	0.02	0.73	0.30
Female	5.64	3.40	0.13	3.92	1.64
$O - E/\sqrt{E}$					
Male	−1.02	−0.79	0.16	0.85	0.55
Female	2.38	1.84	−0.37	−1.98	−1.28

13.7 Large residuals

Once you have carried out the X^2 test of independence you will conclude that there is or is not evidence of association between the two variables. If you find that there is no association, then you should report that this is so, and you can then summarize the table in terms of the overall row or column proportions, using confidence intervals to give an idea of the precision of the estimates. You may also quote the p-value of the test and summarize the data in terms of the percentages.

If there is evidence of association between the two variables in the table, as in the cases in Sections 13.5 and 13.6, then you need to make some attempt to describe the nature of the association. In your interpretation you need to be careful that you do not make any causal links but only speak in terms of associations, as the test is only a test of association. There are two common ways of trying to describe the nature of the association. The test statistic, X^2, is a sum of components, $(O - E)^2/E$, and as large values of X^2 provide evidence of association then looking for the cells which contribute most to this sum will give information on the nature of the association. The second way is to examine the row or column percentages in the table (see Section 13.4.2).

A rough rule of thumb for deciding if a component of X^2 is large is to use the cut-off value of 4. This arises because in large samples the quantities $(O - E)/\sqrt{E}$, known as the X^2 residuals, are approximately normally distributed. Thus we would expect to find that 95% of these residuals would lie in the range ±2. Consequently, we expect 95% of the components of X^2, $(O - E)^2/E$, to lie between 0 and 4. Thus, 4 is a rough rule of thumb for deciding if a component of X^2 is large enough to comment on. Alternatively, X^2 residuals bigger than ±2 can be used. The advantage of the residual is that you can see right away if the observed is larger than the expected (positive residual) or if the observed is smaller than the expected (negative residual).

In Table 13.10 there is one component of X^2 which is much larger than the cut-off of 4. It is 5.64, corresponding to households with a woman as head and with 1 or more bedrooms below standard. Under the hypothesis of no association between bedroom standard and gender the expected number of households is 8.9, and 16 were observed. So the high value of X^2 arises because there are more households with a woman as head with below standard accommodation than is expected. The most important reason for the association between bedroom standard and the gender of the head of the household is that there are many more households with a woman as head living in poor accommodation than we would have expected.

Other cells with large X^2 components are Female, Standard and Female, 2 Above, although neither is over the rough rule-of-thumb cut-off value of 4. The observed and expected values are 62 and 49.1, 32 and 45.3, respectively, and the conclusion is that the association is a result of an over-representation of households with a female head in the poorer standard accommodation and an under-representation among households with above standard accommodation. This is exactly the same interpretation as obtained when looking at the row and column percentages in Section 13.4.1.

13.8 Sample size

A final point about the interpretation of X^2 tests is concerned with the sample size. Let us take the data in Table 13.10 and consider the conclusions that we would come to if they were reduced by a factor of 2 or 3. First halving the sample size, we get

	1+ below standard	Standard	1 above standard	2 above standard	3+ above standard
Male	20	126	204	129	22
Female	8	31	36	16	2

resulting in $X^2 = 8.9$. Reducing the original sample size by two-thirds gives

	1+ below standard	Standard	1 above standard	2 above standard	3+ above standard
Male	14	84	136	86	15
Female	5	21	24	11	1

and now $X^2 = 5.6$.

Although the row and column proportions are exactly the same when the data are reduced, X^2 grows smaller with the size of the sample. Admittedly, the X^2 test becomes inapplicable as some of the expected values become too small. The point is that the statistical significance of the X^2 test, like that of any statistical test, is related to the sample size, through the power of the test. If the sample size is small then the X^2 test has little power to detect any association other than large association. If, on the other hand, the sample size is extremely large then very small departures from equality of proportions will appear to be statistically significant.

With a table where the true row proportions are

Column	A	B	C	D	E
Row 1	0.20	0.20	0.20	0.20	0.20
Row 2	0.18	0.19	0.20	0.21	0.22

with equal numbers in the two rows, the X^2 values on 4 degrees of freedom and the p-values for different sample sizes are

Sample size	100	1000	5000	10 000
X^2	0.1	1.3	6.3	12.5
p-value	0.99	0.87	0.18	0.01

We can see that even when the differences in the row proportions are unlikely to be of any major importance, if the sample size is large enough then you will find that you conclude a statistically significant association.

This investigation illustrates two important things. Firstly, even if there is a true association which is large, the X^2 test may not give a statistically significant result if the sample size is too small. Secondly, if the sample size is very large, associations which are of a very small size such that they are virtually meaningless may be reported as statistically significant. Generally market research and opinion poll surveys are of the order of 1000–3000 individuals, and this sample size achieves some balance between the two extremes.

14

Correlation

- Knowing how to interpret the correlation coefficient and when it should be used
- Knowing how to calculate a confidence interval for the correlation
- Appreciating the effect of sample size on the precision of the estimated correlation
- Knowing how to test if there is evidence of a significant correlation
- Being fully aware of the problem of the interpretation of a significant correlation coefficient – the inference of causality and the effect of spurious correlations
- Knowing when the product moment correlation is not an appropriate measure of association
- Knowing how to calculate and interpret Spearman's rank correlation coefficient.

14.1 Introduction

We have previously looked at the relationship between two quantitative variables through a graphical procedure known as the scatter plot (Chapter 3). We have also introduced the sample correlation coefficient, which is a measure of the strength of the linear association between two variables (Chapter 6). In this chapter we will introduce and use the statistical inferential properties of the correlation coefficient. Before you work through this chapter you should be familiar with the material in Chapter 6.

For two variables X and Y, if we denote the sets of observations in the sample by (x_i, y_i), $i = 1, \ldots , n$, then the sample correlation coefficient is given by

$$r_{xy} = \frac{s_{xy}}{s_x s_y},$$

where s_{xy} is the sample covariance between x and y, and s_x and s_y are the sample standard deviations of the two variables. The sample correlation coefficient will always lie in the range $-1 \leq r_{xy} \leq 1$; $r_{xy} = 0$ indicates that there is no linear association between x and y, while if $r_{xy} = \pm 1$ then all the points will lie on a straight line with positive or negative slope. If the correlation is positive, then when the value of one variable is large the value of the other is also large, and when one is small then the other is also small. If the correlation is negative, then when the value of one variable is large the value of the other is small, and vice versa.

The data presented in Table 14.1 show figures for the economic activity and average dwelling price in the 13 regions of the United Kingdom. With a correlation analysis of

Table 14.1 Economic activity and average dwelling price in the regions of the United Kingdom, in 1996

Region	Economic activity rates, spring 1996 (%)		Average dwelling price, 1996 (£ thousands)
	Males	Females	
North East	67.6	49.4	51.3
North West (GOR)	70.8	52.7	58.1
Merseyside	64.2	47.8	59.4
Yorkshire Humber	70.5	52.9	57.4
East Midlands	73.4	55.3	59.8
West Midlands	72.8	53.0	64.6
Eastern	74.8	55.7	74.3
London	73.5	55.3	94.7
South East (GOR)	75.3	56.6	88.3
South West	71.0	53.5	68.4
Wales	66.4	48.3	55.4
Scotland	70.9	52.4	57.5
Northern Ireland	70.3	48.6	47.9

Source: http://www.ons.gov.uk. © Crown Copyright.

these data we are looking at the association between economic activity and house prices. We would expect higher house prices in areas where there is greater economic activity. This could arise because there is a movement of people into these areas in order to work and the competition for housing drives the prices upwards. In areas which are economically depressed there is less demand for housing as people are not moving house so much and so house prices will tend to be lower.

The correlation between the percentage of men in an area who are economically active and the corresponding percentage of women is 0.922. This, in conjunction with the scatter plot in Figure 14.1, indicates a strong linear association between these two variables. The positive association is entirely to be expected. The correlations between the economic activity percentages and the average dwelling prices are 0.630 for men and 0.736 for women. The slightly higher correlation in the case of women might have a socio-economic rationale in that for many households two salaries are required to provide the necessary finance for a house purchase. However, we have not taken into account any sampling variation inherent in any statistical estimation and so we do not know if there is a significant difference between these two correlations. If there is no difference then the above socio-economic theory is not supported by these data. In the next section we will look at the estimation of the confidence interval in more detail.

14.2 Estimate and confidence interval

14.2.1 Fisher's *z* transformation

The sampling distribution properties of the correlation coefficient are worked out under the assumption of a statistical model for the set of observations. This is the same as in any other situation. In this case, the common assumption we make is that there is a random

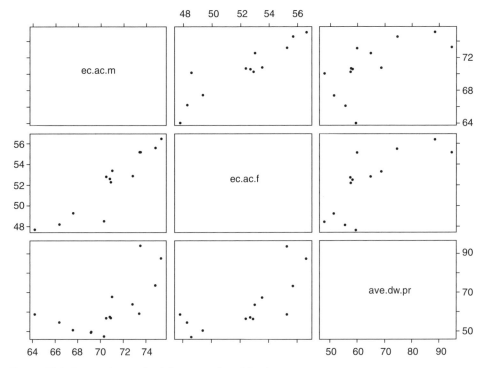

Figure 14.1 Scatter plot matrix of the economic activity data.

sample of n observations from a population. Over the population the two variables (X, Y) follow what is known as a **bivariate normal distribution** with means μ_X and μ_Y, standard deviations σ_X and σ_Y and a population correlation coefficient ρ_{XY}. Essentially this means that both X and Y have to have normal distributions. We have already come across the sample mean and sample standard deviation as estimators of μ_X and σ_X, and the sample correlation coefficient r_{XY} as an estimate of the population correlation coefficient.

As with any estimator, the correlation coefficient has a sampling distribution associated with the use of sample data to estimate a population quantity. Throughout Chapter 11 we have seen that a confidence interval is of the form

$$\text{Estimate} \pm t \text{ value} \times \text{Standard error of the estimate},$$

which means that in order to calculate a confidence interval for the correlation coefficient we need to find its standard error. This is not easy as the correlation coefficient is a ratio of sums of squares and products. For this reason the confidence intervals for the correlation coefficient are often omitted from textbooks.

However, it is not necessary to know the derivation of the standard error in order to calculate the confidence interval. In Chapter 6 we came across the idea of transforming variables to achieve a straight-line relationship, and throughout the book we have transformed variables to achieve a more normal distribution. Now, we will look at a transformation of a statistic, the correlation coefficient, to achieve normality and hence calculate the confidence interval using the well-known procedure above.

The famous statistician R. A. Fisher introduced the z transformation for the correlation coefficient. This is given by

$$z = \frac{1}{2} \log_e \left(\frac{1+r}{1-r} \right),$$

which is defined for $-1 < r < 1$, i.e. it cannot be used if $r = 1$ or $r = -1$. This transformation is required because the correlation coefficient is bounded by -1 and $+1$ and so cannot follow a normal distribution. By taking the logarithm of the ratio we end up with a random variable which can take any value within a given range and has a normal distribution. The standard deviation of z is

$$\sigma_z = \frac{1}{\sqrt{n-3}};$$

we can calculate confidence intervals for z based upon the standard normal distribution and then convert back to r using the back-transformation

$$r = \frac{e^{2z} - 1}{e^{2z} + 1},$$

where e^x is the exponential function.

There is no doubt that this is a complicated procedure, much more so than the confidence intervals for the sample mean and proportion. However, it is really the only feasible way of calculating the confidence interval for a correlation coefficient. This procedure is not pursued much, probably because of its complexity, but if it were used more frequently then there would be less over-interpretation of sample correlations.

In the data in Table 14.1 there are 13 regions and so the standard error of z is given by

$$\sigma_z = \frac{1}{\sqrt{10}} = 0.316.$$

First of all, we will look at the correlation between the employment participation rate for women and the average dwelling price, which was 0.736. The z transformation of this value is

$$z = \frac{1}{2} \log_e \left(\frac{1+r}{1-r} \right) = \frac{1}{2} \log_e \left(\frac{1 + 0.736}{1 - 0.736} \right) = \frac{1}{2} \log_e (6.565) = 0.941.$$

The 95% confidence interval for the correlation coefficient is obtained from

$$z \pm z_{0.025} \sigma_z.$$

Note that we have z used in two different places in this formula, firstly as the transformation of the correlation coefficient, and secondly as the percentage point of the standard normal distribution used to calculate the confidence widths. By this time I hope that you are familiar with the use of z for the standard normal distribution and that the double use of z here will not be too confusing. It is unfortunate that this is the standard statistical terminology for the z transformation of the correlation coefficient.

Substituting the values into the above equation gives

$$z \pm z_{0.025} \sigma_z = 0.941 \pm 1.96 \times 0.316 = (0.321, 1.561).$$

These are the confidence limits for the z transformation and we now need to use the back-transformation to find the limits of the confidence interval for the correlation coefficient:

$$r_L = \frac{e^{2z} - 1}{e^{2z} + 1} = \frac{e^{2 \times 0.321} - 1}{e^{2 \times 0.321} + 1} = \frac{e^{0.642} - 1}{e^{0.642} + 1} = \frac{1.900 - 1}{1.900 + 1} = 0.310,$$

$$r_U = \frac{e^{2z} - 1}{e^{2z} + 1} = \frac{e^{2 \times 1.561} - 1}{e^{2 \times 1.561} + 1} = \frac{e^{3.122} - 1}{e^{3.122} + 1} = \frac{22.678 - 1}{22.678 + 1} = 0.916.$$

The sample correlation coefficient is 0.736 and the 95% confidence interval is (0.310, 0.916). There are two things to note about this interval. Firstly, it is not symmetrical about the estimate of 0.736. This is a consequence of the transformation and the fact that the correlation coefficient is bounded by -1 and 1. If the interval were symmetric then the upper limit could go above 1, and this would not be sensible. Secondly, the interval is wide, covering virtually all of the values of r indicating a moderate to strong positive association.

The correlation between the percentage of men in employment and the average dwelling price is 0.630, and if you go through the 95% confidence interval calculations in the same way as above then you will find that the limits are 0.121 and 0.877. This interval is also very wide and spans much the same range as the confidence interval for women. This means that there is very little evidence to suggest that there is a higher correlation for women than for men and the differences in the sample correlations can just be ascribed to sampling variation.

14.2.2 Sample size and random sampling

The sample size is small at only 13 observations. If we had had exactly the same value for the correlation but from a sample of size 130, then the 95% confidence interval for men would have been (0.51, 0.72), which is reasonably symmetric about the observed correlation of 0.63. With a sample of size 260, the 95% confidence interval would have been (0.55, 0.70); with 1300 observations the interval would have been (0.60, 0.66). In order to estimate the correlation coefficient precisely a very large number of observations will be required. This is generally over 1000 for moderately strong correlations and is of the same order of magnitude as the sample sizes required to estimate percentages or proportions to the same precision of ± 0.03.

If the correlation is very strong, by which we mean around ± 0.9, the confidence interval will be narrower. For the correlation between the economic activity rates for men and women, which is 0.92, the 95% confidence interval from the 13 observations is (0.75, 0.98), narrower than the other two correlations we have calculated. With 130 observations the interval would have been (0.91, 0.93). For strong correlations you do not require as large a sample to achieve a high precision as you do for weak correlations. This is exactly the same situation as with percentages – you need larger samples to estimate percentages around 50% than percentages around 10% or 90%.

With regard to the data used in the example, the assumption of a bivariate normal distribution may be reasonable for each of the pairs of variables that we have looked at. However, it is difficult to justify that we have a random sample from a population, as our data obviously refer to the populations of the 13 regions of the United Kingdom and we have not done any sampling. However, the use of the random sampling idea is still valid for a couple of reasons. Firstly, the measurements themselves – the economic activity and the average dwelling price – are not population values but are, in fact, both based upon surveys. The dwelling price is based upon building society figures and the economic activity on labour force surveys. These surveys incorporate random sampling. Secondly, recall that we have used data from spring 1996 for economic activity and 1996 for average dwelling price. We could have used other seasons in 1996 for the economic activity and other years for both. We would expect the same associations with slightly different data, and so again we have sampling of a sort.

You will often find that it is hard to justify the assumption of a random sample, but it is still better to calculate a confidence interval to give you some idea of precision even if you suspect that there is no random sample at all. There will always be some random variation, and the confidence interval helps to put the precision of the estimate into perspective.

14.3 Testing

14.3.1 *t* test for a correlation

When using the correlation coefficient, the most common question asked by researchers in business and marketing is: is there any association? The correlation coefficient measures linear association, and so the question comes down to a more specific one: is there any linear association between the two variables? From this it is only a short step to the question: is there any evidence that the correlation coefficient is significantly different from zero? The null hypothesis is H_0: $\rho = 0$, and the alternative is normally a two-sided H_1: $\rho \neq 0$. This hypothesis is, in fact, very easy to test and comes down to a t test.

As in the previous section we assume that the set of n observations are a random sample from a bivariate normal distribution. If the null hypothesis is true and $\rho = 0$ then the test statistic is

$$t = \sqrt{n - 2} \, \frac{r}{\sqrt{1 - r^2}},$$

which follows a t distribution on $n - 2$ degrees of freedom. This is a particularly easy test to carry out, much easier than calculating the confidence interval, and is one reason for the popularity of testing correlations rather than interpreting what they mean in terms of the confidence intervals.

It would clearly be of interest to test if there was evidence that there was a linear association between the economic activity rates and the average dwelling price. For men the sample correlation is 0.630 and with 13 regions the test statistic is

$$t = \sqrt{13 - 2} \, \frac{0.630}{\sqrt{1 - 0.630^2}} = \sqrt{11} \, \frac{0.630}{0.777} = 2.69.$$

From the tables of the t distribution on 11 degrees of freedom we can see that $t_{0.025} = 2.20$ and $t_{0.01} = 2.72$. This means that the p-value is $0.02 < p < 0.05$, as we have a two-sided test. There is evidence at the 5% level of significance that the correlation between the economic activity rates and average dwelling price is not equal to zero. This implies that there is a significant linear association.

If we do the test for the correlation between economic activity (women) and average dwelling price, we find that the p-value is $p < 0.01$. It should not surprise you that the p-value is smaller, as the correlation was larger. However, it is important to note that these two tests are not independent of each other. Both of them involve the average dwelling price and so the correlations cannot be independent, particularly when you consider that the economic activity rates for men and women are very highly correlated.

14.3.2 Multiple testing and interpretation of correlation tests

The same problems of multiple testing arise when testing many correlations, but in this instance the problem is more acute as you know that the tests are not independent of each other. It is quite common to see a correlation matrix such as the one in the upper triangle of Table 14.2 which is based upon data discussed in Chapter 6. The full correlation matrix is given on the lower triangle, but only the correlations which are significant at the 5% level are included in the upper triangle. This is done to introduce some clarity into the table. However, there are 28 correlations tested, and you would expect some of them to be significant just by chance even if there was no linear association.

Table 14.2 Correlation matrix of housing data*

	FCAT	CTOQ	CTOU	FCAV	CTOS	CTOW	AHLO	AHLS
FCAT	–	–0.48	0.66	0.81	–0.47	0.54	0.38	0.40
CTOQ	–0.48	–	–0.83	–0.64	0.90	–0.89		
CTOU	0.66	–0.83	–	0.63	–0.79	0.91		
FCAV	0.81	–0.64	0.63	–	–0.64	0.62	0.48	0.54
CTOS	–0.47	0.90	–0.79	–0.64	–	–0.85		
CTOW	0.54	–0.89	0.91	0.62	–0.85	–		
AHLO	0.38	–0.08	–0.08	0.48	–0.08	–0.06	–	0.88
AHLS	0.40	–0.08	–0.07	0.54	–0.11	–0.04	0.88	–

*All correlations are shown below the diagonal; above the diagonal only correlations significant at the 5% level are given.

The multiple testing problem is not the only point about this correlation matrix. The main point is that you know from the definition of the variables that some of them are going to be related and its is pointless carrying out a test of H_0: $\rho = 0$. The second important point is that the scatter plot shows curved relationships in some cases and so the correlation coefficient is not valid as it only measures linear association.

You should not calculate a correlation matrix and do a lot of tests on the individual correlations. Nor should you use a correlation matrix without first looking at all the scatter plots.

14.4 Interpretation: causality and spurious correlations

Correlation coefficients are one of the most over-interpreted sample statistics. All the coefficient does is to measure the strength of a linear association between two variables. As we have previously discussed, there is absolutely no mention of causality in the interpretation. A high correlation coefficient does not mean that one variable depends on the other to such an extent that changing one variable will bring about a consequent change in the other. All of the material in this section is also relevant for the interpretation of X^2 tests of independence (Chapter 13) as they are used to assess the association between two qualitative variables. As with the correlation coefficient, a significant X^2 test does not imply a causal association.

The strong correlation of 0.92 between the economic activity rates for men and women over the 13 regions does not any imply causal relationship. If the rates for men change this does not mean that the rates for women will change as a consequence. The most likely interpretation is that both of these rates increase as a function of a number of other economic variables. These might include the availability of jobs in the area, the productivity in the region, the level of investment in the region. All of these three might be considered as causal in the sense that increasing the level of investment may lead to greater availability of jobs which leads to higher economic activity rates. This may in turn lead to a further increase in investment, and so you have a circular causal relationship. This discussion is here to show that causality requires knowledge of the economic situation over and above the statistical demonstration of an association. The statistical correlation does not imply causality.

In the same way, the existence of significantly positive associations between the economic activity rates for both men and women and the average dwelling price does not prove that any of the speculative economic theories discussed in Section 14.1 are true. All that we know is that a positive correlation is consistent with these theories. Agreement with the theory does not mean that the theory is correct.

If there were no evidence of a positive correlation, leading to a non-significant test, this would suggest that the theory was not justified. However, in order to be more certain about this conclusion we would need to be convinced that the test was sufficiently powerful. This is not the case here as there are only 13 observations and a non-significant result from a small study is not sufficient to refute the theory.

We have not discussed power with respect to the correlation test as we did with tests on means and proportions (Section 12.8). However, the same principle will apply, and large samples are needed to give rise to sufficiently high power. Samples of size 13 are not large enough to inspire confidence in the power of the test. Bearing in mind the samples of hundreds and thousands required for a high precision in the estimate of the correlation coefficient (Section 14.2), we would also need samples of this size to have confidence in the power of the test.

The interpretation of correlation coefficients is made easier in the presence of an underlying physical plausibility. In the examples in this chapter, for all of the high correlations there are reasonably clear economic reasons. Also, there are good economic reasons why eight of the correlations in Table 14.2 were not significantly different from zero. These were the correlations between AHLO and AHLS on one hand and CTOQ, CTOU, CTOS, and CTOW on the other. AHLO and ALHS relate to building society commitments and advances on new buildings, while CTOQ, CTOU, CTOS and CTOW relate to local authority and housing association starts and completions. Building societies lend money for the purchase of private housing, and there is no reason to expect an association with housing starts and completions in the non-private housing sector.

The final difficulty in the interpretation of correlations is the existence of high correlations between two variables which have no obvious physical association. This is known as a **spurious** correlation, and one way in which this can occur is if each of the two variables is associated with a third variable. This means that when interpreting the correlation coefficient you should always have in mind whether or not there is likely to be a physical relationship between the two variables. The most common third variable to which the other variables are related is time.

The data in Table 14.3 (Tame and Elliott n.d.) are concerned with cigarette consumption and advertising in the United Kingdom from 1970 until 1983. They were published by FOREST, an organization founded in 1979 to defend the rights of adults who smoke tobacco.

Table 14.3 Tobacco consumption and advertising

Year	Consumption of cigarettes (millions)	Advertising index	% Low tar
1970	127 000	100.0	NA
1971	122 400	101.1	NA
1972	130 500	114.0	5
1973	137 400	109.7	9
1974	137 000	85.0	10
1975	132 600	76.3	14
1976	130 600	120.4	21
1977	125 900	105.4	22
1978	125 200	71.0	19
1979	124 300	82.8	24
1980	121 500	114.0	28
1981	110 300	93.8	39
1982	102 000	105.9	49
1983	NA	NA	53

The correlation matrix is presented in Table 14.4. You can see a very high negative correlation of -0.95 between consumption of cigarettes and the percentage of cigarettes sold which are low tar. It is very difficult to believe this is a causal relationship – that increasing the consumption of cigarettes leads to a reduction in the percentage of low tar cigarettes sold or increasing the percentage of low tar cigarettes sold leads to a reduction in consumption. It is more likely that this correlation is spurious, arising as the pattern of consumption of cigarettes has changed over time. The correlation of consumption with time is -0.89, indicating a reduction in consumption over the period of observation. There is a positive correlation of 0.94 between time and the percentage of cigarettes consumed which are low tar, indicating an increase in the consumption of low tar cigarettes over time. The scatter plots are not shown, but they indicate that the relationships are approximately linear.

Table 14.4 Tobacco correlations

	Year	Consumption	Advertising	PercLowTar
Year	1.000	−0.888	−0.095	0.943
Consumption	−0.888	1.000	−0.069	−0.952
Advertising	−0.095	−0.069	1.000	0.077
PercLowTar	0.943	−0.952	0.077	1.000

The authors' interpretation of the correlation between cigarette consumption and advertising is given below. Note that regression analysis is closely related to correlation analysis and will be discussed in Chapter 15. The essential point as far as they are concerned is that the correlation between consumption and advertising is −0.07, which is not significantly different from zero, and they use this figure to support their argument that advertising expenditure has no influence on cigarette consumption and so advertising on cigarettes should not be banned:

> The results of statistical analysis in fact permit only the most agnostic conclusions about the influence of advertising on consumption. This is borne out by 'regression analysis', a standard method of statistical analysis for measuring the effect of one variable on another. Regression shows an inverse relationship between advertising and consumption, that is, as advertising rises consumption of cigarettes falls. This is not what one would expect. However, a regression of this data shows that only 0.005% of the variations in consumption can be statistically explained by changes in advertising expenditure.
>
> Statistically little can be confidently said, judged from the data herein, about the effect of advertising on tobacco consumption. Therefore, in practice, the impact of advertising on cigarette consumption has been even less significant than economic theory suggests.

There is nothing statistically wrong with the first conclusions of the authors. However, the correlation between advertising and time is −0.09, and so there is no association between the advertising index and time. If consumption is associated with time and advertising is not associated with time, then it is no surprise that consumption is not associated with advertising. All this goes to show is that the interpretation of correlations is very difficult when you try to attach a causal interpretation – particularly when other important information has been omitted, such as the ban on advertising on television, advertising campaigns encouraging people to stop smoking, and health warnings on cigarette packets and on billboard adverts.

The lesson of spurious correlations is that you should not go overboard with the interpretation of a high correlation unless you have a plausible economic, business or marketing reason why there should be an association. If you find an unexpected correlation between variables which you did not think would be related then you should think carefully about possible reasons for this. You should bear in mind the existence of a spurious correlation, and the fact that you have probably been doing lots of significance tests means that from time to time you must expect to get significant correlations even if there is no association. Similarly, the absence of a correlation does not mean that there is absolutely no influence of one variable on another.

14.5 Rank correlation

We conclude this chapter with a discussion of what to do to estimate and test for an association when the two variables do not come from a bivariate normal distribution. As mentioned in Chapter 6, we can try to find a transformation. An alternative approach, mentioned in Chapter 12, is to adopt a non-parametric approach and use a method based upon the ranks of the data. There are two methods, **Kendall's** τ and **Spearman's correlation**. Of these the latter is the simplest and the only one we will use.

We illustrate the Spearman rank correlation with reference to the correlation between the housing starts for housing associations (CTOQ) and the housing starts for local authorities, new towns and government departments (CTOU) which are part of the correlation matrix in Table 14.2 and also in Chapter 00. The histograms in Figure 14.2 show clearly that neither of the variables has a normal distribution. The scatter plot shows that there is not a straight-line relationship between the two variables – it is a curve. Finally, the scatter plot of the ranks reveals an approximate straight-line relationship with a negative slope. This implies that if we take the correlation of the ranks we will have a measure of association between the two variables.

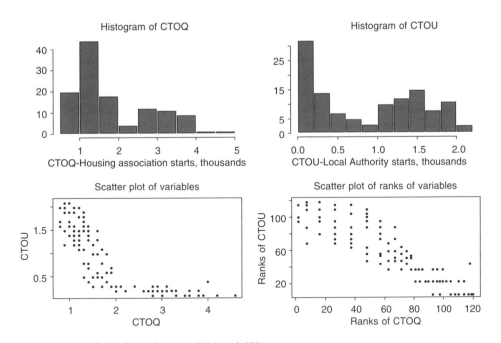

Figure 14.2 Rank correlation between CTOQ and CTOU.

The Spearman rank correlation is denoted r_S and is simply calculated by replacing each observation by its rank and then calculating the correlation based on the ranks. This correlation is interpreted in much the same way as the Pearson correlation coefficient. It is bounded above and below by $-1 \le r_S \le 1$, with $r_S = 0$ indicating no association while $r_S = 1$ indicates perfect positive association.

You should note that the Spearman rank correlation measures the linear association between the ranks of the variables. The ranks contain only ordinal information, so the Spearman rank correlation is a measure of ordinal association between the two variables. In contrast, the Pearson correlation coefficient is a measure of linear association between the two variables.

An appropriate test statistic for the Spearman rank correlation can be enumerated for any sample size, but it is complicated and we will only use an approximation

$$z = \frac{r_S}{1/\sqrt{n-1}} = r_S \sqrt{n-1} .$$

In large samples this statistic follows a standard normal distribution, assuming that there is no association between the two variables, i.e. the null hypothesis is true. For the purposes of using this approximation samples of greater than 20 are sufficiently large.

The Spearman rank correlation for the data in Figure 14.2 is -0.903. There are 120 pairs of observations and the test statistic is $z = -9.85$, which has a p-value which is almost zero. This provides very strong evidence of an ordinal association between these two variables. These data were collected from 1985 until 1994 and came at a time when the Conservative government in the United Kingdom adopted a policy of selling off local authority houses to the tenants and encouraging housing associations to build houses which could then be bought by the tenants, while at the same time removing the incentive for local authorities to build houses which they might then have to sell to the tenants at low prices. In this way the statistical association is a confirmation of the policy effects, and this example shows you how you need to tie in the statistical association with the economic policies to get a full interpretation of the data.

The only technical issue with the Spearman rank correlation is in the calculation of the ranks of the observations. We will illustrate this with a small example using the two of the variables in Table 14.1, reproduced in Table 14.5. To calculate the ranks of the data all you do is assign a rank of 1 to the smallest observation, 2 to the next smallest, and so on. For the average dwelling price in Table 14.5 this is easily done, and the ranks range from 1 in Northern Ireland, where the houses are cheapest on average, to 13 in London, with

Table 14.5 Calculating Spearman's correlation

Region	Economic activity, women	Average dwelling price	Rank economic activity	Rank average dwelling price
North East	49.4	51.3	4	2
North West (GOR)	52.7	58.1	6	6
Merseyside	47.8	59.4	1	7
Yorkshire Humber	52.9	57.4	7	4
East Midlands	55.3	59.8	10.5	8
West Midlands	53.0	64.6	8	9
Eastern	55.7	74.3	12	11
London	55.3	94.7	10.5	13
South East (GOR)	56.6	88.3	13	12
South West	53.5	68.4	9	10
Wales	48.3	55.4	2	3
Scotland	52.4	57.5	5	5
Northern Ireland	48.6	47.9	3	1

the highest average. For the economic activity for women a slight modification is necessary because the rates are exactly the same in East Midlands and London. These two regions are the tenth and eleventh highest in terms of the economic activity, and as the rates are the same we just use the average of the ranks. Thus East Midlands and London are both given a rank of $(10 + 11)/2 = 10.5$, and we carry on with a rank of 12 for East and 13 for South East.

The Spearman rank correlation between the economic participation rates for women and the average dwelling price is 0.806. The test statistic is $z = 2.79$ and the p-value associated with the null hypothesis that there is no association is 0.005. There is evidence of a strong ordinal association between economic activity and average dwelling price. In section 14.3 the Pearson correlation coefficient was found to be 0.730, with associated p-value less than 0.01. In this case we have the same statistical interpretation from the use of the Spearman and Pearson correlations. This will generally be the case when the assumptions which are necessary for the use of the Pearson coefficient are valid – linearity and normality.

In summary, the Pearson correlation is the usual one to use. It is a measure of linear association and should only be used when the scatter plot shows a linear association and the variables follow normal distributions. If either of these assumptions is not valid then you can use the Spearman rank correlation. This is a more general measure of association in that you do not need to have a straight-line relationship. It is really a measure of ordinal association, or linear association of the ranks. In both cases you need to be wary of the interpretation of the correlation and must not automatically ascribe a causal relationship to any significant correlation. Always seek an interpretation of any correlation in terms of the underlying economic or business models. If you cannot do so then it is possible that you have spurious correlation arising because of the influence of other variables.

15

Linear regression

- Understanding the random and systematic components of a linear regression model
- Being aware of the assumptions of this model—straight-line relationship, normal distribution, independent errors with constant variance
- Knowing how to calculate the intercept and slope of the model as well as the residual standard error
- Appreciating the relationship between the slope of the model and the correlation coefficient
- Understanding the meaning of the concept of the proportion of variance explained (R^2)
- Understanding the importance of the residuals as a way of checking the validity of the assumptions of the model
- Knowing how to calculate confidence intervals for the estimated parameters
- Knowing how to calculate confidence intervals for the fitted values and predicted single values and when it is appropriate to use these predictions
- Being aware of the relationship between the linear regression model and the capital asset pricing model in finance
- Knowing how to bring in qualitative explanatory variable through the use of dummy variables
- Appreciating the relationship of the linear regression model with one dummy variable to the pooled two-sample t test
- Appreciating the extension of the model to cases where there are many explanatory variables (multiple regression)

15.1 Introduction

Throughout this book I have tried to emphasize that any statistical analysis has, in the background, the idea of a statistical model. This is described in terms of a probability distribution, often the normal distribution, and the parameters of the distribution, the mean and standard deviation in the case of the normal distribution. We use the data to estimate the parameters and then use our knowledge of sampling distributions, which are derived from the probability model or from large-sample theory, in order to provide confidence intervals and carry out significance tests.

In this chapter we develop this idea further to estimate the parameters of a statistical model for the association between two variables. We have already met this model briefly in Chapter 6, where we used the sample data to estimate the slope and intercept of the model. In this chapter we will focus on the statistical inference properties of the model. This will involve the usual ideas of confidence intervals for parameters of the model and significance tests for hypothesized values of these parameters. Furthermore, as one of the main uses of these models is in prediction we will look at confidence intervals for predictions.

This is the most advanced modelling that we will look at, and in subsequent sections we will see that some of the procedures and tests that we have looked at previously can also be written within this framework. We will also discuss how the model can be extended into more elaborate models.

With any statistical model the inference and estimation are based upon the idea that certain assumptions are valid for the data in the study. We have previously looked at the use of histograms and normal plots for assessing if the observations come from a normal distribution. In this chapter we will show how you can estimate the random errors and use them to assess the validity of the model. This is a crucial aspect of the use of any statistical model. The model is based upon assumptions, and if these are shown to be false then any predictions and conclusions drawn from the model will not have much support and so the model is more or less useless.

15.1.1 Data set: earnings per share and gearing

Throughout this analysis we will use two sets of data. This first is a small set using data taken from the Biz/Ed web site (www.bizednet.bris.ac.uk) and listed in Table 15.1. In this analysis we will investigate if the earnings per share are linearly related to gearing using the data from 1996 and then carry out a test using the data from 1997. We will derive an equation relating the earnings per share to gearing and then use this equation for prediction. The aim of the analysis is primarily to illustrate the model, but the substantive financial part is to try to predict how the earnings per share change with respect to the gearing.

Both gearing and earnings per share are financial ratios used in the financial analysis of company accounts. Gearing is defined as the ratio of long-term liabilities to shareholders' funds and is a measure of the financial risk of a company based on how the company

Table 15.1 Earnings per share and gearing for 11 companies

Source	1996		1997	
	EPS	G	EPS	G
Manchester Utd	18.40	28.60	29.80	7.15
Sheffield Utd	− 0.50	4.35	− 20.00	11.47
Hi-Tec	2.50	2.37	4.50	0.61
Yates Brothers	14.00	0.09	21.50	24.50
Glaxo	56.70	126.83	52.00	97.41
Tate & Lyle	40.50	78.17	22.60	76.44
Zeneca	67.80	22.26	77.00	23.27
Cadbury-Schweppes	34.10	47.61	68.70	34.08
Harry Ramsden's	10.60	0.88	12.10	20.36
Unilever	21.50	41.46	44.70	23.31
Kwik-Save	−14.61	0.00	27.75	2.33

finances itself. If gearing is 50% then for every £100 the shareholders have invested, the company has raised £50 from-long term loans. Shareholders' funds carry no risk to the company, but long-term loans do carry risk. A low gearing means less risk, but it may also mean that the company is not making enough long-term investments. The earnings per share are defined as the ratio of the profit of the company to the number of shares, that is, the amount of money each share has earned for the shareholder. If this ratio is high then the company has good profits relative to the shareholder investment.

15.1.2 Data set: supermarket location

The second set of data is a much larger one, which we encountered in Chapter 1. The problem is to develop a statistical model which can be used to predict turnover for a store based upon a number of other variables which may be related to the turnover. If such a model can be developed then the supermarket chain will be in a better position to decide upon the location of new stores and also to monitor the performance of existing stores, relative to the others. Illustrative data from five of the 153 supermarkets from a chain are presented in Table 15.2.

Table 15.2 Supermarket location data[1]

	Turnover[2]	Area[3]	Competitor[4]	Percentage in social class ABC[5]	Catchment area population[6]
1	9176	100	NC	42.2	15 182
2	8670	221	C	7.1	16 712
3	9317	95	NC	34.6	16 925
4	6992	157	C	36.1	11 723
5	7562	295	NC	57.5	12 323
Mean	8429.8	165.1	[NC% = 45.1]	35.3	14 885.7
Std. dev.	1942.4	87.3	[C% = 54.9]	16.3	31 71.2

[1] The numerical values in these data are all fictitious (see Chapter 1). The setting is real and the data are similar to the observed data.
[2] Average weekly turnover over a period of one year (£).
[3] Selling area in the store (m^2).
[4] The presence of a direct competitor (C) or not (NC) in the store's catchment area, defined to be the area within a 15–minute drive of the store. For a town like Berwick on Tweed this would encompass the whole town and some of the surrounding countryside. For a store in Leeds, it would be a small part of Leeds centred on the store.
[5] The percentage of households in social class ABC in the catchment area.
[6] The population in the catchment area.

The company specializes in small local supermarkets and aims to attract shoppers living close to the store or calling in on their way back from work. Its main market is among families shopping a little every day as opposed to those who drive to supermarkets out of town each week. The stores generally only have one manager and a few assistants, and tend to be located in relatively more deprived areas of towns.

15.2 Linear regression model

15.2.1 Systematic and random components

The scatter plot of the earnings per share and gearing data is shown in Figure 15.1. This shows that there is a reasonable straight-line relationship between earnings per share and gearing. The higher the gearing the higher the earnings per share, and vice versa. The line in the graph is the fitted straight line through these points and the vertical lines represent the deviations from the model. These deviations are known as the **residuals**. The straight line represents the systematic part of the model, while the residuals come from the random part of the model.

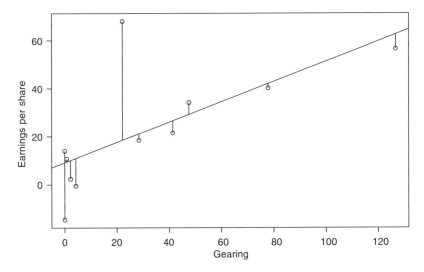

Figure 15.1 Scatter plot of earnings per share and gearing.

If we let y represent the earnings per share and x the gearing, then we write the statistical model as

$$y_i = \beta_0 + \beta_1 x_i + \varepsilon_i, \quad i = 1, \ldots, n,$$

where there are n pairs of observations. The systematic part of the model, $\beta_0 + \beta_1 x_i$, describes the relationship between the **dependent** or **response variable**, y, and the **independent** or **explanatory variable**, x, while the random part, ε_i, represents the variation of the observations about the systematic part of the model. These random errors are assumed to have a mean of zero, $E[\varepsilon_i] = 0$, and a constant variance, $Var[\varepsilon_i] = \sigma^2$, and to follow a normal distribution.

We have already met the systematic part in Chapter 6, where we also used the estimates for the parameters of the systematic part of the model to draw the lines on scatter plots. As the expected value of the error term is zero, the systematic part of the model represents the relationship between the expected value of y and x:

$$E[y_i \mid x_i] = E[\beta_0 + \beta_1 x_i + \varepsilon_i] = E[\beta_0 + \beta_1 x_i] + E[\varepsilon_i] = \beta_0 + \beta_1 x_i.$$

This is the expected value of y given x. As we treat x as a fixed value, there is no random variation in x, and so $E[\beta_0 + \beta_1 x_i] = \beta_0 + \beta_1 x_i$. This uses the rules for expected values in Section 9.1. The two parameters of the systematic part are the intercept, β_0, and the slope, β_1. The intercept represents the expected value of y when $x = 0$. The slope measures the rate of change in the expected value of y for a unit change in x. Generally, the slope is the more important of the two parameters and is the one that we shall pay more attention to.

Unlike the calculation of the correlation coefficient (see Chapter 14), the choice of which variable is the response variable and which is the explanatory variable is important. The independent variable is the one which is thought to cause changes in the response variable, or is thought to have an independent effect on the response variable. If there is a time order for the two variables, such as spending money on advertising and then measuring the subsequent retail sales, then the response variable is the variable which comes second, retail sales, and the explanatory (independent) variable is the one which comes first, advertising spend. The response variable depends upon the explanatory variable, not necessarily the other way round. In the supermarket example the turnover is the dependent variable because we think of the value of the turnover as being influenced by how large the store is, how large the catchment area is, what its social composition is, and whether the store has competition locally, which are explanatory variables. In the earning per share example, we think of gearing as being the independent variable as the earnings per share may be influenced by the financial risk of the company. In some cases it is hard to decide which variable is the response and which the explanatory variable, and a final clue is that if you are trying to predict the value of one of the variables then this variable should be the response variable.

The parameters of the linear regression model, β_0 and β_1, are estimated by a method known as **least-squares estimation**. When fitting a model to data we want the model to explain as much of the variation in the data as possible. The random part models the variation about the systematic part and we want to make this as small as possible. All of the error terms, ε_i, contribute to the variation, and a measure of the total variation is

$$\sum_{i=1}^{n} \varepsilon_i^2 = \sum_{i=1}^{n} (y_i - \beta_0 - \beta_1 x_i)^2.$$

This is known as the **sum of squares of the errors**. There are other measures of total variation that we could use, such as the sum of the absolute values of the errors, $\sum |\varepsilon_i|$, but with the sum of squares it is possible to derive equations for the estimates of β_0 and β_1. You should note that we cannot use the simple sum of the errors, $\sum \varepsilon_i$, as this will always be zero. You can see this from the vertical lines in Figure 15.1, where the four above the line will contribute a positive error and the seven below will contribute a negative error such that the overall sum will be zero.

15.2.2 Parameter estimates

The sum of squares of the errors, $\sum_{i=1}^{n} (y_i - \beta_0 - \beta_1 x_i)^2$, is a function of two unknown parameters, β_0 and β_1, because the data (x_i, y_i) are known values. We estimate the parameters as the values of β_0 and β_1 which make this sum of squares as small as possible. This is a mathematical problem of minimizing a function and is accomplished by differentiating the sum of squares with respect to the two parameters and solving the two equations. This gives us the following estimates, where the 'hat' on the parameters is used to denote that they are estimates, as opposed to the theoretical values in the model:

$$\hat{\beta}_1 = \frac{s_{xy}}{s_x^2}, \qquad \hat{\beta}_0 = \bar{y} - \hat{\beta}_1 \bar{x},$$

where

$$s_{xy} = \frac{1}{n-1}\sum(x - \bar{x})(y - \bar{y}), \qquad s_x^2 = \frac{1}{n-1}\sum(x - \bar{x})^2.$$

We have already used these estimates in Chapter 6.

The random part of the model is the error terms, ε_i, and this is where the probability modelling comes in. We assume that the errors are all independent of each other, meaning that the value of ε_1 has no influence on the value of ε_2, and so on. We also assume that the errors are normally distributed with a mean of zero. Finally, we assume that the variance of the ε_i is the same for all observations, and is denoted σ^2. This variance is a third parameter which we need to estimate.

With the estimated values for the slope and intercept we can calculate the **fitted values** corresponding to each of the observations. These are given by

$$\hat{y}_i = \hat{\beta}_0 + \hat{\beta}_1 x_i.$$

They represent the points along the line in Figure 15.1 corresponding to each of the 11 x-values. We can estimate the errors, ε_i, using the difference between the observed values, y_i, and the fitted values, $\hat{y}_i$. These differences are known as the **residuals**:

$$e_i = y_i - \hat{y}_i = y_i - \hat{\beta}_0 - \hat{\beta}_1 x_i.$$

These residuals can be positive or negative and, as mentioned above, their sum $\sum e_i = 0$. As the residuals, e_i, are estimates of the errors, ε_i, we can substitute the residuals into the sum of squares of the errors, $\sum \varepsilon_i^2$, to get the **residual sum of squares**, $\sum e_i^2$. From this we get our estimate of the variance of the errors in the model:

$$\hat{\sigma}^2 = \frac{1}{n-2}\sum e_i^2 = \frac{1}{n-2}\sum(y_i - \hat{y}_i)^2.$$

This is just like the estimate of the variance of any set of observations, except that the mean of the residuals ($\bar{e}$) is zero and so does not appear in the formula, and the divisor is $n - 2$ rather than $n - 1$. This occurs because two parameters have been estimated in the fixed part of the model, namely the estimates of β_0 and β_1, whereas in the case of a single sample of observations only the mean is estimated in the fixed part of the model.

15.2.3 Regression calculations: earnings per share

We will now illustrate the calculation of these quantities using the earning per share and gearing data. In practice, you will not need to worry about these too much as there are many computer programs which do the calculations. One thing to note in going through

Table 15.3 Calculations for the regression of earnings per share on gearing

	G (x)	EPS (y)	G2 (x^2)	EG (xy)	E2 (y^2)
Manchester Utd	28.60	18.40	817.9600	526.2400	338.5600
Sheffield Utd	4.35	−0.50	18.9225	−2.1750	0.2500
Hi-Tec	2.37	2.50	5.6169	5.9250	6.2500
Yates Brothers	0.09	14.00	0.0081	1.2600	196.0000
Glaxo	126.83	56.70	16 085.8489	7 191.2610	3 214.8900
Tate & Lyle	78.17	40.50	6 110.5489	3 165.8850	1 640.2500
Zeneca	22.26	67.80	495.5076	1 509.2280	4 596.8400
Cadbury-Schweppes	47.61	34.10	2 266.7121	1 623.5010	1 162.8100
Harry Ramsden's	0.88	10.60	0.7744	9.3280	112.3600
Unilever	41.46	21.50	1 718.9316	891.3900	462.2500
Kwik-Save	0.00	−14.61	0.0000	0.0000	213.4521
Totals	352.62	250.99	27 520.8310	14 921.8430	11 943.9121

the calculations is that many decimal places are recorded in the intermediate calculations to avoid any problems with rounding errors. The final equation is usually only quoted to one more significant figure than the original data.

From the totals in Table 15.3 we can calculate the means, variances and standard deviations

$$\bar{x} = 32.056\,364, \quad \bar{y} = 22.817\,273;$$

$$s_x^2 = 1621.712, \quad s_y^2 = 621.7005, \quad s_{xy} = 687.601\,63.$$

This means that the slope and intercept are estimated as

$$\hat{\beta}_1 = \frac{s_{xy}}{s_x^2} = \frac{687.601\,63}{1621.712} = 0.423\,997\,5,$$

$$\hat{\beta}_0 = \bar{y} - \hat{\beta}_1\bar{x} = 22.817\,273 - 0.423\,997\,5 \times 32.056\,364 = 9.225\,455.$$

The fitted line is

$$\hat{y}_i = \hat{\beta}_0 + \hat{\beta}_1 x_i = 9.23 + 0.42x_i.$$

The fitted values and residuals are shown in Table 15.4. The fitted values have been obtained by substituting the values for x in the equation for the fitted line, and the residuals are the difference between the observed y-values and the fitted y-values. The residuals are positive when the observed value is above the fitted line, and negative when the observed value is below the fitted line (see also Figure 15.1). The sum of the residuals is zero and the sum of squares of the residuals is $\sum e_i^2 = 3301.5913$. This means that the estimated error variance, also known as the variance of the residuals, is

$$\hat{\sigma}^2 = \frac{1}{n-2} \sum e_i^2 = \frac{3301.5913}{9} = 366.843\,473.$$

Thus the estimated error standard deviation is $\hat{\sigma} = 19.15$. This is also known as the **residual standard error**.

Table 15.4 Fitted values and residuals

	x(G)	y(EPS)	$\hat{y}$	e
Manchester Utd	28.60	18.40	21.35	–2.95
Sheffield Utd	4.35	–0.50	11.07	–11.57
Hi-Tec	2.37	2.50	10.23	–7.73
Yates Brothers	0.09	14.00	9.26	4.74
Glaxo	126.83	56.70	63.00	–6.30
Tate and Lyle	78.17	40.50	42.37	–1.87
Zeneca	22.26	67.80	18.66	49.14
Cadbury-Schweppes	47.61	34.10	29.41	4.69
Harry Ramsden's	0.88	10.60	9.60	1.00
Unilever	41.46	21.50	26.80	–5.30
Kwik-Save	0.00	–14.61	9.23	–23.84

15.3 Relationship of the slope to the correlation coefficient

15.3.1 Proportion of variance explained by the model

The slope of the linear regression line is given by $\hat{\beta}_1 = s_{xy}/s_x^2$, and this is very similar to the equation for the sample correlation coefficient, which is $r_{xy} = s_{xy}/s_x s_y$. These two quantities are related through the equations

$$\hat{\beta}_1 = r_{xy}\frac{s_y}{s_x}, \qquad r_{xy} = \hat{\beta}_1\frac{s_x}{s_y}.$$

This means that exactly the same information is contained in the slope of the line as in the correlation coefficient. The only differences between the two are in the interpretation and the scale. The slope measures the rate of change of the y variable with respect to x, and so its units are the ratio of the y units to the x units. If the y variable is in metres and the x variable is in seconds, then the units of the slope are metres per second. If the x and y variable are both in the same units, for example both in pounds sterling, then the slope is dimensionless. The correlation coefficient is a dimensionless coefficient between –1 and 1 measuring the amount of linear association. There are no upper and lower bounds on the slope.

There is a second interpretation of the correlation coefficient wiÉ in linear egression models, and this is related to the **percentage of variation explained by the model**. It also links two of the variances we have used in the model. The sample variance of the dependent variable is

$$s_y^2 = \frac{1}{n-1}\sum(y-\bar{y})^2,$$

and this measures the variability in the dependent variable without paying any attention at all to the relationship between y and x. The residual variance,

$$\hat{\sigma}^2 = \frac{1}{n-2}\sum(y_i - \hat{y}_i)^2,$$

measures the variation in the y observations about the straight-line model. The two key components of these variances are the **total sum of squares**, $\sum(y - \bar{y})^2$, and the **residual sum of squares**, $\sum(y - \hat{y})^2$. The ratio of these two quantities is related to the correlation coefficient in that

$$1 - r_{xy}^2 = \frac{\sum(y - \hat{y})^2}{\sum(y - \bar{y})^2} = \frac{\text{Residual sum of squares}}{\text{Total sum of squares}}$$

is the proportion of the total variation which is associated with the residuals. This proportion *cannot* be explained by the straight-line model. This means that

$$r_{xy}^2 = 1 - \frac{\sum(y - \hat{y})^2}{\sum(y - \bar{y})^2} = 1 - \frac{\text{Residual sum of squares}}{\text{Total sum of squares}}$$

is the proportion of variance which *can* be explained by the model. This term is usually referred to as **R-squared**, or the **coefficient of determination**, and is normally expressed as a percentage.

15.3.2 Interpretation of R^2

In the earnings per share and gearing data the correlation coefficient is $r_{xy} = 0.685$, and so $r_{xy}^2 = 0.469$. We can interpret this by saying that 47% of the variation in the earnings per share among the 11 companies is explained by the straight-line relationship with gearing. Alternatively, we can say that gearing explains 47% of the variation in the earnings per share. With this relationship we have managed to explain almost half of the variability in the earnings per share values.

For the supermarket data, the percentages of variance in turnover explained by the three quantitative explanatory variables are

Shop area	24.7%
Percentage of social class ABC in catchment area	10.3%
Catchment area population	17.7%
All three	55.3%

These figures are much lower than for the earnings per share data set and, in finance and business examples, are much more realistic. The area of the shop is the best single predictor of the turnover, but it only accounts for 25% of the total variation in the turnover figures over all the shops.

Although we derived the R^2 measure from the correlation coefficient, the former is actually more general and measures the relationship of the residual sum of squares to the total sum of squares. These two quantities do not depend upon the number of explanatory variables in the model, and so we can include all three quantitative variables in the supermarket data. When we do so we find that collectively these three variables explain 55% of the variation in the turnover figures, which is better.

R^2 tells us the percentage of variation explained by the model. As it is derived from the correlation coefficient, which must lie between −1 and 1, R^2 has to be bounded below by

0% and above by 100%. It can only be 0% if the correlation is zero and there is no linear association. It can only be 100% if all the points lie on a straight line and the residual sum of squares is zero.

Clearly the higher the value of R^2 the better the model, and the closer the points lie to the straight line. It is difficult to give general guidance as to reasonable values for R^2. It may be that even with small values, less than 25%, the model will still give you important information about the relationship between the explanatory variable and the dependent variable. If the model is to be used for prediction of future values of the dependent variable then you would want to have a high value of R^2, of the order of 60–70% at least.

15.4 Confidence intervals for the estimates

The main aim of a linear regression analysis is to estimate some quantities. Often this will include the slope of the relationship and predictions for a number of values of the explanatory variable. As in Chapter 11, whenever a quantity is estimated we need to find a way of conveying the precision of the estimate. This has been achieved in earlier chapters through the use of confidence intervals, and this is what is required here.

The 95% confidence intervals used in Chapter 11 were all of the form

$$\text{Estimate} \pm t_{0.025} \times \text{Standard error}$$

These were based upon the t distribution as the normal distribution was the probability model and the standard error had to be estimated from the data. This will be the framework here as we have exactly the same situation of a normal distribution for the errors and the error variance, σ^2, is unknown and estimated by the residual variance, $\hat{\sigma}^2$.

In previous chapters it was possible to derive the variance of the estimators or to show you by simulation that the formulae are valid. In this chapter this approach will not always be adopted as some of the mathematics is a little tricky. Quite often I will just quote the formula and then go on to show you how it is used and interpreted.

15.4.1 Slope

The estimate of the slope in the linear regression model is

$$\hat{\beta}_1 = \frac{s_{xy}}{s_x^2}.$$

If you substitute for s_{xy} and s_x^2, cancel out the two $n-1$ terms, and note that $\sum(x-\bar{x})\bar{y} = 0$, then you can show that this is equal to

$$\hat{\beta}_1 = \frac{\sum(x-\bar{x})y}{\sum(x-\bar{x})^2}$$

This is a linear combination of y-values which we have assumed are independent and normally distributed with a variance of σ^2, as the x-values are assumed to be fixed constants. This means that we know how to work out the expected value and variance of $\hat{\beta}_1$ using the same methods as in Chapter 9. After a bit of algebra you can show that

$$E[\hat{\beta}_1] = \beta_1,$$

$$\text{Var}[\hat{\beta}_1] = \frac{\sigma^2}{\sum(x - \bar{x})^2} = \frac{\sigma^2}{(n-1)s_x^2}.$$

This means that $\hat{\beta}_1$ is an unbiased estimate of the true slope, β_1. You may recall from Chapters 10 and 11 that this is an important quantity of any sample estimate. The estimated standard error of $\hat{\beta}_1$ is the square root of the variance, with $\hat{\sigma}$ replacing σ, and is

$$\text{SE}(\hat{\beta}_1) = \frac{\hat{\sigma}}{\sqrt{\sum(x - \bar{x})^2}}.$$

which means that the $100(1 - \alpha)\%$ confidence interval is

$$\hat{\beta}_1 \pm t_{\alpha/2}\,\text{SE}(\hat{\beta}_1),$$

where the degrees of freedom for the t distribution are $n - 2$. This also leads us to a method of testing hypotheses about the slope. To test that the slope is equal to a specified value, $H_0: \beta_1 = \beta_{10}$, where β_{10} is the specified value of the slope, we use a t test with

$$t = \frac{\hat{\beta}_1 - \beta_{10}}{\text{SE}(\hat{\beta}_1)}$$

This will follow a t distribution on $n - 2$ degrees of freedom if the null hypothesis is true.

The most common choice for the specified value is $\beta_{10} = 0$, but in finance when investigating the relationship between individual shares and the market it is common to test if the slope is equal to 1. Other values are possible, depending upon the context.

The reason why the test of $H_0: \beta_1 = 0$ is very important is that it permits us to choose between two very different models. If the alternative hypothesis is true then we have $H_1 :$ $\beta_1 \neq 0$ and the slope is non-zero. This means that the linear regression model is

$$y = \beta_0 + \beta_1 x,$$

indicating that y depends upon x. On the other hand, if the null hypothesis is true then the slope is zero and the model becomes

$$y = \beta_0,$$

which means that y does not depend upon x. Thus the test of $H_0: \beta_1 = 0$ against $H_1: \beta_1 \neq 0$ is a test of the linear association between y and x.

15.4.2 Slope in the gearing and earnings per share example

In the gearing and earnings per share example, the slope is estimated as $\hat{\beta}_1 = 0.423\,997\,5$. We calculate the sum of squares of the x observations from $\sum(x - \bar{x})^2 = (n-1)s_x^2 = 10 \times 1621.712 = 16217.12$. This leads to the standard error of the slope,

$$\text{SE}(\hat{\beta}_1) = \frac{\hat{\sigma}}{\sqrt{\sum(x - \bar{x})^2}} = \frac{19.15}{\sqrt{16217.12}} = 0.1504.$$

There are $n = 11$ companies so the t distribution has 9 degrees of freedom. The value we need for the 95% confidence interval is 2.262. The 95% confidence interval for the slope of the linear regression of earnings per share on gearing is

$$0.4240 \pm 2.262 \times 0.1504 = (0.084, 0.764).$$

For every 1% increase in the gearing ratio, the earnings per share are predicted to increase by 0.42 pence. The 95% confidence interval ranges from 0.08 pence to 0.76 pence and, relative to the slope, is quite wide. One way of interpreting this confidence interval is to say that all slopes from as low as 0.08 to as high as 0.76 are consistent with the data.

As the 95% confidence interval does not include zero it is clear that a slope of 0 is not consistent with these data. In practice, there is seldom any need to formally test the null hypothesis, $H_0: \beta_1 = 0$, as the main aim is usually one of estimation. However, we illustrate the calculations and interpretation. Our test statistic is

$$t = \frac{\hat{\beta}_1 - \beta_{10}}{SE(\hat{\beta}_1)} = \frac{0.4240}{0.1504} = 2.819.$$

The p-value is 0.02, and as this is less than 0.05 we say that there is evidence at the 5% level that the slope of the relationship between earnings per share and gearing is not equal to zero. This implies that earnings per share are linearly related to gearing.

15.4.3 Relationship between testing the slope and testing the correlation

As there is a formal relationship between the slope of the linear regression model and the correlation coefficient, the test that the slope is equal to zero is exactly the same as the test that the sample correlation coefficient is equal to zero. The correlation between earnings per share and gearing is 0.6848, and the t statistic for testing that the correlation is zero is, from Chapter 14,

$$t = \sqrt{n - 2} \; \frac{r}{\sqrt{1 - r^2}} = 2.819.$$

This is exactly the same as the test that the slope is equal to zero. Consequently, there is no need to test the correlation and the slope in the same analysis, and testing one is equivalent to testing the other. If the correlation is significantly different from zero then the slope will also be significantly different from zero.

15.4.4 Intercept

The intercept in the linear model, which is the expected value of y when x takes the value zero, is estimated using

$$\hat{\beta}_0 = \bar{y} - \hat{\beta}_1 \bar{x}.$$

In order to work out the variance of $\hat{\beta}_0$ we need to use the variance of $\bar{y}$, the variance of $\hat{\beta}_1$ and the covariance between them. After some tricky algebra we get

$$E[\hat{\beta}_0] = \beta_0,$$

$$\text{Var}[\hat{\beta}_0] = \sigma^2 \frac{\sum x^2}{n \sum (x - \bar{x})^2}.$$

These results show you that $\hat{\beta}_0$ is an unbiased estimate of β_0 and that the variance of $\hat{\beta}_0$ is a function not only of the error variance, σ^2, and the sample size, n, but also the sums of squares of the x variables.

You can carry out hypothesis tests on the intercept and calculate confidence intervals in the same fashion as for the slope. These procedures are not as commonly used as for the slope because in most circumstances the intercept is a less important parameter of the model than the slope.

15.4.5 Intercept in the earning per share example

The intercept is estimated as $\hat{\beta}_0 = 9.225\,455$ and the standard error of the intercept is estimated using

$$\text{SE}(\hat{\beta}_0) = \hat{\sigma} \sqrt{\frac{\sum x^2}{n \sum (x - \bar{x})^2}} = 19.15 \sqrt{\frac{27\,520.8310}{11 \times 16\,217.12}} = 7.522.$$

Thus the 95% confidence interval for the intercept is

$$9.225 \pm 2.262 \times 7.522 = 9.225 \pm 17.015 = (-7.79, 26.24).$$

As you can see this is a very wide interval, ranging from –8 pence per share to 26 pence per share. The range of the observed earnings per share in Table 15.1 is from –14 pence to 67 pence, and you can see that the confidence interval for the intercept spans a large part of this range. Thus the prediction of the linear model is not very precise.

The intercept is the predicted value of the earnings per share when the gearing is zero. This may be important in financial analysis but is probably no more important than the predicted value when gearing is 5% or 10% or 30%. In most cases the intercept is just one of the many predicted values along the line, and in many circumstances is not important at all.

15.4.6 Interpretation of intercept: fixed and variable costs

Within management accounting there is one area in which the intercept is an important parameter of the model. This is in the analysis of fixed and variable costs. The data in Figure 15.2 illustrate the relationship between the overhead costs of a factory and the number of units produced. Within a management accounting analysis, the intercept of the

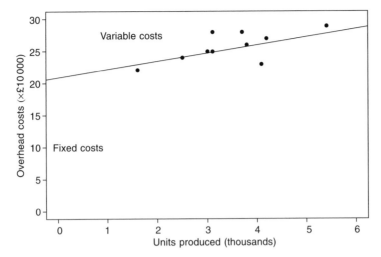

Figure 15.2 Overhead costs and units produced.

model is an estimate of the overhead costs associated with the production of no units and so gives an estimate of the fixed costs of the production process. The slope estimates the relationship between the overhead costs and the number of units produced and so is concerned with the variable costs of production.

In the graph the intercept is estimated as 20.9, with a standard error of 2.1. There are 12 observations, representing quarterly data from three consecutive years, and the 95% confidence interval for the fixed costs is (16.2, 25.6) in units of £10000. Thus the fixed costs are £209000, with a 95% confidence interval of (£162000, £256000). In this analysis the variable costs represent a much smaller percentage of the total overhead costs. The slope of the linear regression is 1.29 in units of £10000 per 1000 items produced or £12.9 per item. Thus the predicted variable costs for 1000 items are £12900.

When we are estimating the intercept in this case we have to assume that the linear relationship observed between the x-values of 1.6 and 5.4 in Figure 15.2 will continue right down to zero. This is an untested assumption. The confidence interval for the intercept in Figure 15.2 is very wide, and part of the reason for this is the large distance between $x = 0$ and the rest of the x-values.

15.5 Centring

As the intercept is estimated as $\hat{\beta}_0 = \bar{y} - \hat{\beta}_1\bar{x}$, the fitted line is

$$\hat{y}_i = \hat{\beta}_0 + \hat{\beta}_1 x_i = \bar{y} - \hat{\beta}_1\bar{x} + \hat{\beta}_1 x_i.$$

If we substitute $\bar{x}$ for x_i then we obtain the fitted value corresponding to the mean of the x-values. This is

$$\hat{y}_i = \bar{y} - \hat{\beta}_1\bar{x} + \hat{\beta}_1\bar{x} = \bar{y}.$$

Thus the mean of the y-values is the fitted value corresponding to the mean of the x-values, and so the straight line will always pass through the point $(\bar{x}, \bar{y})$.

As the intercept seldom has an important interpretation and as the value $x = 0$ is frequently far away from the observed x-values it is prudent to centre the x-values about their mean before fitting the linear regression model. This will have no effect on the estimate of the slope, no effect on the residual standard error, and no effect on the fitted values and predictions. However, it will have an effect on the estimate of the intercept and does mean that the intercept will have an interpretation.

The estimate of the intercept is

$$\hat{\beta}_0 = \bar{y} - \hat{\beta}_1 \bar{x};$$

thus if the x-values are centred on their mean so that $\bar{x} = 0$ then the estimated intercept will become

$$\hat{\beta}_0 = \bar{y}.$$

Furthermore, the variance of the intercept, based upon x-values which are centred, will be

$$\text{Var}[\hat{\beta}_0] = \sigma^2 \frac{\sum x^2}{n \sum (x - \bar{x})^2} = \sigma^2 \frac{x^2}{n \sum (x)^2} = \frac{\sigma^2}{n},$$

again because $\bar{x} = 0$ when the x-values are centred on their mean.

15.5.1 Example: Earnings per share

We illustrate the effects of centring using the data in Table 15.5. The explanatory variable is now the fourth column, in which the mean (32.06) has been subtracted from all the x-values to centre them. The comparison of the results from the centred and uncentred versions of the same model is shown in Table 15.6. You can see that exactly the same

Table 15.5 Centring gearing

Source	EPS (y)	G (x)	(x − x̄)
Manchester Utd	18.40	28.60	−3.46
Sheffield Utd	−0.50	4.35	−27.71
Hi-Tec	2.50	2.37	−29.69
Yates Brothers	14.00	0.09	−31.97
Glaxo	56.70	126.83	94.77
Tate and Lyle	40.50	78.17	46.11
Zeneca	67.80	22.26	−9.80
Cadbury-Schweppes	34.10	47.61	15.55
Harry Ramsden's	10.60	0.88	−31.18
Unilever	21.50	41.46	9.40
Kwik-Save	−14.61	0.00	−32.06

$\bar{x} = 32.06$.

Table 15.6 Comparison of centred and uncentred estimates

Parameter	Uncentred		Centred	
	Estimate	Standard error	Estimate	Standard error
$\hat{\beta}_0$	9.2255	7.5229	22.8173	5.7749
$\hat{\beta}_1$	0.4240	0.1504	0.4240	0.1504
$\hat{\sigma}$	19.15	–	19.15	–
r_{xy}^2	0.4689	–	0.4689	–

estimates are obtained for all the parameters except the intercept. In the centred version the intercept is interpreted as the predicted earnings per share for companies with an average value for the gearing. This is the average earnings per share for the 11 companies in the sample and so has some meaning with regard to the interpretation of the data.

15.5.2 Example: Supermarket location

In many cases the estimated value of the intercept is nonsensical in terms of the data under consideration. In the supermarket data the estimated model is

$$\text{Turnover} = 6603 + 11.1 \times \text{Area.}$$

This means that if a shop has zero area then the predicted turnover is 6603, which is ludicrous. The mean shop area is 165 m^2 and if we centre Area on this value we get the following fitted model:

$$\text{Turnover} = 8429 + 11.1(\text{Area} - 165),$$

which is exactly the same line as before except that the intercept corresponds to the estimated turnover for a shop with an area of 165 m^2. It is not necessary to always centre on the mean and sometimes you can get a better interpretation if you centre the observations on another value in the centre of the distribution of the explanatory variable. In this case we might choose a shop of 100 m^2, 150 m^2 or 200 m^2. The regression lines are exactly the same:

$$\text{Turnover} = 7710 + 11.1(\text{Area} - 100),$$

$$\text{Turnover} = 8263 + 11.1(\text{Area} - 150),$$

$$\text{Turnover} = 8816 + 11.1(\text{Area} - 200).$$

The only difference is that the intercepts refer to turnover at different areas.

15.6 Case study: capital asset pricing model

The capital asset pricing model is widely used in finance and investment analysis as a means of managing the risk associated with portfolios of investments. A portfolio is just a collection of stocks or shares. According to this model the risk of any asset is split between systematic and specific risk. The risk of any asset is compared to the risk of the stock market.

The systematic risk is the risk of holding the market portfolio. Over time the values of stock market shares will change and, as the market moves, each individual asset is more or less affected. The systematic risk of an asset is associated with the risk of the whole stock market. The specific risk of an asset is the risk which is unique to that asset.

A simplified version of the capital asset pricing model can be written as

$$R_p = R_f + \beta(R_m - R_f) + \varepsilon,$$

where R_p is the risk of the asset, R_f is the risk-free rate, R_m is the risk of the whole market and ε is the specific risk of the asset. This is very similar to a linear regression model, and the systematic risk β is estimated by regressing the returns for the asset on the returns for the market. If the risk-free rate is zero then the model corresponds to a linear regression through the origin, i.e. with the intercept equal to zero. The same is true if $\beta = 1$.

The risk associated with the market, obtained by regressing R_m on itself, will obviously give a value of $\beta = 1$. Assets which have a β greater than one tend to have a greater return than the market when the market has a positive return and a greater loss than the market when the market is losing money. Details of the capital asset pricing model can be found in many financial text books and also in World Wide Web resources such as www.contingencyanalysis.com/_frame/framesglossarycapitalassetpricingmodel.htm.

We will illustrate the use of the linear regression model in connection with the capital asset pricing model using data of monthly returns in the New York Stock Exchange from 1959 until 1992 (Table 15.7). The data consist of the monthly returns from Treasury bills (the risk-free rate), the return on the New York Stock Exchange (the market) and the returns for five different industry sectors – primary industry, manufacturing industry, transport, trade and finance – which are the portfolios in this analysis.

The results of fitting the models are shown in Table 15.8 for each of the five sectors. In the primary industry the intercept is estimated as 0.0009 and the p-value for testing if this is significantly different from zero is 0.58, implying that there is no evidence to suggest that the intercept is not zero. The slope (β) for this model is estimated as 0.94,

Table 15.7 Extract of monthly returns on the New York Stock Exchange

	Treasury bills	NYSE	Primary	Manufacturing	Transport	Trade	Finance
1959							
April	0.0021	0.0367	−0.0057	0.0516	0.0128	0.0021	0.0451
May	0.0006	0.0164	0.0067	0.0303	−0.0104	0.0240	−0.0309
June	0.0013	−0.0014	−0.0268	0.0026	−0.0149	0.0157	0.0025
July	0.0006	0.0340	0.0484	0.0396	0.0179	0.0182	0.0274
1992							
April	0.0007	0.0200	0.0567	0.0260	0.0399	−0.0218	0.0185
May	0.0013	0.0039	0.0531	0.0001	−0.0023	0.0062	0.0116
June	0.0005	−0.0187	−0.0301	−0.0300	0.0004	−0.0166	0.0048
July	0.0009	0.0389	0.0432	0.0350	0.0569	0.0368	0.0362

Source: Normandin and St-Amour (1998). Data can be downloaded from
http://ideas.uqam.on/ideas/data/jaejapmet.html.

Table 15.8 Results of fitting linear regression models to the New York Stock Exchange data

	Estimate	St. error	$t*$	p	Confidence limits	
					Lower	Upper
Primary						
Intercept	0.0009	0.0017	0.5486	0.5836	−0.0023	0.0042
Slope	0.9335	0.0372	−1.7848	0.0751	0.8605	1.0065
Manufacturing						
Intercept	0.0001	0.0003	0.3843	0.7010	−0.0005	0.0008
Slope	1.0345	0.0078	4.4241	0.0000	1.0192	1.0498
Transport						
Intercept	0.0011	0.0010	1.0748	0.2831	−0.0009	0.0030
Slope	0.7195	0.0224	−12.5423	0.0000	0.6756	0.7633
Trade						
Intercept	0.0011	0.0014	0.7900	0.4300	−0.0016	0.0038
Slope	1.1082	0.0309	3.5058	0.0005	1.0477	1.1687
Finance						
Intercept	0.0002	0.0008	0.2382	0.8118	−0.0014	0.0017
Slope	1.0616	0.0178	3.4546	0.0006	1.0267	1.0966

*The t statistic for the intercept corresponds to testing the hypothesis that the intercept is equal to zero. The t statistic for the slope corresponds to the hypothesis that the slope is equal to 1, which is the slope of the market.

which is less than 1. The test statistic for testing if the slope is different from 1 is $t = -1.78$, with $p = 0.08$. At the 5% significance level we would say that there is no evidence to reject the null hypothesis that the slope is equal to zero. This implies that the primary industries have the same systematic risk as the market.

In all the other industry sectors the evidence from the estimates and their standard errors is that the intercept is zero. However, the slope is different from 1 in all of them. For the transport sector it is well below 1 ($t = -12.5$, $p < 0.0001$), while for the others it is greater than 1. This implies that over the period of 33 years when the market has boomed the transport sector has not given as great a return, but when the market has gone down the transport sector has not gone down as much. The other three sectors have provided a better return when the market has risen, but have sustained bigger losses when the market has fallen.

This type of information is used by investment analysts to balance up a portfolio of stocks and shares with a lower risk. For example, you might combine investments in companies with different slopes (betas in finance). If you can find companies with negative slopes then these are companies with a high return when the market is falling, and they can help to offset the losses incurred by companies with betas much greater than 1.

The example illustrates how statistical methods are used within the context of finance through a specific model. In this instance the model is the simplest one. There is currently considerable use of complex statistical models in finance. A vast quantity of data are available from stock exchanges and within companies on buying and selling transactions. This has led to the development and use of statistical and computational techniques such as data mining, classification, and complex regression models. All of these techniques have basic assumptions and principles which are similar to those discussed in this chapter and in the book in general.

15.7 Residuals and outliers: checking the validity of the model

Within the linear regression model

$$y_i = \beta_0 + \beta_1 x_i + \varepsilon_i, \qquad i = 1, \ldots, n,$$

the distributional assumptions are all concerned with the errors, ε_i, which are assumed to be independent of each other and normally distributed with a mean of zero and a variance of σ^2. The residuals from the fitted model

$$e_i = y_i - \hat{y}_i = y_i - \hat{\beta}_0 - \hat{\beta}_1 x_i$$

are the estimates of these errors, and it is these residuals that are used to check the validity of the assumptions.

15.7.1 Normality and outliers

The easiest assumption to check is that of normality, and a plot of the residuals will do this. If all the residuals lie on an approximate straight line then we can be quite happy with the assumption of normality. If there is a curved relationship then this suggests that the normal distribution is not valid and that we have to try a transformation of the y variable, as discussed in Chapter 6.

The normal plot of the residuals from the regression of turnover on shop area for the supermarket data is shown in Figure 15.3. All of the points lie very close to a straight line and so we conclude that the assumption of a normal distribution for the errors is reasonable.

If the points do not lie on a straight line then this suggests that something is wrong with the assumption of a normal distribution. It may also suggest that there are outliers

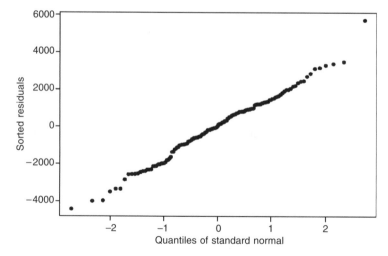

Figure 15.3 Normal plot of residual from supermarket data.

in the data. We discussed outliers in Chapter 3. These are data points which lie far away from the rest of the observations. Within the context of linear regression an outlier will be identified as a point whose residual is far away from the rest. As the residuals are distributed about zero, an outlier will be a very large or very small value.

One way to investigate if there are any outliers is through the use of a boxplot of the residuals. Another way is from the normal plot above, where we are looking for points which are not on or near the straight line. In Figure 15.3 there is only one point, the large residual of £5700, which is not on the straight line. In these data we have 153 supermarkets and one outlier is not really unexpected, nor will it cause a problem for the analysis.

Examples of some common types of residual plot are shown in Figure 15.4 for situations where the assumptions are not valid. In Figure 15.4a there is clearly one outlier with a large positive residual. The correct procedure here is to go back to the data and check that the values of x and y have been entered correctly. If the data are correct then you may find that there is some physical reason why this data point is not like the rest. For example, the measurement may have been taken under different circumstances. In such cases it is permissible to omit this point from the modelling analysis, but you must state clearly that you have done so in your report.

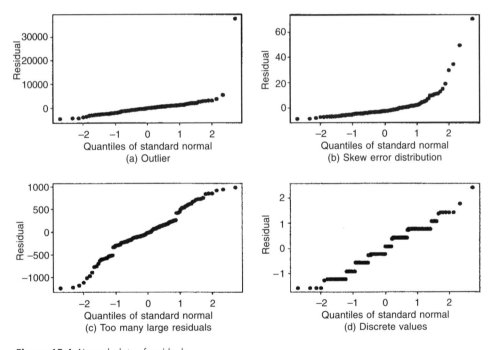

Figure 15.4 Normal plots of residuals.

In Figure 15.4b there is a curved relationship between the residuals and the expected values. This implies that the errors come from a skew distribution rather than a symmetric one. In some cases it is possible to correct this by taking a transformation of the response variable, for example by using the logarithmic or square root transformation.

The first two graphs are the most common and the third (Figure 15.4c) is relatively rare. This is the sort of plot you get if there are too many large residuals relative to what you would expect. The general shape of the curve resembles an 'S'. In this case the model is wrong, but there is no quick way to correct it. The last type of plot (Figure 15.4d) crops up when you are carrying out a regression analysis using response variables and explanatory variables which are discrete with a small number of values. An example would be regressing a grade of examination, measured on a five-point scale, say, to grade of effort, measured on a four-point scale. The linear regression model cannot be appropriate here as the response variable is not quantitative but ordinal.

15.7.2 Constant variance

The second assumption to check is that the variance of the errors is constant. This is known as **homoscedasticity**. If the variance is not constant we say that there is **heteroscedasticity**. The most common departure from a constant variance is to find that the variance depends upon the y-value or, equivalently, upon the x-value. The error variance is estimated from the residuals, so plotting the residuals against the fitted values, or the x-variable, will give you information on whether or not the error variance is constant. It will also give information about outliers.

The plot of residuals against fitted values for the supermarket data is shown in Figure 15.5. The two horizontal lines are drawn at $\pm 1.96\hat{\sigma}$, as we would expect that 95% of the residuals will lie within this band. There are seven points outside these limits, and as we have 153 shops we expect $153 \times 0.05 = 7.6$ points outside the limits. In this plot all of the points lie within a broad band and there is consequently no evidence that the residuals tend to be larger or smaller over any one range of fitted values. This confirms that the assumption of a constant error variance is reasonable.

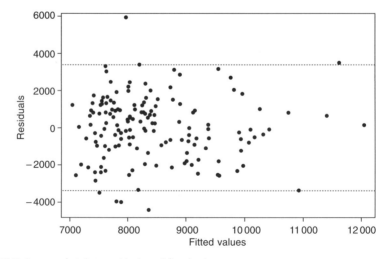

Figure 15.5 Supermarket data: residuals and fitted values.

Some plots of residuals against fitted values where the assumptions are not valid are shown in Figure 15.6. In Figure 15.6a there is one outlier with a large residual, and you should identify this point and check that the data are correct. Figure 15.6b shows that at low fitted values the residuals are spread over a small range while at larger fitted values the range of the residuals increases. This is consistent with the error variance increasing with x or with y. This is a common event and you can try to minimize the effect by taking a transformation of the response variable such as $\log(y)$. The less common situation of variance decreasing with increasing fitted value is illustrated in Figure 15.6c. Figure 15.6d shows a curved relationship between the residuals and the fitted values. This indicates that there is a curved relationship between y and x which has not been taken into account in the linear model. One way to correct this is to try to transform the x variable or to include x^2 terms in the model as well as simply x.

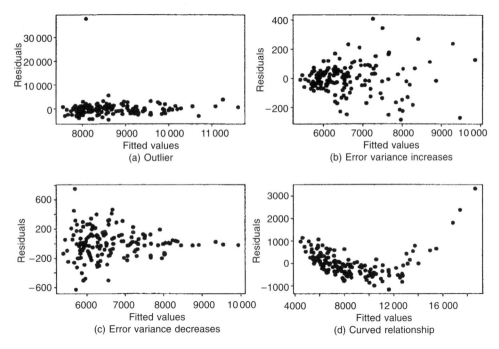

Figure 15.6 Common residual and fitted value plots.

15.7.3 Independence of errors

A third assumption to check is that the errors are independent of each other. This assumption will generally hold in the kinds of data you are likely to want to analyse in business settings, but it is a very difficult one to check in practice. The most common situation where the errors may not be independent is when observations are collected sequentially in time. Data of this type are very common in finance and economics.

The influence of the time effects may mean that if you have a positive residual in one period then you are also likely to have a positive residual in the adjacent periods. This is known as **positive serial correlation**. You can get **negative serial correlation** where positive residuals are followed by negative residuals and vice versa. A time series plot of the residuals may exhibit some non-independence through a cyclical pattern or trend. If the plot reveals a random scatter of points, then we would conclude that there was no evidence of serial correlation.

Neither the earnings per share nor the supermarket data sets are suitable for an investigation of correlation among the residuals, as neither set was collected sequentially over time. A time series plot of the residuals from a regression of the monthly returns for the finance and services industry sector in the United States against the New York Stock Exchange return is shown in Figure 15.7. These data were previously discussed in Section 15.6 in connection with the capital asset pricing model. This plot exhibits some mild serial correlation among the residuals in that there is some evidence of cyclical behaviour, particularly from about 1975 onwards, though it is also visible during the 1960s. There is also evidence of short periods of increased variability in the residuals. In financial terms such periods are known as periods of **high volatility**. The most obvious instance of this is in the mid-1970s, but there are also periods of high volatility in the early 1980s.

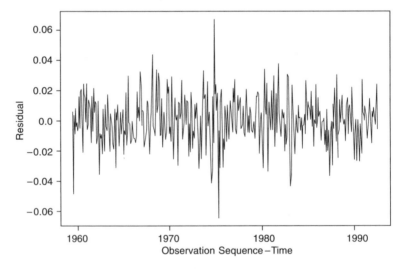

Figure 15.7 Serial correlation: times series plot of residuals.

15.7.4 Summary

All of the suggestions for checking and improving the model involve quite tricky modelling decisions, and these are best done in consultation with statisticians. The plots are there for you to see when the assumptions made in the simple linear regression model are not true so that you can then go and get appropriate help for your analysis. Statistical modelling is a tricky affair and requires a considerable amount of judgement. This is easier if you have had a lot of experience, but very hard for novices.

15.8 Confidence intervals for a predicted mean and predicted single value

15.8.1 Predicting the response variable

The most common reason for fitting a linear regression model is to estimate the effect of the explanatory variable on the dependent or response variable. This is concerned with providing an estimate of the slope of the relationship and an interpretation of this slope in the context of the analysis. So in the supermarket example we find that the slope of the regression of Turnover on Area is 11.1 with a standard error of 1.6. This means that a shop with an area of 100 m^2 more than another shop is expected to have a Turnover of £1110 more, with a 95% confidence interval of (£790, £1430).

When we discussed correlation we really had no concept of a response variable and an explanatory variable, and all we were doing was looking to measure the linear association. Although the mathematics of linear regression and correlation are very similar, there is a big difference in interpretation. When fitting a linear regression model we are thinking of some form of dependent relationship where changes in one variable are associated with changes in another. So in a sense we have the idea of the explanatory variable coming before the response variable. Although we know from a marketing and selling point of view that there is no direct causal relationship between the size of a shop and its turnover (increasing the size of the shop need not necessarily increase the turnover) we do expect that there is going to be a dependent relationship (larger shops are likely to have larger turnovers).

As linear regression models are based upon the notion of an explanatory variable which is thought of as influencing the response variable we come to the second main use of these models, which is in predicting the value of the response variable. With these predictions it is then possible to use the statistical model to inform business decisions. Obviously such predictions are going to be based on assumptions, and we will look at them in detail at the end of this section.

In the supermarket example we have an equation which can be used to predict the turnover from the area based upon the results from the existing shops. If a new site becomes available, the managers of the chain of supermarkets can use the model to predict the turnover for the new store, based upon its (known) area. They can then put this together with their other cost data on overheads and salaries to estimate how profitable the store is likely to be. On the basis of this they can then decide whether or not to expand into the new premises.

15.8.2 Predicted mean value and confidence interval

Although there was much discussion in Section 15.5 about centring and the interpretation of the intercept, this does not matter for prediction. The predictions of the fitted model will be exactly the same whether or not you centre the explanatory variable.

An important aspect of prediction is in providing a confidence interval for the fitted value, and in order to calculate this we need to have the variance of the fitted value. The fitted value corresponding to x_p is

$$\hat{y}_p = \hat{\beta}_0 + \hat{\beta}_1 x_p,$$

and to derive the variance of this we need to use the variance of the estimated intercept, the variance of the estimated slope and the covariance between them. We did this type of calculation in Chapters 9 and 10 and the principle is exactly the same, but the algebra here is harder and the variance is

$$V[\hat{y}_p] = \sigma^2 \left(\frac{1}{n} + \frac{(x_p - \bar{x})^2}{\sum(x - \bar{x})^2} \right).$$

The important thing to notice about this equation for the variance is that it is a quadratic function of the distance x_p is from the mean of x. If $x_p = \bar{x}$, then the predicted value is $\hat{y}_p = \bar{y}$ and the variance is equal to σ^2/n, which is the variance of the sample mean. This variance will then increase the further x_p goes from $\bar{x}$.

The confidence interval uses the estimated value of the standard error, which is

$$SE(\hat{y}_p) = \hat{\sigma} \sqrt{ \left(\frac{1}{n} + \frac{(x_p - \bar{x})^2}{\sum(x - \bar{x})^2} \right) }.$$

The $100(1 - \alpha)\%$ confidence interval is

$$\hat{y}_p \pm t_{\alpha/2} \, SE(\hat{y}_P),$$

where the t distribution has $n - 2$ degrees of freedom.

15.8.3 Example: Earnings per share

In the earnings per share and gearing example the fitted line is

$$\hat{y}_i = \hat{\beta}_0 + \hat{\beta}_1 x_i = 9.23 + 0.424 x_i,$$

and the estimated error standard deviation is $\hat{\sigma} = 19.15$. Furthermore, $\bar{x} = 32.06$, $\sum(x - \bar{x})^2 = 16\,217.12$ and there are $n = 11$ companies. In this example we predict the earnings per share for a company with a gearing of 50%. The predicted value is

$$\hat{y}_p = 9.23 + 0.424 \times 50 = 30.43 \text{ pence per share.}$$

The estimated standard error of this predicted mean value is

$$SE(\hat{y}_p) = 19.15 \sqrt{ \left(\frac{1}{11} + \frac{(50 - 32.06)^2}{16217.12} \right) } = 19.15 \sqrt{0.110\,76} = 6.37.$$

The 95% confidence interval for the predicted mean value is therefore

$$30.43 \pm 2.262 \times 6.37 = 30.43 \pm 14.42 = (16.0, 44.8) \text{ pence per share.}$$

Based upon the linear regression model for earnings per share and gearing we would say that for companies with a gearing of 50% the predicted mean earnings per share is 30.4 pence and the 95% confidence interval is (16, 45) pence.

15.8.4 Interpretation of predicted mean value

This confidence interval gives you a very clear idea of the precision of the predicted mean value. If it is narrow then the mean is predicted with a high precision. If the interval is wide in relation to the range of the response variables then the prediction does not have a very high precision. In this case the observed range of the earnings per share is from −14.61 to 67.80 and you can see that the 95% confidence interval is wide relative to this. Consequently, based upon 11 companies, you would not have very reliable predictions. If there were data from 55 companies, with exactly the same fitted line and residual standard error then the 95% confidence interval would be (30.4, 36.4) pence, which is much more precise.

As in all other areas of statistical estimation, the precision of the estimate depends upon how many observations you have. The more observations, the greater the precision. Furthermore, in predictions from the linear regression model the precision also depends upon the spread of the explanatory variables. The greater the observed range of the x variables, the greater will be the precision. While you can see that this is true from the equation for the standard error of the predicted mean value, it is also a matter of common sense. To get information on the relationship between y and x you would want to collect information over as wide a range of x as possible and not just restrict yourself to a small part of the possible x-values.

The 95% confidence intervals for all predicted means on the fitted line are shown in Figure 15.8. There are two things to notice here. The first is that the confidence intervals are curved, and the second is that they fan out at either end of the line. This means that the precision of the estimated means is higher in the centre of the range of x-values. Specifically, the smallest confidence interval corresponds to predictions made at the mean of the explanatory variable. Just as important is the fact that the estimates corresponding to x-values at the extremes of the range of x, i.e. near the maximum and minimum values of x, will not have a very high precision.

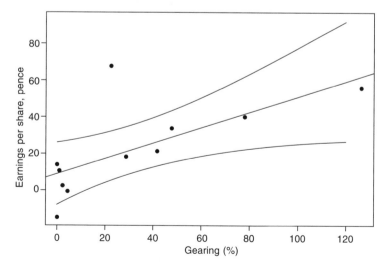

Figure 15.8 Ninety-five per cent confidence interval for predicted means.

The implication of this curved relationship is that any predictions are best done using values around the centre of the distribution of the x-values. If you have to make predictions based upon x-values close to the maximum or minimum, then you must bear in mind that they will have a low precision.

The 95% confidence intervals and the fitted lines for the predicted mean values of turnover for the three separate regressions in the supermarket data are shown in Figure 15.9. Compared to the confidence interval curves in Figure 15.8, the intervals are much narrower relative to the spread of the y observations. This occurs because there are many more observations in this data set, 153 compared to 11, which means that the predicted mean turnovers are obtained with a high degree of precision. For example, the predicted mean turnover for shops with 200m^2 of selling area is £8800 and the 95% confidence interval for this mean is (£8500, £9100). This is small relative to the range of turnovers of £3800 to £15 000.

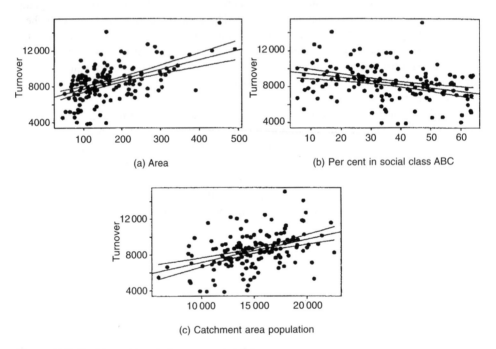

(a) Area

(b) Per cent in social class ABC

(c) Catchment area population

Figure 15.9 Confidence intervals for supermarket data.

15.8.5 Predicted single value and confidence interval

In Figure 15.9 there are a large number of observed points which lie outside the confidence intervals. These confidence intervals refer to the fitted values, which are mean values. So for the earnings per share the interval of (16, 45) pence is the confidence interval for the mean earnings per share of all companies with a gearing of 50%. This is important. It is not for a single company, but for the mean of all conceivable companies with a gearing of 50%.

Similarly, in the supermarket example, the interval of (£8500, £9100) is the 95% confidence interval for the mean of all shops with a selling area of 200 m². The reason for developing a regression model in this situation is to make predictions for the turnover of a single shop, not for the mean of all shops. Thus, not only do we have to take into account the standard error of the predicted mean but we also have to take into account the variation of the observations about the mean.

The only difference between predicting a mean value and predicting a single observation is that in the latter case we need to take into account the error variance. The predicted value of the response variable from a single value x_p is the same as the fitted value,

$$\hat{y}_p = \hat{\beta}_0 + \hat{\beta}_1 x_p.$$

So the prediction of the single value is exactly the same as the predicted mean of all observations with a value of x_p.

When we work out the variance of this single value we have to take into account the error that is associated with this single value and use

$$\text{Var}[\hat{y}_s] = \text{Var}[\hat{y}_p + \varepsilon_s] = \text{Var}[\hat{y}_p] + \text{Var}[\varepsilon_s] = \sigma^2 \left(\frac{1}{n} + \frac{(x_p - \bar{x})^2}{\sum (x - \bar{x})^2} \right) + \sigma^2.$$

The variance of a predicted single value is just the variance of the predicted mean plus the error variance. The estimated standard error of the predicted single value is

$$\text{SE}(\hat{y}_s) = \hat{\sigma} \sqrt{\left(1 + \frac{1}{n} + \frac{(x_p - \bar{x})^2}{\sum (x - \bar{x})^2} \right)}.$$

This will always be greater than the standard error of the predicted mean. The confidence interval for a single predicted value derived from this standard error is often referred to as a **prediction interval**, partly to distinguish it from the confidence interval for the fitted mean value.

15.8.6 Prediction intervals for the earnings per share and supermarket data

The predict earnings per share for a company with a gearing of 50% is

$$\hat{y}_s = 9.23 + 0.424 \times 50 = 30.43 \text{ pence per share.}$$

The estimated standard error of this single predicted value is

$$\text{SE}(\hat{y}_s) = 19.15 \sqrt{\left(1 + \frac{1}{11} + \frac{(50 - 32.06)^2}{162\,17.12} \right)} = 19.15 \sqrt{1.110\,76} = 20.18 \text{ pence.}$$

The 95% prediction interval for the predicted single value is

$$30.43 \pm 2.262 \times 20.18 = 30.43 \pm 45.65 = (-15.2, 76.1) \text{ pence per share.}$$

This is extraordinarily wide because we are providing a confidence interval for a single value. This means that we have to take into account not only the sampling variation in the estimation of the parameters of the model but also the error variance which is added on. This means that SE($\hat{y}_s$) will always be larger than SE($\hat{y}_p$), often very much larger, and the only way to reduce the size of SE($\hat{y}_s$) is by reducing the residual standard error, $\hat{\sigma}$. The only feasible way of doing this is by finding a better model, with a lower value of $\hat{\sigma}$. This can often be achieved by including in the model more explanatory variables, which are each related to the response variable. We discuss this in Section 15.10.

The prediction intervals for the prediction of a single shop from the supermarket data are shown in Figure 15.10. Although these are curves as in Figure 15.9, this cannot be seen, as the term which dominates is the residual standard error which is £1690 in Figure 15.10a, £1845 in Figure 15.10b and £1768 in Figure 15.10c. You can see clearly that while the confidence interval for the predicted mean turnover for all shops with a specified area is narrow, the prediction interval for a single shop is wide. It is this latter interval which is needed by the managers of the supermarket chain.

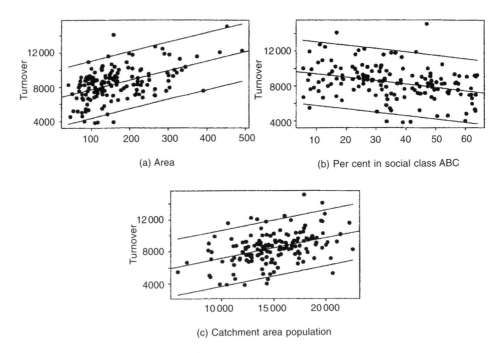

(a) Area

(b) Per cent in social class ABC

(c) Catchment area population

Figure 15.10 Single-value prediction intervals.

15.8.7 Assumptions involved in making predictions

When predicting an observation, either a single observation or the mean response, you are making an estimation of the unknown. This is not an easy thing to do, but using a statistical model, which has been validated, is the best way of going about it. This process is not without its drawbacks, but provided you are aware of them and do not treat your predictions as gospel truth then you can use them to inform your business decisions.

There are a number of implicit assumptions made when predicting, and before discussing them we assume that you have developed a model which is valid according to the assumptions listed in Section 15.2. This means that you are sure that there is a linear relationship, that the errors are normally distributed, that they are independent, and that the error variance is constant.

The data that you have for developing a prediction model have been collected at some time in the past. The predictions are to be made at some time in the future. An implicit assumption is that the relationship that you know to have existed in the past will still exist in the future. Effectively this means that we are assuming that the intercept and slope will be unchanged, the residual variance the same, and the linear model the same. None of these assumptions can be checked.

This problem is illustrated in Table 15.9, where we first of all look at the relationship between earnings per share and gearing for the 11 companies in 1996 and again for the same 11 companies in 1997. In 1996 there is a significant relationship between these two quantities with slope 0.42 and 95% confidence interval (0.08, 0.76). In 1997 there is no evidence of a significant association and the slope is estimated as 0.31 (−0.10, 0.72). Part of the problem here is that there are only 11 companies, but you can see that to predict earnings per share from gearing in 1997 using a model developed in 1996 would not be correct as the model is not stable.

Table 15.9 Investigation of stability of regression coefficients

			Confidence limits	
	Estimate	Std error	Lower	Upper
Earnings per share and gearing				
1996				
Intercept	9.23	7.52	−7.79	26.24
Slope	0.42	0.15	0.08	0.76
$\hat{\sigma}$	19.15			
1997				
Intercept	20.95	10.49	−2.77	44.67
Slope	0.31	0.21	−0.16	0.79
$\hat{\sigma}$	28.16			
New York Stock Exchange finance				
1959–1965				
Intercept	−0.0001	0.0012	−0.0025	0.0022
Slope	1.0833	0.0284	1.0276	1.1389
$\hat{\sigma}$	0.0172			
1966–1992				
Intercept	0.0006	0.0010	−0.0014	0.0026
Slope	1.0416	0.0221	0.9983	1.0849
$\hat{\sigma}$	0.0142			

However, stability of the regression model is seen in the second example, where we use the relationship between the returns in the financial sector and the stock exchange returns for the New York Stock Exchange data. The estimates for the first half of the series are very similar to those for the second half. In this example we have a much larger set of data and so the effects of sampling variation on the fitted models are much less.

A second major issue with prediction is concerned with the prediction of observations outside the observed range of values for the explanatory variable. We can illustrate this problem with reference to Figure 15.10, and consider what turnover would be predicted for a shop with selling area of 1000m². There is no problem with the algebra, and if we substitute into the equation we find that the predicted turnover is

$$\text{Turnover} = 6603 + 11.1 \times 1000 = £17\,700.$$

The problem is that a shop of 1000 m² is much larger than any of the other existing shops in the sample. The largest has an area of 491m².

The danger with predictions outside the range of observed x-values is that we do not know that the linear relationship will hold. All that we have been able to do in fitting the model and checking its validity is to establish that there is a linear relationship within the observed values of x. We no nothing outside this range and so predictions outside this range are based upon faith that the model will be valid. Again there is no way of testing this assumption. You should treat such predictions accordingly and not place much reliance on them.

A final point about predictions is that if the R^2 of the model is low then the model does not really account for very much of the variation in the response variable. Consequently, predictions based on the model will not be of much use as there is a great deal of unexplained variation. If a model only explains 20% of the variation then it is not going to be much use for prediction, though it may be useful for understanding the relationships among the variables. A model with R^2 in excess of 70% is normally going to be good for predictions because a large part of the variability is explained by the model and the residual variation is quite small.

15.9 Qualitative explanatory variables

15.9.1 Dummy variables

So far we have confined ourselves to relationships between two quantitative variables. In the supermarket data one of the possible explanatory variables for the turnover is the presence of a direct competitor in the neighbourhood. This is a qualitative variable taking only two values – yes or no. Such values can be included as explanatory variables in a regression analysis using a technique known as **dummy variables**. They cannot be included as response variables in a linear regression analysis as the assumptions of a normal distribution and constant variance cannot be valid.

First of all, we consider the problem of a qualitative variable which takes only two values. With dummy variable coding we introduce a new variable which takes only two values, 0 or 1, and we let 0 represent one of the two levels of the qualitative variable and 1 the other. It does not matter which level is assigned to 0 and which to 1. This is illustrated in Table 15.10 where we have chosen to let 0 in the dummy variable represent 'No competitor in the neighbourhood' and 1 to represent 'Competitor in the neighbourhood'. In the linear regression we use Turnover as the response variable and the competitor dummy variable as the explanatory variable.

The fitted model can then be written as

Table 15.10 Dummy variable coding

	Turnover (y)	Area	Competitor	Competitor dummy variable (d)	Percentage in social class ABC	Catchment area population
1	9176	100	No competitor	0	42.2	15182
2	8670	221	Competitor	1	7.1	16712
3	9317	95	No competitor	0	34.6	16925
4	6992	157	Competitor	1	36.1	11723
5	7562	295	No competitor	0	57.5	12323

$$\hat{y} = \hat{\beta}_0 + \hat{\beta}_1 d.$$

As d takes only two values there are going to be only two fitted values:

$$\hat{y} = \begin{cases} \hat{\beta}_0 & \text{when } d = 0, \\ \hat{\beta}_0 + \hat{\beta}_1 & \text{when } d = 1. \end{cases}$$

This provides us with an interpretation of the parameters of the model. $\hat{\beta}_0$ is the predicted mean value of the response variable when the dummy variable takes the value zero. $\hat{\beta}_0 + \hat{\beta}_1$ is the predicted mean value when the dummy variable is one. Thus the difference between these two fitted values is $\hat{\beta}_1$, which then represents the effect of level 1 compared to level 0 of the dummy variable.

15.9.2 Supermarket example: effect of competitor on turnover

The estimates for the regression of turnover on the dummy variable representing the presence of a competitor are shown in Table 15.11. As the intercept $\hat{\beta}_0$ represents the mean value of the turnover when the dummy variable takes the value 0 we know that the estimated turnover when there are no competitors is £8956.30.

Table 15.11 Dummy variable regression

Parameter	Estimate	Std error	t	p
$\hat{\beta}_0$	8956.30	227.38	39.39	0.0000
$\hat{\beta}_1$	−958.94	306.87	− 3.12	0.0021
$\hat{\sigma}$	1888.75			

The slope ($\hat{\beta}_1$) represents the effect of a competitor in the neighbourhood compared to no competitor in the neighbourhood. As the estimated effect is negative at −£959, there is a lower predicted mean turnover when there is a competitor. This is what you would expect. The predicted mean turnover when there is a competitor present is 8956.30 + (−958.94) = £7997.36.

The *t*-values are used for hypothesis testing. The one corresponding to $\hat{\beta}_0$ is used to test H_0: $\beta_0 = 0$, i.e. the mean turnover in shops with no competitor is equal to zero. This is not a particularly interesting hypothesis to test because we expect turnover to be much greater than zero in any shop.

The hypothesis about β_1 is much more interesting. If $\beta_1 = 0$ then there is no difference between the mean turnover in shops with a competitor and in shops without a competitor. If $\beta_1 \neq 0$ then there is going to be a difference. The test statistic is −3.12 with $p = 0.002$, and as this is less that 0.05 there is strong evidence that the mean turnover when there is no competitor is different than when there is a competitor.

The parameter β_1 gives the difference in mean turnover between the group of shops with no competitor and the group with a competitor. This is estimated as −£959, with a 95% confidence interval of (−£1565, −£353).

15.9.3 Relationship to pooled *t* test

Although this model has been fitted within the framework of a linear regression it is very similar, in principle, to the types of models discussed in Sections 12.4 and 11.4. There are two groups of observations – shops with a competitor present and shops without a competitor. The regression coefficient $\hat{\beta}_1$ gives the difference in the means, $\hat{\beta}_0$ being the mean of one group and $\hat{\beta}_0 + \hat{\beta}_1$ being the mean of the other group.

The means and standard deviations of the turnover from the two groups of observations are presented in Table 15.12. A comparison of these with the results of the linear regression reveals exactly the same values for the means. Furthermore, if you calculate the pooled standard deviation using the data in Table 15.12 you find that it is 1888.75, which is identical to the residual standard error from the regression model. Also the pooled two-sample *t* statistic based upon the data in Table 15.12 is −3.12 which is identical to the *t* statistic for the test of the hypothesis that the regression coefficient $\beta_1 = 0$. Finally, the confidence interval for the difference between the two based upon data in Table 15.12 is (−1565.26, −352.63), which is exactly the same as the confidence interval for β_1.

Table 15.12 Two group summary statistics

	n	Mean	Standard deviation
No competitor	69	8956.30	1989.505
Competitor	84	7997.36	1802.018

15.9.4 Summary

The purpose of this section has been to show you how to include qualitative variables into the linear regression model and to show you that when the qualitative variable has only two levels the linear regression analysis is exactly the same as when estimating the differences between two means and testing for differences between two means. The advantage of the linear regression approach is that it is easy to extend it in two ways. Firstly, we can include more explanatory variables in the model and so we can look at the joint effects of qualitative and quantitative variables. We will turn to this in Section 15.10.

The second advantage is that we can include qualitative variables which have three or more levels corresponding to differences between three or more groups. This is often discussed in the framework of a topic known as analysis of variance, but it is essentially the same as linear regression. For a qualitative variable with three levels it is necessary to use two dummy variables, if there are four levels use three dummy variables. For three levels these are coded as

Qualitative variable levels	Dummy 1	Dummy 2
A	0	0
B	1	0
C	0	1

The fitted regression model becomes

$$\hat{y} = \hat{\beta}_0 + \hat{\beta}_1 d_1 + \hat{\beta}_2 d_2,$$

so that $\hat{\beta}_0$ is the fitted value for group A ($d_1 = d_2 = 0$), $\hat{\beta}_0 + \hat{\beta}_1$ the fitted value corresponding to group B ($d_1 = 1, d_2 = 0$), and $\hat{\beta}_0 + \hat{\beta}_2$ the fitted value corresponding to group C ($d_1 = 0, d_2 = 1$). Thus $\hat{\beta}_1$ represents the difference between group B and group A, while $\hat{\beta}_2$ represents the difference between group C and group A.

15.10 More than one explanatory variable: multiple regression

In the supermarket data the most important of the quantitative explanatory variables was the selling area of the shop. If we fit this model and then take the residuals and plot them against the other variables in the data – the percentage in social class ABC, the population in the catchment area and the presence or absence of a competitor – then we get the residual plots in Figure 15.11. All of these plots reveal a systematic pattern. In Figure 15.11a the residuals tend to be positive when the percentage in social class ABC is low and negative when the percentage is high. In Figure 15.11b there is an increasing trend in the residuals with increasing population. Figure 15.11c shows predominantly negative residuals when there is a competitor and positive residuals in the absence of a competitor.

These plots all suggest that these variables also play an important part in the prediction of turnover over and above the effect of area. This means that we would like to include them in the regression model, and this is very easy to do. Regression with two or more explanatory variables is referred to as **multiple regression**. The estimation of the parameters of the model is carried out using computer programs but the principles are exactly the same as for regression with just one variable.

The parameter estimates are presented in Table 15.13. They are based upon data where the selling area is centred on 150 m^2, the percentage in social class ABC is centred on

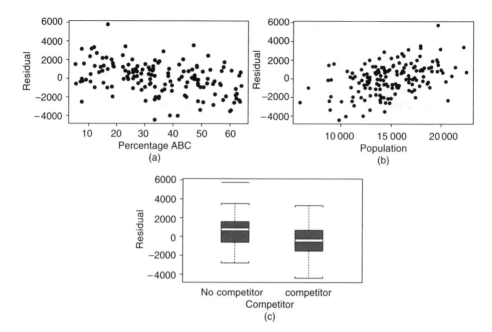

Figure 15.11 Residual plots after fitting area.

Table 15.13 Parameter estimates from multiple regression model

	Estimate	Std error	*t*	*p*
Intercept	9110.844	133.903	68.041	0
Area	11.171	1.024	10.909	0
Percentage ABC	− 53.775	5.562	− 9.669	0
Population	0.296	0.029	10.342	0
Competitor dummy	− 1460.683	180.856	− 8.076	0
Residual standard error	1096			

35% and the population centred on 15 000. This means that the intercept corresponds to the mean turnover in shops with an area of 150 m², located in neighbourhoods of 15 000 where 35% of the population are in social class ABC, and there is no competing shop nearby. The estimated turnover is £9111.

All of the regression coefficients are highly significant, indicating that they have a role to play in the estimation of the turnover of the shop. The scatter plots (Figure 15.9) all suggest that a linear relationship is appropriate for all the variables. We interpret the coefficients by saying that turnover is estimated to increase by £11 for every increase in area of 1m², to decrease by £54 for every increase of 1% in the percentage of the population who are in social class ABC, to increase by £0.30 per person and to decrease by £1461 if there is a competing shop in the neighbourhood.

The estimates in Table 15.14 are obtained from the four separate linear regressions with only one explanatory variable. There are two things to notice in the comparison of this table with Table 15.13. The first is that the residual standard error is much lower in

Table 15.14 Four separate regressions

		Estimate	Std error
Area	Intercept	8263.136	138.722
	Slope	11.064	1.571
	Residual standard error	1691	
Percentage ABC	Intercept	8439.491	149.234
	Slope	−38.320	9.199
	Residual standard error	1846	
Population	Intercept	8459.232	143.072
	Slope	0.2573	0.0452
	Residual standard error	1769	
Competitor	Intercept	8956.304	227.3793
	Slope	−958.947	306.8718
	Residual standard error	1889	

the multiple regression than in any of the single linear regressions. This occurs because we have explained much more in the systematic part of the model with four explanatory variables than with only one explanatory variable. In fact the R^2 for the multiple regression model is 69%, whereas the most it was for any of the simple linear regressions was 25%.

The second point is that the estimates of the slopes are different. In two cases there is just a little difference – Area and Population – but in the other two there is more of a difference. The reasons are quite complex – but, in general, you should expect to find differences, sometimes quite large differences, between the estimated slopes in a series of simple regressions with one explanatory variable and a multiple regression. In the multiple regression the slope for one variable takes into account all the other relationships in the model. The technical way of saying this is that they adjust for the effects of the other variables. In the simple linear regression the other variables are not considered and the slope is unadjusted.

As an example, the unadjusted effect of a competitor is to reduce the turnover by £958. Adjusting for the effects of area, percentage in social class ABC, and population, the effect of a competitor is to reduce the turnover by £1460. The difference arises because of relationships among the explanatory variables. Shops with competitors tend to be in areas with a slightly higher population and with a slightly lower percentage in social class ABC. When you take into account the effects of these two variables the effect of a competitor becomes greater.

You can make predictions from a multiple regression model in the same way as for a simple linear regression model. Again there are formulae for the standard errors and confidence or prediction intervals. As the residual standard error is smaller in the multiple regression you would expect the intervals to be narrower. The fitted values for two shops are shown in Table 15.15 with the corresponding values of the explanatory variables. The standard error for a single predicted shop is still high, as it is dominated by the residual standard error. The standard error of the mean is much lower. Both standard errors are lower than the corresponding values from the use of only area in the model in Section 15.8.

Multiple regression analysis is a complex modelling process and there are potentially many pitfalls. One problem which is particularly prevalent in finance and economics is something known as **multicollinearity**. This arises when the explanatory variables are all

Table 15.15 Fitted values and standard errors

Area	Competitor	Percent ABC	Population	Fitted value	Standard error of mean	Standard error of single value
100	No	30	14 000	8525	149.8	1096.4
300	Yes	40	15 000	9353	181.3	1096.5

very highly correlated with each other. The effect of multicollinearity is to make the regression coefficients unstable and to increase their standard errors. This can mean that the overall R^2 is high but that none of the explanatory variables are significant.

15.11 Summary

Linear regression analysis is among the most important statistical modelling techniques. It is certainly widely used. Part of the reason for this is that it allows you to investigate the relationships among variables and use the patterns in these relationships to predict values of the response variable. Predictions form a great part of decision making in business and industry, and any tools which will assist in this process are potentially very useful.

Regression analysis has many uses in business and finance, and there is a whole branch of economics, known as **econometrics**, which is derived from this modelling approach. Within the realms of quality control one can imagine that the number of defects in a production process is related to the speed at which the process is running. A regression model could be developed to investigate this and to set the production speed at such a speed as to ensure that the number of defects is acceptably low.

The revenue of a company should be linked to the amount it spends upon advertising in the various media. A multiple regression model can be developed to predict the revenue from the amount spent on advertising in television, magazines, newspapers and billboards. This would enable the company to judge the most important advertising media for increasing sales.

In the United Kingdom, the cost of car insurance varies according to the age and sex of the driver, as well as the number of years of driving experience, previous insurance claims and driving convictions. The loading placed on the different categories of driver can be derived from multiple regression models regressing the amount of the claim on the various characteristics of the drivers.

A key feature of regression analysis, and virtually all other statistical analyses, is that they are based upon the concept of a statistical model. This is a mathematical description of a relationship together with a probability description of the random variation in the data. Whenever using such a model it is important to check that the assumptions on which the model is based are correct. This can normally always be done but in circumstances when it is impossible, such as predicting future events, you should be cautious about the predictions and realize that they are only predictions – better than guesses, but not facts.

The more complex the model the more that can go wrong with it and the more sensitive it is to violations in the assumptions that are necessary for the model. Before fitting a linear regression model it is always necessary to do a scatter plot of the data to check that linearity is a reasonable assumption. Then it is necessary to check up on the normality of the residuals and other assumptions. You also need to check on the presence and possible effect of outliers and on other possible explanatory variables. Once you have a model you are happy with then you can interpret it to try to understand the relationships in the data and use the model for predictions, taking into account the error involved in making such predictions.

Appendix A

Table of the standard normal distribution

	0.00	0.01	0.02	0.03	0.04	0.05	0.06	0.07	0.08	0.09
0.0	0.5000	0.5040	0.5080	0.5120	0.5160	0.5199	0.5239	0.5279	0.5319	0.5359
0.1	0.5398	0.5438	0.5478	0.5517	0.5557	0.5596	0.5636	0.5675	0.5714	0.5753
0.2	0.5793	0.5832	0.5871	0.5910	0.5948	0.5987	0.6026	0.6064	0.6103	0.6141
0.3	0.6179	0.6217	0.6255	0.6293	0.6331	0.6368	0.6406	0.6443	0.6480	0.6517
0.4	0.6554	0.6591	0.6628	0.6664	0.6700	0.6736	0.6772	0.6808	0.6844	0.6879
0.5	0.6915	0.6950	0.6985	0.7019	0.7054	0.7088	0.7123	0.7157	0.7190	0.7224
0.6	0.7257	0.7291	0.7324	0.7357	0.7389	0.7422	0.7454	0.7486	0.7517	0.7549
0.7	0.7580	0.7611	0.7642	0.7673	0.7704	0.7734	0.7764	0.7794	0.7823	0.7852
0.8	0.7881	0.7910	0.7939	0.7967	0.7995	0.8023	0.8051	0.8078	0.8106	0.8133
0.9	0.8159	0.8186	0.8212	0.8238	0.8264	0.8289	0.8315	0.8340	0.8365	0.8389
1.0	0.8413	0.8438	0.8461	0.8485	0.8508	0.8531	0.8554	0.8577	0.8599	0.8621
1.1	0.8643	0.8665	0.8686	0.8708	0.8729	0.8749	0.8770	0.8790	0.8810	0.8830
1.2	0.8849	0.8869	0.8888	0.8907	0.8925	0.8944	0.8962	0.8980	0.8997	0.9015
1.3	0.9032	0.9049	0.9066	0.9082	0.9099	0.9115	0.9131	0.9147	0.9162	0.9177
1.4	0.9192	0.9207	0.9222	0.9236	0.9251	0.9265	0.9279	0.9292	0.9306	0.9319
1.5	0.9332	0.9345	0.9357	0.9370	0.9382	0.9394	0.9406	0.9418	0.9429	0.9441
1.6	0.9452	0.9463	0.9474	0.9484	0.9495	0.9505	0.9515	0.9525	0.9535	0.9545
1.7	0.9554	0.9564	0.9573	0.9582	0.9591	0.9599	0.9608	0.9616	0.9625	0.9633
1.8	0.9641	0.9649	0.9656	0.9664	0.9671	0.9678	0.9686	0.9693	0.9699	0.9706
1.9	0.9713	0.9719	0.9726	0.9732	0.9738	0.9744	0.9750	0.9756	0.9761	0.9767
2.0	0.9772	0.9778	0.9783	0.9788	0.9793	0.9798	0.9803	0.9808	0.9812	0.9817
2.1	0.9821	0.9826	0.9830	0.9834	0.9838	0.9842	0.9846	0.9850	0.9854	0.9857
2.2	0.9861	0.9864	0.9868	0.9871	0.9875	0.9878	0.9881	0.9884	0.9887	0.9890
2.3	0.9893	0.9896	0.9898	0.9901	0.9904	0.9906	0.9909	0.9911	0.9913	0.9916
2.4	0.9918	0.9920	0.9922	0.9925	0.9927	0.9929	0.9931	0.9932	0.9934	0.9936
2.5	0.9938	0.9940	0.9941	0.9943	0.9945	0.9946	0.9948	0.9949	0.9951	0.9952
2.6	0.9953	0.9955	0.9956	0.9957	0.9959	0.9960	0.9961	0.9962	0.9963	0.9964
2.7	0.9965	0.9966	0.9967	0.9968	0.9969	0.9970	0.9971	0.9972	0.9973	0.9974
2.8	0.9974	0.9975	0.9976	0.9977	0.9977	0.9978	0.9979	0.9979	0.9980	0.9981
2.9	0.9981	0.9982	0.9982	0.9983	0.9984	0.9984	0.9985	0.9985	0.9986	0.9986
3.0	0.9987	0.9987	0.9987	0.9988	0.9988	0.9989	0.9989	0.9989	0.9990	0.9990
3.1	0.9990	0.9991	0.9991	0.9991	0.9992	0.9992	0.9992	0.9992	0.9993	0.9993
3.2	0.9993	0.9993	0.9994	0.9994	0.9994	0.9994	0.9994	0.9995	0.9995	0.9995
3.3	0.9995	0.9995	0.9995	0.9996	0.9996	0.9996	0.9996	0.9996	0.9996	0.9997
3.4	0.9997	0.9997	0.9997	0.9997	0.9997	0.9997	0.9997	0.9997	0.9997	0.9998
3.5	0.9998	0.9998	0.9998	0.9998	0.9998	0.9998	0.9998	0.9998	0.9998	0.9998
3.6	0.9998	0.9998	0.9999	0.9999	0.9999	0.9999	0.9999	0.9999	0.9999	0.9999
3.7	0.9999	0.9999	0.9999	0.9999	0.9999	0.9999	0.9999	0.9999	0.9999	0.9999
3.8	0.9999	0.9999	0.9999	0.9999	0.9999	0.9999	0.9999	0.9999	0.9999	0.9999
3.9	1.0000	1.0000	1.0000	1.0000	1.0000	1.0000	1.0000	1.0000	1.0000	1.0000

Appendix B

Percentage points of the *t* distribution

Degrees of freedom	Probability in upper tail					
	0.2	0.1	0.05	0.025	0.01	0.005
1	1.3764	3.0777	6.3138	12.7062	31.8205	63.6567
2	1.0607	1.8856	2.9200	4.3027	6.9646	9.9248
3	0.9785	1.6377	2.3534	3.1824	4.5407	5.8409
4	0.9410	1.5332	2.1318	2.7764	3.7469	4.6041
5	0.9195	1.4759	2.0150	2.5706	3.3649	4.0321
6	0.9057	1.4398	1.9432	2.4469	3.1427	3.7074
7	0.8960	1.4149	1.8946	2.3646	2.9980	3.4995
8	0.8889	1.3968	1.8595	2.3060	2.8965	3.3554
9	0.8834	1.3830	1.8331	2.2622	2.8214	3.2498
10	0.8791	1.3722	1.8125	2.2281	2.7638	3.1693
11	0.8755	1.3634	1.7959	2.2010	2.7181	3.1058
12	0.8726	1.3562	1.7823	2.1788	2.6810	3.0545
13	0.8702	1.3502	1.7709	2.1604	2.6503	3.0123
14	0.8681	1.3450	1.7613	2.1448	2.6245	2.9768
15	0.8662	1.3406	1.7531	2.1314	2.6025	2.9467
16	0.8647	1.3368	1.7459	2.1199	2.5835	2.9208
17	0.8633	1.3334	1.7396	2.1098	2.5669	2.8982
18	0.8620	1.3304	1.7341	2.1009	2.5524	2.8784
19	0.8610	1.3277	1.7291	2.0930	2.5395	2.8609
20	0.8600	1.3253	1.7247	2.0860	2.5280	2.8453
21	0.8591	1.3232	1.7207	2.0796	2.5176	2.8314
22	0.8583	1.3212	1.7171	2.0739	2.5083	2.8188
23	0.8575	1.3195	1.7139	2.0687	2.4999	2.8073
24	0.8569	1.3178	1.7109	2.0639	2.4922	2.7969
25	0.8562	1.3163	1.7081	2.0595	2.4851	2.7874
26	0.8557	1.3150	1.7056	2.0555	2.4786	2.7787
27	0.8551	1.3137	1.7033	2.0518	2.4727	2.7707
28	0.8546	1.3125	1.7011	2.0484	2.4671	2.7633
29	0.8542	1.3114	1.6991	2.0452	2.4620	2.7564
30	0.8538	1.3104	1.6973	2.0423	2.4573	2.7500
Normal	0.8416	1.2816	1.6449	1.9600	2.3263	2.5758

Appendix C

Percentage points of the χ^2 distribution

df	0.01	0.025	0.05	0.1	0.5	0.9	0.95	0.975	0.99
1	0.0002	0.0010	0.0039	0.0158	0.4549	2.7055	3.8415	5.0239	6.6349
2	0.0201	0.0506	0.1026	0.2107	1.3863	4.6052	5.9915	7.3778	9.2103
3	0.1148	0.2158	0.3518	0.5844	2.3660	6.2514	7.8147	9.3484	11.3449
4	0.2971	0.4844	0.7107	1.0636	3.3567	7.7794	9.4877	11.1433	13.2767
5	0.5543	0.8312	1.1455	1.6103	4.3515	9.2364	11.0705	12.8325	15.0863
6	0.8721	1.2373	1.6354	2.2041	5.3481	10.6446	12.5916	14.4494	16.8119
7	1.2390	1.6899	2.1673	2.8331	6.3458	12.0170	14.0671	16.0128	18.4753
8	1.6465	2.1797	2.7326	3.4895	7.3441	13.3616	15.5073	17.5345	20.0902
9	2.0879	2.7004	3.3251	4.1682	8.3428	14.6837	16.9190	19.0228	21.6660
10	2.5582	3.2470	3.9403	4.8652	9.3418	15.9872	18.3070	20.4832	23.2093
11	3.0535	3.8157	4.5748	5.5778	10.3410	17.2750	19.6751	21.9200	24.7250
12	3.5706	4.4038	5.2260	6.3038	11.3403	18.5493	21.0261	23.3367	26.2170
13	4.1069	5.0088	5.8919	7.0415	12.3398	19.8119	22.3620	24.7356	27.6882
14	4.6604	5.6287	6.5706	7.7895	13.3393	21.0641	23.6848	26.1189	29.1412
15	5.2293	6.2621	7.2609	8.5468	14.3389	22.3071	24.9958	27.4884	30.5779
16	5.8122	6.9077	7.9616	9.3122	15.3385	23.5418	26.2962	28.8454	31.9999
17	6.4078	7.5642	8.6718	10.0852	16.3382	24.7690	27.5871	30.1910	33.4087
18	7.0149	8.2307	9.3905	10.8649	17.3379	25.9894	28.8693	31.5264	34.8053
19	7.6327	8.9065	10.1170	11.6509	18.3377	27.2036	30.1435	32.8523	36.1909
20	8.2604	9.5908	10.8508	12.4426	19.3374	28.4120	31.4104	34.1696	37.5662
21	8.8972	10.2829	11.5913	13.2396	20.3372	29.6151	32.6706	35.4789	38.9322
22	9.5425	10.9823	12.3380	14.0415	21.3370	30.8133	33.9244	36.7807	40.2894
23	10.1957	11.6886	13.0905	14.8480	22.3369	32.0069	35.1725	38.0756	41.6384
24	10.8564	12.4012	13.8484	15.6587	23.3367	33.1962	36.4150	39.3641	42.9798
25	11.5240	13.1197	14.6114	16.4734	24.3366	34.3816	37.6525	40.6465	44.3141
26	12.1981	13.8439	15.3792	17.2919	25.3365	35.5632	38.8851	41.9232	45.6417
27	12.8785	14.5734	16.1514	18.1139	26.3363	36.7412	40.1133	43.1945	46.9629
28	13.5647	15.3079	16.9279	18.9392	27.3362	37.9159	41.3371	44.4608	48.2782
29	14.2565	16.0471	17.7084	19.7677	28.3361	39.0875	42.5570	45.7223	49.5879
30	14.9535	16.7908	18.4927	20.5992	29.3360	40.2560	43.7730	46.9792	50.8922

References

Jones, G. and Martin, C. (1977) *The Social Context of Spending in Youth*. Briefing No. 11, June. Centre for Educational Sociology, Edinburgh University.

Kerrison, S. and Macfarlane, A. (eds) (2000) *Official Health Statistics: An Unofficial Guide*. London: Arnold.

Normandin, M. and St-Amour, P. (1998) Substitution, risk aversion, taste shocks and equity premia. *Journal of Applied Econometrics* **13**, 265–281.

Lackman, L.C. (1987) The impact of capital adequacy constraints on bank portfolios. *Journal of Business Finance and Accounting* **13**, 587–596.

Moody's (1983) Moody's Banks.

O'Muircheartaigh, C. and Lynn, P. (1997) The 1997 UK pre-election polls. *Journal of the Royal Statistical Society Series A* **160**, 381–385.

Office for National Statistics (1998a) *Social Trends 28*. London: The Stationery Office.

Office for National Statistics (1998b) *Annual Abstract of Statistics: 1998*. London: The Stationery Office.

Romaniuk, H., Skinner, C.J. and Cooper, P.J. (1999) Modelling consumers' use of products. *Journal of the Royal Statistical Society Series A* **162**, 407–421.

Sillitoe, K. and White, P.H. (1992) Ethnic group and the British Census: the search for a question. *Journal of the Royal Statistical Society Series A* **155**, 141–163.

Skuse, D.H., James, R.S., Bishop, D.V., Coppin, B., Dalton, P., Aamodt-Leeper, G., Bacarese-Hamilton, M., Creswell, C., McGurk. R. and Jacobs, P.A. (1997) Evidence from Turner's syndrome of an imprinted X-linked locus affecting cognitive function. *Nature* **387**(6634), 705–708.

Stray, S., Naude, P. and Wegner, T. (1994) Statistics in management education. *British Journal of Management* **5**, 73–82.

Tame, Chris R. and Elliott, Nick (n.d.) Up in smoke: The economics, ethics and politics of tobacco advertising bans. www.forest-on-smoking.org.uk/publics/advertising.htm.

Tukey, J. (1977) *Exploratory Data Analysis*. Reading, MA: Addison Wesley.

Wright, D. (1997) Football standings and measurement levels. *The Statistician* **46**, 105–110.

Index